Biogeography

SCIENCES

Ecosystems and Environment
Field Directors – Françoise Gaill and Dominique Joly

Biodiversity, Subject Head – Fabienne Aujard

Biogeography

An Integrative Approach of the Evolution of Living

Coordinated by
Eric Guilbert

iSTE

WILEY

First published 2021 in Great Britain and the United States by ISTE Ltd and John Wiley & Sons, Inc.

ISTE Ltd
27-37 St George's Road
London SW19 4EU
UK

www.iste.co.uk

John Wiley & Sons, Inc.
111 River Street
Hoboken, NJ 07030
USA

www.wiley.com

Library of Congress Control Number: 2021941650

British Library Cataloguing-in-Publication Data
A CIP record for this book is available from the British Library
ISBN 978-1-78945-060-6

ERC code:
PE10 Earth System Science
PE10_13 Physical geography
LS8 Ecology, Evolution and Environmental Biology
LS8_1 Ecosystem and community ecology, macroecology

MIX
Paper from responsible sources
FSC® C013604

Contents

Preface

Eric Guilbert
UMR7179 MECADEV, National Museum of Natural History, Paris, France

I am pretty sure that most of the scientists working on evolution or the ecology of living organisms did not start as biogeographers … However, when interested in understanding how living organisms have evolved, and how they are organized in relation to their environment, linked to biotic and abiotic variables, biologists naturally arrive at biogeography. Biogeography is the main approach when embracing the history of living species as a whole. It is a huge world, and it is only growing. After a quick search online (Lavoisier.fr), I found 1,291 books on Biogeography written since 1985. Another search, on the Web of Science (wcs.webofknowledge.com), shows 36,567 papers (written between 1957 and 2021) with "biogeography" in the title (Figure P.1). Among these, 25% are associated with "ecology" and 19% with "evolutionary biology". However, not only ecology and evolutionary biology are linked to biogeography, but a wide range of disciplines, from geography (9%) to genetics (8%) or conservation (7%).

This is why biogeography involves a wide variety of different methods. Which do we use for what purpose? The questions in biogeography are very diverse, and thus the approaches used may be very different. In the second overview of biogeography, Dawson *et al.* (2016) said the most common term in the 521 abstracts of the 7th biennial meeting of the International Biogeographic Society was "distribution". In terms of space, "region" was the most common and in terms of time, "history" or "historical" was the most common, followed by "future". The word "species" was mentioned in 85% of the abstracts. Biogeography still remains the study of the distribution of species in space and time, despite a widening range

Biogeography,
coordinated by Eric Guilbert.

of topics. The future of species is one of the major interests, with the history of species distribution.

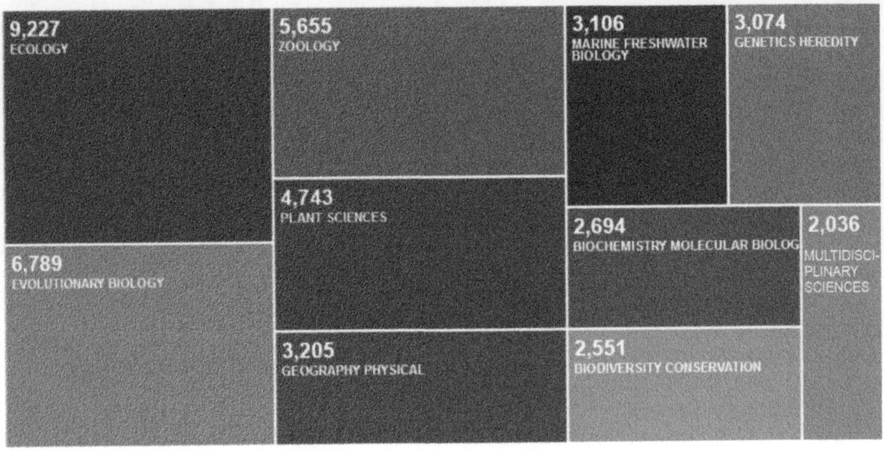

Figure P.1. *Representation of the number of publications according to different keywords. Source: Web of Science (wcs.webofknowledge.com), March 19, 2021*

If terrestrial plant distribution is the main historical beginning of biogeography studies, with the *Essai sur la géographie des plantes*, written by von Humboldt and Bonpland (1805), the concept was quickly broadened to include other living organisms; see Wallace's *The Geographical Distribution of Animals* (1876). After that, progress in taxonomy, systematics and phylogenetics made these studies even broader. New techniques, such as DNA sequencing, ecological niche modeling and many others, have allowed biogeographical approaches to be widened. Dealing with the distribution of *Nothofagus* (the iconic Southern beech) is not the same as dealing with the propagation of the Ebola virus. *Nothofagus* has been a key group in biogeographical studies on plants for over 170 years (Cook and Crisp 2005). Despite a huge amount of literature on the subject, the evolution of *Nothofagus* remains controversial (Hill *et al.* 2015). On the other hand, the Ebola virus was totally unknown before the 1970s (Pourrut *et al.* 2005) and, today, drivers that shape the epidemic are much better known, thanks to improvements in approaches and methods (see Chapter 12).

Not only do approaches differ according to the biology of organisms but they also differ according to the environment. Dealing with the biogeography of freshwater fish is not the same as dealing with the biogeography of marine fish. If freshwater habitats can be considered as islands of water in the middle of the land (see Chapter 8), oceans are much less fragmented and more stable environments (see

Chapter 9). How should bacterial distribution in the soil be considered? Drivers of community assemblages of bacteria are specific (Fierer *et al.* 2007). In addition, taxonomic recognition of bacteria involves molecular tools (see Chapter 7). Is the biogeography of water beetles studied in the same way as the biogeography of cave beetles? The drivers of their distribution may not be the same (see, for example, Arribas *et al.* 2012; Faille *et al.* 2014), even if both can be considered as in an insular environment.

A lot of very good books on biogeography have been edited; see, for example, the fifth edition of *Biogeography* (Lomolino *et al.* 2017), or *Conservation Biogeography* (Ladle *et al.* 2011). Most provide the very bases of biogeography, theories and methods, a wide range of approaches, historical and original cases with nice illustrations. And yet, new studies and methodological novelties are coming out every year, and the number and variety make biogeography so attractive and exciting!

In this book, we have chosen to present an overview of biogeography through different specialists, disciplines, particular living groups or ecosystems and challenging topics, trying to cover a wide range of the current studies in such a broad and multidisciplinary science. The biogeography of terrestrial plants and animals is the very basis of the discipline, yet many topics in these two groups remain unstudied. Many books already cover biogeographical studies on plants and animals and these will not be considered in this book.

After a particular overview of the history of biogeography in Chapter 1, in Chapter 2 we will have an intensive study of the challenges and perspectives of the main analytical approaches used in biogeography, an always-changing world. We will then investigate different approaches, such as phylogeography, dealing with the geographical distribution in gene lineages within species or between closely related species (Chapter 3). Another approach we will look at, in Chapter 4, is geophysical, where the geology and climate conditions are put before biotic processes. We will study different ecosystems, such as islands, in Chapter 5. Since MacArthur and Wilson (1967), how can we miss the famous case of island biogeography! Caves (Chapter 6) are also an extreme ecosystem where species develop specific adaptive skills. Soil bacteria (Chapter 7) is almost an unknown world, a world in which much is still to be done in terms of biogeography, and it deserves a specific approach. In the same way, fungi play an essential role in ecosystem processes, and remain largely unstudied. In Chapter 8, we will also explore the biogeography of fungi, reacting differently according to environmental conditions. In Chapters 9 and 10, we will deal with two different environments, that of freshwater biogeography (Chapter 9) and that of marine biogeography (Chapter 10). One is like islands of water in the land, while the other is open areas, where terrestrial approaches are not always applicable. Finally, we will focus on particular approaches that are challenging today and may be of greater importance in the future, such as the biogeography of diseases (Chapter 11), a very current field of research, climate change (Chapter 12) and conservation (Chapter 13), two fields that are also closely linked to the human impact on the distribution of species in space and time.

Biogeography is not only a discipline that has been questioning the evolution of species and ecology since naturalists started exploring the world. Understanding patterns and processes of the distribution of species in space and time may provide solutions to the challenges humanity has faced since the era called the Anthropocene and its consequences, such as the biodiversity crisis and global warming.

July 2021

P.1. References

Arribas, P., Velasco, J., Abellan, P., Sanchez-Fernandez, D., Andujar, C., Calosi, P., Millan, A., Ribera, I., Bilton, D.T. (2012). Dispersal ability rather than ecological tolerance drives differences in range size between lentic and lotic water beetles (Coleoptera: Hydrophilidae). *Journal of Biogeography*, 39, 984–994.

Cook, L.G. and Crisp, M.D. (2005). Not so ancient: The extant crown group of Nothofagus represents a post-Gondwanan radiation. *Proceedings of the Royal Society B: Biological Sciences*, 272(1580), 2535–2544.

Dawson, M.N., Axmacher, J.C., Beierkuhnlein, C., Blois, J., Bradley, B.A., Cord, A.F., Dengler, J., He, K.A., Heaney, L.R., Jansson, R., Mahecha, M.D., Myers, C., Nogués-Bravo, D., Papadopoulou, A., Reu, B., Rodríguez-Sánchez, F., Steinbauer, M.J., Stigall, A., Tuanmu, M.-N., Gavin, D.G. (2016). A second horizon scan of biogeography: Golden Ages, Midas touches, and the Red Queen. *Frontiers of Biogeography*, 8(4), 1–30.

Faille, A., Andújar, C., Fadrique, F., Ribera, I. (2014). Late Miocene origin of a Ibero-Maghrebian clade of ground beetles with multiple colonisations of the subterranean environment. *Journal of Biogeography*, 41, 1979–1990.

Fierer, N., Bradford, M.A., Jackson, R.B. (2007). Toward an ecological classification of soil bacteria. *Ecology*, 88(6), 1354–1364.

Hill, R.S., Jordan, G.J., Macphail, M.K. (2015). Why we should retain Nothofagus *sensu lato*. *Australian Systematic Botany*, 28(3), 190.

von Humboldt, A. and Bonpland, A. (1805). *Essai sur la géographie des plantes ; accompagné d'un tableau physique des régions équinocoxiales*. Levrault, Schoell & Co., Paris.

Ladle, R.J. and Whittaker, R.J. (2011). *Conservation Biogeography.* Wiley-Blackwell Press, Oxford.

Lomolino, M.V., Riddle, B.R., Whittaker, R.J. (2017). *Biogeography*, 5th edition. Oxford University Press, Sunderland, MA.

MacArthur, R.H. and Wilson, E.O. (1967). *The Theory of Island Biogeography.* Princeton University Press, Princeton, NJ.

Pourrut, X., Kumulungui, B., Wittmann, T., Moussavou, G., Délicat, A., Yaba, P., Nkoghe, D., González, J.-P., Leroy, E.M. (2005). The natural history of Ebola virus in Africa. *Microbes and Invection.* 7, 1005–1014.

Wallace, A.R. (1876). *The Geographical Distribution of Animals: With a Study of the Relations of Living and Extinct Faunas as Elucidating the Past Changes of the Earth's Surface.* Harper & Brothers, New York, NY.

1

Origins of Biogeography: A Personal Perspective

Malte C. EBACH

University of New South Wales and The Australian Museum, Sydney, Australia

1.1. Introduction: a history of scientific practice

In the 1810 Preface of his *Theory of Colours*, Johann Wolfgang von Goethe wrote, "The history of an individual displays his [or her] character, so it may here be well affirmed that the history of science is science itself" (Goethe, cited in Duck and Petry 2016, p. xxxv). What Goethe meant is that the past practices of scientists are instances of scientific practice regardless of age. In other words, scientists do not need historians of science to interpret past scientific practice. Scientific practice, no matter how old, will always remain a part of science.

Where, then, does history fit in? Many historians and philosophers of science discuss scientific ideas rather than practice. The reason is that historians and philosophers of science *do not* engage in scientific practice and are therefore not always able to interpret what we do due to a lack of training or experience or both. British biologist Peter Medawar discussed this in 1968: "What scientists do has never been the subject of a scientific, that is, ethological inquiry… It is no use looking to scientific 'papers', for they not merely conceal but actively misrepresent the reasoning that goes into the work they describe" (Medawar 1968, p. 151).

Can historians and philosophers of science trust what scientists say as opposed to what they do? Understanding what scientists do is perhaps a much better way to

Biogeography,
coordinated by Eric GUILBERT.

understand the scientific process, rather than believing what they say. Take the science of systematics or taxonomy as an example. Edgar Anderson noted that it is "difficult to write about the taxonomic method [because] in its broadest aspects it has never been described. Taxonomists are more like artists than like art critics; they practice their trade and don't discuss it" (Anderson, cited in Haas 1954, p. 65). Indeed. What of biogeographical practice? In the case of taxonomy, the result is a list of diagnostic characteristics, names and photographic plates. In biogeography, the results are usually maps. Maps are representative of classifications and are an ideal starting point to understand the early biogeographic method. Before we look at the 18th and 19th biogeographical practices, it is important to understand what biogeography is and how it may be defined[1].

1.1.1. *What is biogeography?*

The Oxford English Dictionary defines biogeography as "the branch of biology that deals with the geographical distribution of plants and animals. Also: the characteristics of an area or organism in this respect" (OED 2021). The definition is extremely broad, and present-day biogeographers work on a myriad of topics with very different aims and methods. In this sense, biogeography is multidisciplinary, namely the geographical aspect of any field of study that deals with natural objects, be it animals, plants, bacteria fungi, viruses, humans and abiogenesis. It has even made it into nursing pedagogy. In any case, biogeography is a field of enquiry that is dependent on the questions, aims and methods of a particular field. Ecological biogeography, for instance, endeavors to answer ecological questions using methods in ecology. Biogeography is not an independent science with its own unique methods, aims or goals, and attempts to unify it as an "integrative biogeography" have not been successful. Rather than unifying, integrative biogeography discarded a whole suite of approaches and goals in favor of others. We only need to look at the 18th century origins of plant and animal geographies to see that attempts at unification had posed theoretical as well as methodological problems, leading to a split by the end of the 19th century and an abrupt hiatus by the beginning of the 20th century.

1.2. A history of phyto- and zoogeographical classification

1.2.1. *Terminology*

The goal of finding a phyto- and zoogeographical classification united 18th and 19th century plant and animal geographies. It is important to note that the term

1. For a detailed account of the history of 18th and 19th century biogeographies, see Ebach (2015).

biogeography only appeared in the latter half of the 19th century; Jordan (1883) coined it in German, and Merriam (1892) in English; however, neither author defined the term. The above OED definition of biogeography may have first appeared in Nelson (1978), 90 years after it was first coined[2]. Regardless, it would be historically incorrect to use the term biogeography for any theory, aims or methods used by 18th and 19th century plant and animal geographers. The term biogeography gained popularity in the latter half of the 20th century. I will use the terms plant and animal geographies, botanical geography, and phytogeography and zoogeography to suit the parlance of the time.

1.2.2. *How classification works*

Without a classification, it is impossible to scientifically study any natural object. Organisms not only need names, but they need categories to convey meaning. Finding myself in the middle of the North American forest, my friend warned me from afar, "Watch out for the poison ivy!" I knew what ivy was, but how could you tell normal ivy from poison ivy? He could list a set of characteristics that diagnose poison ivy. I am not a botanist so these characteristics would be meaningless; moreover, I would need to touch the poison ivy in order to identity it. "It looks green!" came the reply. My friend was not a botanist either, so I was to avoid a green type of ivy, which I assumed was of a climbing variety. This basic form of communication is helpful to avoid danger or useful to discriminate an edible berry from the one which is toxic. Communication is an essential part of scientific practice as well as its goal. The first question any taxonomists would ask is "What is this?" followed by "What are its characteristics?" and then "What is it most like?" The same is true when we deal with the distributions of animals and plants. Kangaroos do not come from "over there". They have a particular place in the world, which is determined through a hierarchical classification:

Earth

  Southern Hemisphere

    Australia

2. I avoid using the phrase "Father of". Coining a term does not justify ownership of the whole field. Herman Jordan and Clinton Hart Merriam used the term only once. Jordan casually refers to the term as though it was already in use, and Merriam uses it to describe his "Bio-geographic map".

There was also an understanding of a basic classification with the poison ivy:

Plant

Ivy

Green

While this is a basic classifications, it works in communicating information and meaning. In order for the classification to work, biogeographers need to be able to communicate plant and animal distributions clearly, either as a list, table or map. A distribution map simply lists where certain organisms occur. Zimmermann (1777) published the first modern distribution map (Figure 1.1) in which the global distributions of terrestrial quadrupeds are named in the areas where they occur. Note that the hierarchy is purely geographical:

Old World

Europe

Africa

Asia

New World

North America

South America

Australasia

Zimmermann's map was accompanied by a 600-odd page treatise that described the distributions in four chapters:

Chapter I: Animals dispersed throughout the world and their degeneration

Chapter II: Introduction

Part One. Quadrupeds of both the Old and New World

Part the Latter: Quadrupeds of the Old World

Chapter III. Quadrupeds of the New World

Chapter IV. In which the animals are generally treated by the dispersion across the surface, whose consequences are added in the history of the planet (Zimmermann 1777, p. xxiv)

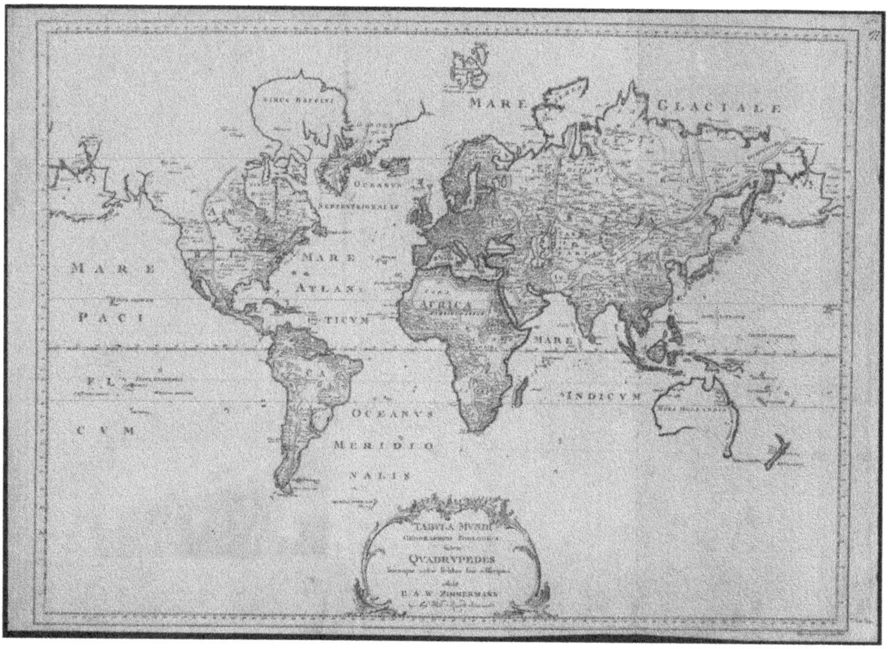

Figure 1.1. *Zimmermann's Tabula mundi geographico zoologica sistems quadrupedes hucusque notos sedibus suis adscriptos, second edition of 1783. Source: National Library of Australia. For a color version of this figure, see www.iste.co.uk/guilbert/biogeography.zip*

Although Zimmermann's zoogeographical classification is basic, it does serve a purpose – to explain distribution and dispersal of quadrupeds. Compare the first of these zoogeographical classifications with a more recent bioregionalization by Morrone (2015, Figure 1.2):

Holarctic kingdom

Nearctic region

Palearctic region

Holotropical kingdom

Neotropical region

Ethiopian region

Oriental region

Austral kingdom

Cape region

Andean region

Australian region

Antarctic region

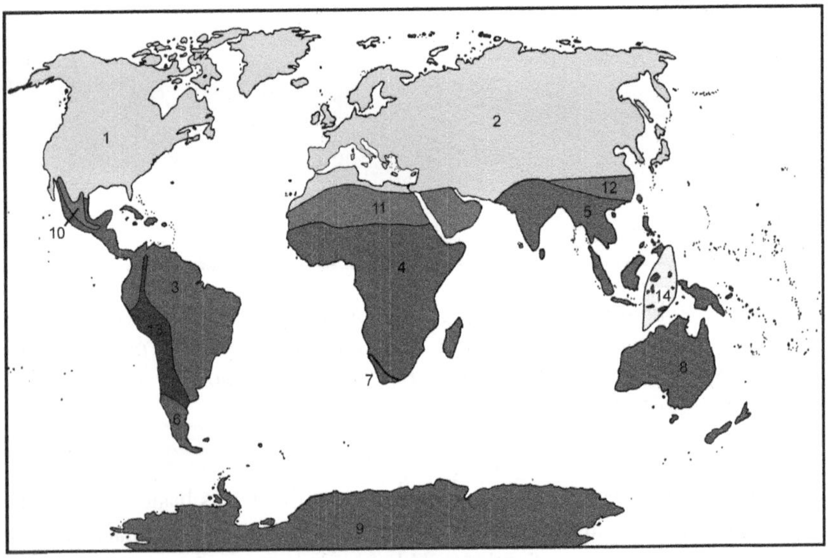

Figure 1.2. *The biogeographic regions of Morrone (2015). Areas in yellow are part of the Holarctic kingdom: 1. Nearctic region; 2. Palearctic region. Areas in red are part of the Holotropical kingdom: 3. Neotropical region, 4. Ethiopian region, 5. Oriental region. Areas in blue are part of the Austral kingdom: 6. Andean region 7. Cape region, 8. Australian region, 9. Antarctic region. Areas in orange and purple are transition zones: 10. Mexican, 11. Saharo-Arabian, 12. Chinese, 13. Andean, 14. Indo-Malayan (Wallacea) (Escalante and Morrone 2020, p. 12, Figure 1.2) For a color version of this figure, see www.iste.co.uk/guilbert/biogeography.zip*

The areas of Morrone (2015) serve a different purpose, namely to be able to study the relationships of these areas. In fact, Zimmermann's and Morrone approaches are completely different methodologically, theoretically and historically; yet, both need a classification in order to be able to communicate their ideas to other plant and animal geographers. In other words, it is impossible to do animal and plant geography without a classification. Classifications, however, do not necessarily overlap as in the case of Zimmermann and Morrone. The classification of Zimmermann is based on the geographical regions of the world in the 18th century (note that Antarctica is missing in the former), while that of Morrone has its historical roots in both the zoogeographic Sclater–Wallacean and phytogeographic Humboldtian tradition. But where Zimmermann and Morrone do overlap is that they use the distributions of named species.

One of the many benefits of doing plant and animal geographies, both in the 18th and 21st centuries, is that plant and animal distributions are available in the form of a database. Zimmermann had access to the many travelogues of explorers of his day such as James Cook, Louis Lahontan and Jan Struys, whereas Morrone had access to recent biogeographical classification (i.e. regionalizations) that had used publicly available digital distribution databases such as GBIF. Practically, these two approaches are the same. They are time and cost-effective (neither had to go into the field and collect and describe new species) and they are quick to access (both had access to libraries of one sort or another). Many plant and animal area classifications during the 18th and 19th were collated via the literature (e.g. Stromeyer 1800; Pritchard 1826). But what if you had the funds to go to an area that was poorly understood and under-collected? What if you had the funds to collect plant specimens and other data? How would you classify the natural history of an area? Alexander von Humboldt faced this problem during his journey to New Granada in present-day Colombia during the late 18th century and solved it in a most ingenious way.

1.2.3. *Botanical geography versus the geography of plants*

The *Essai sur la géographie des plantes* (Humboldt and Bonpland [1805] 1807) and the 30 volume *Le voyage aux régions équinoxiales du Nouveau Continent, fait en 1799, 1800, 1801, 1802, 1803 et 1804* (Humboldt and Bonpland 1814–1829) were the result of a voyage that involved a phenomenal amount of collecting specimens and detailed observations and data collection of astronomical, geological and atmospheric phenomena. In fact, Humboldt amassed enough data to keep himself and Aimé Bonpland occupied until his death in 1859. The problem with collecting a large amount of data is synthesizing it into something useful. The 3,500 species descriptions that would take 15 years to complete was simply too long for someone to make a constructive classification of a fairly unknown area. Humboldt devised something new: a classification based on vegetation. Rather than divide an area based on species distributions, Humboldt used vegetation types, such as "(1) the scitaminales form (*Musa, Pothos* and *Dracontium*); (2) the palms; (3) the tree-ferns", and justifies them:

> These divisions based on physiognomy have almost nothing in common with those made by botanists who have hitherto classified them according to very different principles. Only the outlines characterizing the aspect of vegetation and the similarities of impressions are used by the person contemplating nature, whereas descriptive botany [taxonomy] classifies plants according to the resemblance of their smallest but most essential parts [...]. The absolute beauty of these shapes, their harmony, and the contrast arising from their being together, all this makes what is called the character of nature in various regions (Humboldt and Bonpland 2009, pp. 73–74).

The notion that plant forms can be used to classify entire regions is central to Humboldtian plant geography. Without needing to know what the species are and to which taxonomic groups they belong, botanists can simply observe overall plant forms in order to classify the vegetation type. Compare this with another attempt at botanical geography using what Humboldt calls "very different principles".

In the same year, Augustin Pyramus de Candolle published the third edition of *Flore Française* (de Lamarck and de Candolle 1805), which was accompanied by an unusual map (Figure 1.3). The *Carte Botanique de la France* and a text explaining its function (de Candolle 1805) were "to show the general distribution of plants in France ... The map should be considered more of an attempt to apply a specific methodology rather than an attempt to show the complete plant geography of France'' (de Candolle, cited in Ebach and Goujet 2006, p. 763).

The contrast between Humboldt's and de Candolle's attempt is striking. Whereas Humboldt proposes to use plant form, de Candolle opts to use:

> 1) Temperature, as determined by distance from the equator, height above sea level and southern or northerly exposure.
>
> 2) The mode of watering, which is more or less the quantity of water that reaches the plant, the manner by which water is filtered through

the soil and the matter that is dissolved in the water, which may or may not be harmful to the growth of the plant.

3) The degree of soil tenacity or mobility (de Candolle, cited in Ebach and Goujet 2006, p. 768).

Figure 1.3. *Carte Botanique de France, pour la 3e Edition de la Flore française par A.G. Dezauche fils Ingénieur Hydrogéologue de la Marine an 13 (1805) "Botanical map of France for the 3rd Edition of Flore française by A.G. Dezauche the son, Marine Hydrological Engineer on the 13th year of the Revolution (1805)" (see Ebach and Goujet 2006, Figure 1.1). For a color version of this figure, see www.iste.co.uk/guilbert/biogeography.zip*

The method proposed by de Candolle did not catch on, and by 1820 de Candolle had chosen to use plant distributions instead, dividing the world into 20 regions:

(1) Boreal Asia, Europe, and America; (2) Europe south of the boreal region and north of the Mediterranean; (3) Siberia; (4) the

> Mediterranean area; (5) eastern Europe to the Black and Caspian Seas; (6) India; (7) China, Indochina, and Japan; (8) Australia; (9) south Africa; (10) east Africa; (11) tropical west Africa; (12) Canary Islands; (13) northern United States; (14) northwest coast of North America; (15) the Antilles; (16) Mexico; (17) tropical America; (18) Chile; (19) southern Brazil and Argentina; (20) Tierra del Fuego (de Candolle, cited in Nelson 1978, pp. 283–284).

Again, de Candolle's plant regions failed to find acceptance. Humboldtian Joakim Frederik Schouw dismissed the regions:

> Candolle compares 20 floras, or as he calls them, regions. In his method, which he has developed studying these floras, [Candolle] does not reveal the characteristics that each form takes; it appears that the main basis for the division [of the regions] is current distributions (Schouw 1823, p. 504, my translation).

So too did his son Alphonse de Candolle, who considered "artificial systems", which are a detriment to science "when they are considered to be natural" (de Candolle 1855, pp. 1304–1305). So what then is a natural region?

The Humboldtians believed that both biotic and abiotic factors, such as climate, were vital in recognizing plant forms and plant regions:

> To have an exact acquaintance with these principal forms of vegetation is of the greatest importance to a phyto-geographical division of the globe, as they principally fix the natural physiognomy of different countries. Humboldt is the first who has made such a classification of vegetation, and this must be taken as the foundation of all further inquiry into the subject. It is not until we are somewhat intimately acquainted with the various characteristic forms of plants, that we will be able to recognise the peculiarities of each flora, and to characterise the physiognomy of each country (Meyen 1846, p. 106).

The problem was deciding *which* abiotic factors were crucial. Schouw had a list of factors that he thought were crucial, so did Heer (1835), Meyen (1846), Sendtner (1854), Lorenz (1863), von Marilaun (1863) and Grisebach (1872). Schouw's regions all corresponded to climatic zones (e.g. flora Aplino-arctica) and had a dominant vegetation type (e.g. provincia Cichoriacearum), a practice that the Humboldtians used to define areas. Natural areas (and area classifications) were based on the distributions of vegetation that are driven by climate over time, not dissimilar to the concept of biomes of today. Yet by the end of the 19th century, there was a variety of different area classifications that led Charles Edward Moss to observe that "the subject of ecological plant geography has suffered and still suffers very considerably from a lack of uniformity in the use of its principal terms" (Moss 1910, p. 18).

One idea mentioned by de Candolle (1820) was adopted by the Humboldtians, namely, that of stations and habitations (see Nelson 1978):

> By the term station I mean the special nature of the locality in which each species customarily grows; and by the term habitation, a general indication of the country wherein the plant is native. The term station relates essentially to climate, to the terrain of a given place; the term habitation relates to geographical, and even geological, circumstances … The study of stations is, so to speak, botanical topography; the study of habitations, botanical geography … The confusion of these two classes of ideas is one of the causes that have most retarded the science, and that have prevented it from acquiring exactitude" (de Candolle 1820, p. 383, translated in Nelson 1978, p. 280)[3].

Stations [habitats] and habitations [regions] were the only natural hierarchy that the Humboldtians adopted. The stations were based on physiognomy and the habitations on endemism. Schouw proposed that "at least half the known species are particular [endemic] to a region; 2. that 1/4 of the genera are either fully [endemic] or mostly occur in a region. 3. that single families are either fully [endemic] or mostly occur in a region (Schouw 1823, p. 504, my translation)". These rules were not that dissimilar to those proposed by de Candolle: "(1) every species tends to occupy a certain space, and the determination of the laws that govern species distributions is the study of habitations; as (2) there are more species in the tropics than in high latitudes; (3) the numbers of species of monocots and dicots vary in certain ways; (4) certain numbers of species are recorded for certain countries" (de Candolle 1820, pp. 392–400 in Nelson 1978, p. 281). In this sense, you preserve both classifications: plant taxonomy is used to identify plants and species; genera and families are used to determine the larger habitations; and plant forms and vegetation are used to classify the smaller stations. In *The History of Biology*, Nordenskiöld (1936) summed up plant geography as having taken "two courses", "a systematical, which is ultimately based on Linnæus's observations and theories in

3. Later, Gareth Nelson was to note that "the concepts of station and habitation are important in Candolle's view, for they define two different sciences, which persist into the modern era […]. No matter, the terms as used by Candolle, have modern counterparts: ecological and historical biogeography. Ecological biogeography is the study of stations; historical biogeography, the study of habitations" (Nelson 1978, p. 280, footnote 31, 281).

connexion with the distribution of the plant species, and a morphological [physiognomic], which has its origins in Humboldt's theories on the morphological association of different vegetable types with different countries and forms of landscape". Of the latter, Nordenskiöd states, "œcological plant geography does not investigate the nature of the flora, but of the vegetation. It works not with species, but with plant communities [populations]" (Nordenskiöld 1936, pp. 560–561). I will return to the plight of plant geography at the start of the 20th century later.

1.2.4. *Zoogeography: a search for natural regions*

Animal geography had a later start than plant geography. Although Zimmermann (1778–1783) was the first to consider an animal geography, it was confined to quadrupeds. Unlike the Humboldtians and de Candolle, animal geographers rarely looked at faunal regions, instead preferring to look at taxon-specific distributions. Also, a contemporary of Zimmermann, Johan Christian Fabricius, proposed eight climatic regions "from which the Stations of insects are judged" (Fabricius 1778, p. 154). Zoologists did not adopt the Humboldtian tradition of using "form" and dismissed the climatic regions of Fabricius as arbitrary or artificial:

> This simple statement is enough to convince us that there is a lot of arbitrariness in these divisions (Latreille 1815, pp. 40–41, my translation).

> [Fabricius] … by not attempting to demonstrate the correctness of any one of his divisions, seems to have subsequently abandoned them altogether, since no one, it may be fairly presumed, was more qualified than himself to discover the artificial nature of his theory (Swainson 1835, pp. 10–11).

Similarly, the regions proposed by Pierre André Latreille in 1817, based on latitudinal and longitudinal gradients along climatic zones, were equally dismissed:

> Any division of the globe into climates, by means of equivalent parallels and meridians, wears the appearance of an artificial and arbitrary system, rather than to one according to nature (Kirby and Spence 1828, p. 487).

Entomologists William Swainson, William Kirby and William Spence were insistent on the necessity of defining natural regions, namely "those grand divisions of animal geography pointed out by nature, and immediately recognized by every naturalist" (Swainson 1835, p. 11). Swainson places William Sharp Macleay and Humboldt among those who recognize that natural areas are not "regulated by isothermal lines" (Swainson 1835, p. 12). The famous *Tableau Physique* of Humboldt, depicting mounts Chimborazo and Cotopaxi in the Andes, were drawn in cross-section in order to *highlight* isothermal lines[4] (Figure 1.4). We can imagine that Swainson considered all lines, be they isothermal lines or latitudinal lines, as artificial. Yet, isothermal lines were quite popular with plant and animal geographers in delimiting climatic zones. Given, however, that climate and plant forms define vegetation, we could argue that the Humboldtians would consider such lines to portray natural areas. After all "only those vegetative formations deserve to be recognised as independent plant forms, which conform to the influence of climate" (Grisebach 1866, p. 384).

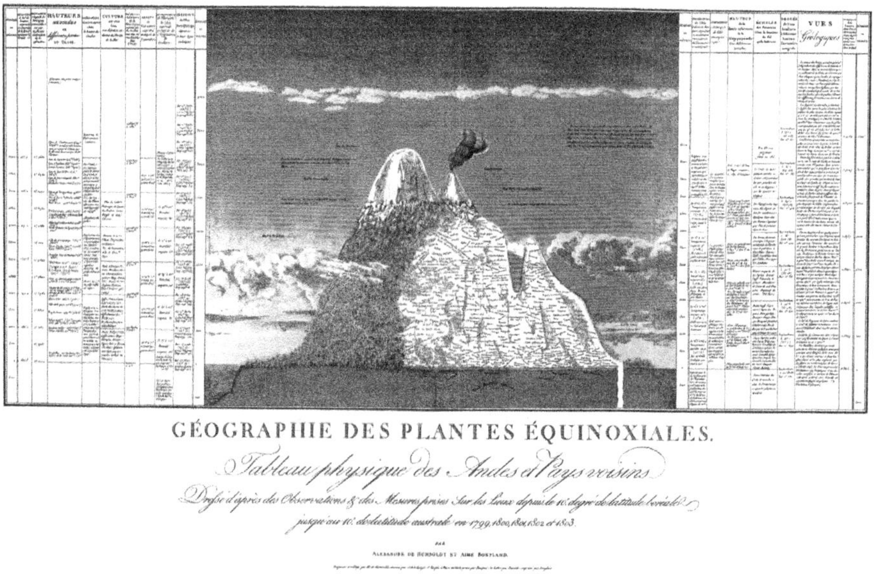

Figure 1.4. *Humboldt's Tableau physique showing a cross-section of Mount Chimborazo and Mount Cotopaxi in the Andes. The full title of the map reads:* Geographie des plantes equinoxiales : tableau physique des Andes et Pays voisins. Dressé d'après des observations et des Mesures prises sur les lieux depuis le 10.degré de l'attitude australe en 1799, 1800, 1801, 1802 et 1803 *(in Humboldt and Bonpland 1807) (source: http://cybergeo.revues.org/docannexe/image/25478/img-7.jpg). For a color version of this figure, see www.iste.co.uk/guilbert/biogeography.zip*

4. The Tableau lacks the actual lines, but instead has a table on either side of the cross-section depicting the temperatures at elevation. Essentially, Humboldt has created a sophisticated isothermal line.

While some zoologists would adopt a more Humboldtian approach to delimiting areas using climate, such as the homoiozoic belts of Forbes (1856) or the life zones of Merrian (1892), others sought to look only at the distribution of species. An earlier attempt at using animal distributions to define animal regions was proposed by Johann Karl Wilhelm Illiger in 1815. Illiger (1815) attempted to summarize the global geographical distribution of mammals by counting the number of species that occurred in each continent. Much of Illiger (1815) are tables listing the names of genera, families and orders and the numbers of species found in each continent. The work is completely taxonomic, devoid of any measurements of temperature or rainfall, and synthetic, as it had collated names of species from the works of others. The body of work is divided into sections or "comparative summaries" based on each of the tables, including a description of the faunal distribution of a list of the taxa. A modern-day biogeographer would be impressed with the volume of data but perplexed with how little Illiger had done with it. German paleontologist Johann Andreas Wagner synthesized Illiger (1815) into four mammalian provinces in a three-part work published between 1844 and 1846 (Wagner 1844–1846). The work also included the first known global biogeographic map (Figure 1.5). Wagner was also the first to use a hierarchical classification of zones, provinces, sub-provinces and regions (Table 1.1). More important, Wagner considered these divisions to be natural. In staying with Illiger's style of only listing the distributions of mammals, Wagner stands out in zoological and botanical geography by ignoring abiotic factors such as climate. Wagner's contemporary Ludwig Schmarda's *Distribution of Animals* (Schmarda 1853), for example, discusses the influence of climate, water and temperature, as well as denoting "tropical forms" and the interaction with vegetation. Schmarda's map of the geographical distribution of animals is possibly the most detailed of any zoogeographical study in the 19th century. The map depicts 21 terrestrial and 10 oceanic areas and the distributions of various taxa, as well as the location of reefs, atolls, the direction of ocean currents, isothermal lines and latitude and longitude. Unlike Wagner's classification, Schmarda's was non-hierarchical and partially based on climate. Unfortunately, Wagner's work was largely ignored and those who did notice it never quite fully appreciated its significance. The nearest zoogeographical work to resemble Wagner's appeared a decade later. Philip Lutley Sclater, who used a similar method to Wagner, namely counting taxa, proposed a hierarchical classification. Sclater, however, offered something new:

> In the Physical Atlases lately published, which have deservedly attracted no small share of attention on the part of the public, too little regard appears to have been paid to the fact that the divisions of the earth's surface usually employed are not always those which we most

natural when their respective Faunæ and Flora are taken into consideration. The world is mapped out into so many portions, according to latitude and longitude, and an attempt is made to give the principal distinguishing characteristics of the Fauna and Flora of each of these divisions; but little or no attention is given to the fact that two or more of these geographical divisions may have much closer relations to each other than to any third, and, due regard being paid to the general aspect of their Zoology and Botany, only form one natural province or kingdom (as it may perhaps be termed), equivalent in value to that third (Sclater 1858, pp. 130–131).

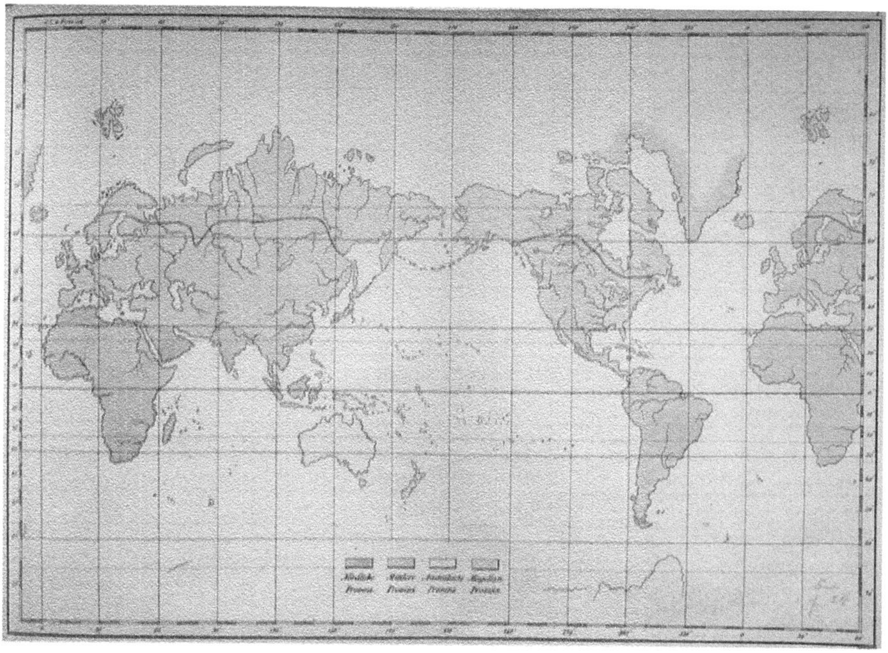

Figure 1.5. *"Representation of the distribution of mammals according to their zones and their provinces. The southern boundary of the northern polar province is indicated by a line of a different color, drawn somewhat further south than the equatorial border of the arctic fox ([Vulpes] lagopus), though not so far in some places as the reindeer may descend there on their summer migrations. The southern polar province is not included in this map, because it is only in the process of discovery and, according to all previous experience, it does not harbour land mammals" (Wagner 1844, 241, Table 1.1). For a color version of this figure, see www.iste.co.uk/guilbert/biogeography.zip*

Zone	Province	Sub-Province	Region
Northern	Polar	Europe	
		Nowaja Semlja	
		Siberia	
		America	
		Greenland	
	Old World	a. Middle Europa	
		b. South Siberia	
		c. Binnenmeerisches Steppes	
		d. Mediterranean Basin	
		e. High Asia	
		f. Japan	
	North America	Temperate North America	
Middle or Tropical	South Asia	–	
	Africa	South Africa	
		West Africa	
		Madagascar	
	Tropical America	I. Western Slopes	a) Coastal region
			b) Western Sierra region
			c) Cordilleran region
		II. Eastern Slopes	d) Pana region
			e) Eastern Sierra region
			f) Forest region
Southern	Australian	Southeast Australia	
		South Australia	
		Southwest Australia	
		Northwest and North Australia	
		Van Diemansland	
	Magellanian	Pamas	
		Patagonian	
		Chilean	
	Southern Polar	–	

Table 1.1. *The hierarchical classification of zones, provinces, sub-provinces and regions listed in Wagner (1844–1846). Note that Wagner considered a third Southern Polar province, but omitted it partly because "we know too little about it and partly since it has no land-animals, and the marine mammals for the most part the same ones are found on the coasts of South America, South Africa, and Australia" (Wagner 1844, p. 86)*

Sclater mentions the relationships of natural areas, a concept that was to appear over a century later (see Hennig 1950, 1966; Brundin 1966), but was never explored further in his 1858 essay. Sclater's regions were adopted by Wallace (1876), who shared similar sentiments regarding arbitrary lines:

> The divisions in use till quite recently were of two kinds; either those ready made by geographers, more especially the quarters or continents of the globe; or those determined by climate and marked out by certain parallels of latitude of by isothermal lines. Either of these methods was better than none at all; [but] it will be evident, that such divisions must have often been very unnatural, and have disguised many of the most important and interesting phenomena which a study of the distribution of animals presents to us … The merit of initiating a more natural system, that of determining zoological regions, not by any arbitrary or a priori consideration but by studying the actual ranges of the more important groups of animals, is due to Mr. Sclater (Wallace 1876, vol. 1, pp. 52–53).

Wallace's revision of Sclater's regions is perhaps the most significant of all the geographical classification of the 19th century as it unified zoogeography under a single classification. Even though there were certain disagreements over terminology (see Ebach 2015), the areas have withstood the test of time, with the same divisions appearing in the 21st century studies (e.g. Holt *et al.* 2013; Morrone 2015). Regardless of its popularity today, the Sclater–Wallacean areas and the whole notion of topographical zoogeography were challenged as "essentially static" and "wrong". "Instead of thinking of fixed regions, it is necessary to think of fluid faunas" (Mayr 1946, p. 5). For the newly developing field of population genetics and the Modern Synthesis, "zoogeography has had a similar fate very much like taxonomy. It was flourishing during the descriptive period of biological sciences. Its prestige, however, declined rapidly" (Mayr 1944, p. 1). Taxonomy apparently had run its course. Long live populations!

1.3. Ecology versus taxonomy: populations not species

The "two courses" of plant classification did not sit well with early ecologists. Linnaean taxonomy and species were considered to be arbitrary and artificial, while vegetations and plant forms were considered to be natural. Ecologists Eugenius Warming, Andreas Schimper, Frederic Clements and Henry Chandler Cowles were incredibly wary of taxonomic description, something Clements (1905, p. 11) decried as “vague descriptive articles” (Hagen 1986). Cowles warned:

> Taxonomy must be scientific. It must require for its devotees a training as rigid as that required by professional workers in morphology, physiology or ecology. Species-making by taxonomic tyros must be abandoned ... These things will not, be endured much longer; a little more and the sinning taxonomists will be cast out into the outer darkness where there shall be wailing and gnashing of teeth (Cowles 1908, pp. 270–271).

Were ecologists truly enraged with taxonomy, or were they after recognition for their newly developing field? Hagen (1986) proposed that early ecologists such as Clements were attempting to legitimize their discipline by distancing it from amateur botany “and to place it on a credible scientific basis” (Hagen 1986, p. 200). One way to draw attention to this is by defining their “new discipline in opposition to what they believed was a moribund, nineteenth-century, natural-history tradition” (Hagen 1986, p. 213). The move towards communities of plant forms, rather than species, was not unique to early 20th century ecologists. Neither was the notion that descriptive taxonomy and zoogeography were static and moribund. Species are essentially hypothetical, in the sense that a set of diagnostic characteristics are given a name and assigned to various types. Organisms that show these diagnostic characteristics are then assigned that name. Plant or life forms, however, were based on the traits of a community or a population. Moreover, populations were observable, that is, quantifiable, rather than abstract in the way species are defined. Populations changed over time, as they evolved new traits, and showed new adaptations. Populations would form faunal elements, thereby relinquishing the need for formal areas. You can have Old World elements in the New World, populations that disperse and establish themselves as new populations in new areas. Populations could also form refuges in cases of changes in climate or landscape, such as advancing ice caps. At the population level, so many more current geographic and climatic events can be used to explain the distribution of populations. Yet, the area classification remained; only the way plant and animal geographers define their areas varied. Mayr, who thought the Sclater–Wallacean areas were static, “descriptive, essentially regional, and non-dynamic” (Mayr 1946, p. 4), proposed a “Classification of faunal elements of the Americas” (Mayr 1946, p. 11), in which the elements were defined as “Old World”, “Holarctic” and “Pantropical”. These are the same static areas Mayr denounced earlier in the same paper. The only difference is that the subregions are treated as "faunal elements", that is, populations, which are assumed to have dispersed between the larger regions.

Plant communities, populations or faunal elements are simply ways in which we describe the smallest unit of classification. We may speak of a population of koalas, but koalas (*Phascolarctos cinereus*) have to be identified through a unique or diagnostic set of characteristics. Koalas are diprotodontids and share the characteristics of other diprotodontids, such as kangaroos. Kangaroos and koalas are marsupials, such as the American opossum, and kangaroos, koalas and opossums are mammals. Without such a classification, we would not be able to identify a population or various individuals within a community. The "two courses", namely vegetation and species (flora) classifications, which were so prevalent in plant geography and classification during the early 20th century, did unify by the 1940s. In 1947, Ronald Good proposed a "classification of the world into floristic units" (Good 1965, pp. 30–32), which was updated by Takhtajan *et al.* (1986). The hierarchical classification, divided up into kingdoms, subkingdoms and regions and subregions, is reminiscent of the Sclater–Wallacean regions (i.e. Neotropical, Indo-Malaysian, Australian). The smaller subregions are based on climate (e.g. Central Deserts) or geopolitical or geographical areas (e.g. Borneo, Mexican Highlands) and seem to resemble plant communities. Good seems to have combined the larger regions with the smaller plant communities into a single classification. In zoogeography, the Scalterian–Wallacean regions prevailed into the 21st century. Early objections to a "static" area classification never challenged how the classification functions, but rather how areas are defined. Ortmann (1902) was very particular on how areas should be defined:

> 1) Any division of the earth's surface into zoogeographical regions which starts exclusively from the present distribution of animals, without considering its origin, must be unsatisfactory, since always only certain cases can be taken in while others remain outside of this scheme.
>
> 2) Considering the geological development of the distribution of animals, we must pronounce it impossible to create any scheme whatever that covers all cases.
>
> 3) Under these circumstances it is incorrect to regard the creation of a scheme of animal distribution as an important feature or purpose of zoogeographical research (Ortmann 1902, pp. 269–270).

There are multiple ways to define areas; however, no one method can claim that it finds *natural* areas; this is simply assumed. Regardless of how we define areas, an area classification is always needed in order to communicate what it is we are trying to convey.

The Sclater–Wallacean areas are still in use today, and various authors using geospatial methods have identified similar classifications to that of Sclater and Wallace (Figure 1.6) using different models (Kreft and Jetz 2010; Proches and Ramdhani 2012; Holt *et al.* 2013; Figures 1.7–1.9). If we look at these three different studies using different data, methods and theories, we find that the same Sclaterian–Wallacean areas keep appearing. While these approaches have different origins in their ideas and methods, the practice of looking at and proposing area classification has not changed since (Zimmermann 1777).

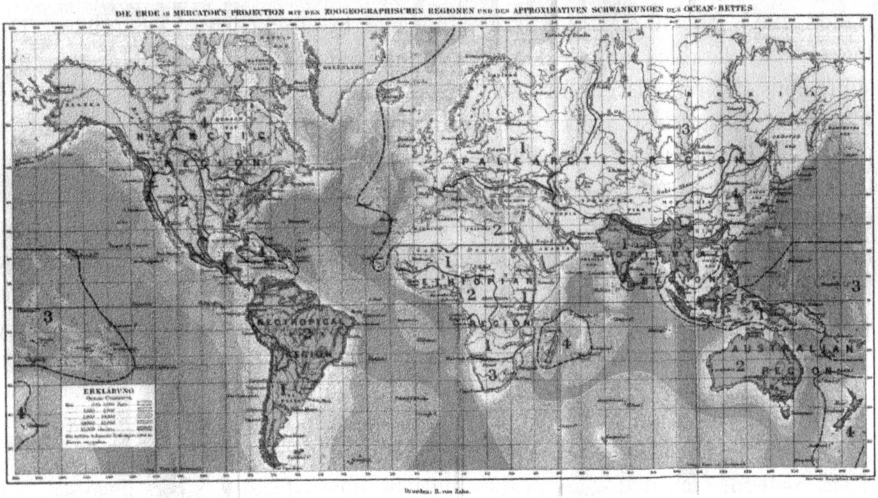

Figure 1.6. *"Map of the World, showing the Zoo-Geographical Regions and the contour of the Ocean-bed" (Wallace 1876, frontispiece). For a color version of this figure, see www.iste.co.uk/guilbert/biogeography.zip*

Let's return to Goethe's observation that "the history of science is science itself". In the case of area classification, we see that biogeographers keep doing the same thing, proposing area classifications, but using very different and independent approaches. If we look at scientific theories and methodologies, we find multiple origins in biogeography. But biogeographic practice, namely area classification, seems to constantly reinvent itself. The history of science is truly science itself.

Figure 1.7. *"The six major biogeographical divisions are highlighted in the dendrogram with large coloured rectangles: orange, Australian; red, Neotropical; brown, African; yellow, Oriental; blue, Palaearctic; green, Nearctic. The first 30 groups in the dendrogram (small rectangles) and in the map are displayed in different colours. Additionally, the first 60 groups are indicated with black boundaries in the map" (Kreft and Jetz 2010, Figure 1.9). For a color version of this figure, see www.iste.co.uk/guilbert/biogeography.zip*

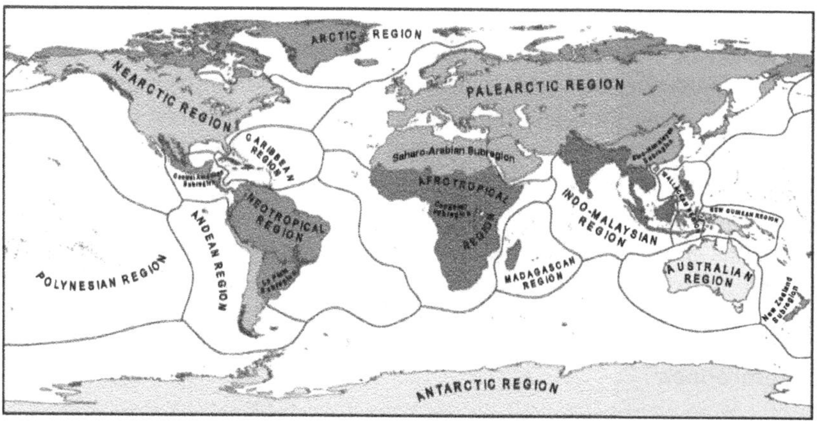

Figure 1.8. *Vertebrate zoogeographical regions and subregions (Proches and Ramdhani 2012, Figure 1.2). For a color version of this figure, see www.iste.co.uk/guilbert/biogeography.zip*

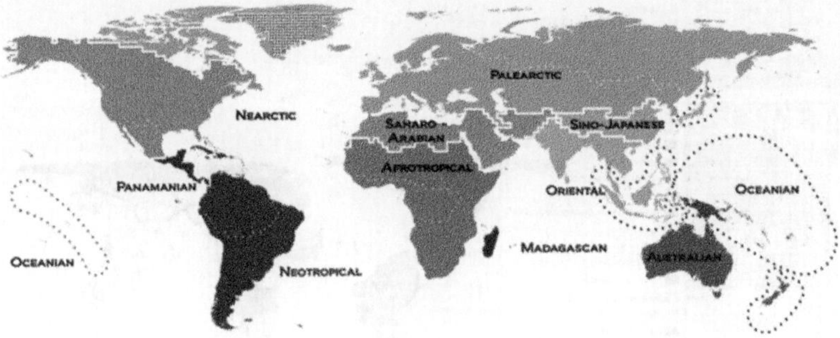

Figure 1.9. *"Map of the terrestrial zoogeographic realms and regions of the world" (Holt* et al. *2013, Figure 1.1). For a color version of this figure, see www.iste.co.uk/guilbert/biogeography.zip*

1.4. Conclusion

It is impossible to practice biogeography without an area classification, whether those areas are simply geopolitical, geographical or based on endemism or taxonomic distribution. The plant and animal geographies of the 18th, 19th and early 20th centuries have shown us that no matter how you divide up the world, you have to call your areas *something*. How plant and animal biogeographers define their areas also depends on the background of the authors. Humboldt, a naturalist, wished to use plant form and vegetation; de Candolle, a taxonomist (in fact, he coined the term taxonomy), used the distributions of species in what was later described as topographical plant geography; Schouw, following in the Humboldtian tradition, used vegetation; Alphonse de Candolle, a systematist, preferred to use endemism; Mayr, an evolutionary biologist, wanted to use populations or elements. Regardless, they were all doing the same thing, area classification, in order to express their theories of the world using a variety of different methods. Yet, area classification seemingly had its detractors, namely those who thought it was static or artificial. These objections were mostly about how areas are defined, rather than area classification *per se*. Area classification will always be practiced as long as there is a study of the distribution of organisms.

1.5. References

Brundin, L. (1966). Transantarctic relationship and their significance, as evidenced by chironomid midges. With a monograph of the subfamilies Podonominae and Aphroteniinae and the austral Heptagyiae. *Kungliga Svenska Vetenskapsakademiens Handlingar*, 11, 1–472.

de Candolle, A.P. (1805). Explication de la carte botanique de la France. In *Flore française, ou descriptions succinctes de toutes les plantes qui croissent naturellement en France : disposées selon une nouvelle méthode d'analyse, et précédées par un exposé des principes élémentaires de la botanique*, 3rd edition, de Lamarck, J-B.P.A. and de Candolle, A.P. (eds). Desray, Paris.

de Candolle, A.P. (1820). Essai élémentaire de géographie botanique. In *Dictionnaire des Sciences Naturelles*, Levrault, F.G. (ed.). Paris.

de Candolle, A.L.P.P. (1855). *Géographie botanique raisonnée.* Masson, Paris.

Clements, F. (1905). *Research Methods in Ecology*. University Publishing Company, Lincoln, NE.

Cowles, H.C. (1908). An ecological aspect of the conception of species. *The American Naturalist*, 42, 265–271.

Duck, M. and Petry, M. (2016). *Goethe's Exposure of Newton's Theory: Apolemic on Newton's Theory of Light and Colour*. Imperial College Press, London.

Ebach, M.C. (2015). *Origins of Biogeography – The Role of Biological Classification in Early Plant and Animal Geography*. Springer, New York, NY.

Ebach, M.C. and Goujet, D. (2006). The first biogeographical map. *Journal of Biogeography*, 33, 761–769.

Escalante, T. and Morrone, J.J. (2020). Evolutionary biogeography and the regionalization of the Neotropics: A perspective from the mammals. *Mastozoología Neotropical*, 27, 4–14.

Fabricius, J.C. (1778). Philosophia entomologica. *Impensis Carol.* Ernest, Bohnii, Hamburg.

Forbes, E. (1856). Map of the distribution of marine life, illustrated chiefly by fishes, molluscs and radiata: Showing also the extent and limits of the homoiozoic belts. In *The Physical Atlas of Natural Phenomena (Plate 31)*, Johnston, A.K. (ed.). William Blackwood and Sons, Edinburgh.

Good, R. (1965). *The Geography of Flowering Plants*. Longmans, Green and Co., New York, NY.

Grisebach, A.H.R. (1866). Der gegenwärtige Standpunkt der Geographie der Pflanzen. *Geographisches Jahrbuch*, 1, 373–402.

Grisebach, A.H.R. (1872). *Die Vegetation der Erde nach ihrer klimatischen Anordnung*. Wilhelm Engelmann, Leipzig.

Hagen, J. (1986). Ecologists and taxonomists: Divergent traditions in twentieth-century plant geography. *Journal of the History of Biology*, 19, 197–214.

Haas, O. (1954). A geneticist on the taxonomic method. *Systematic Zoology*, 3(2), 65.

Heer, O. (1835). *Die Vegetationsverhältnisse des sudöstlichens Theils des Cantons Glarus; ein Versuch, die pflanzengeographischen Erscheinungen der Alpen aus climatischen und Bodenverhältnissen abzuleiten*. Orell, Fiissli, Zurich.

Hennig, W. (1950). *Grundzüge einer Theorie der phylogenetischen Systematik.* Deutsche Zentralverlag, Berlin.

Hennig, W. (1966). *Phylogenetic Systematics.* University of Illinois Press, Urbana, IL.

Holt, B.G., Lessard, J., Borregaard, M.K., Fritz, S.A., Araújo, M.B., Dimitrov, D., Fabre, P.H., Graham, C.M., Graves, G.R., Jønsson, K.A., Nogués-Bravo, D., Wangm Z., Whittaker, R.J., Fjeldså, J., Rahbek, C. (2013). An update of Wallace's zoogeographic regions of the world. *Science*, 339, 74–78.

von Humboldt, A. and Bonpland, A. (1807). *Voyage de Humboldt et Bonpland. Première partie. Physique Générale, et relation historique du voyage. Premier Volume, Contenant Essai sur la Géographie des plantes, accompagné d'un Tableau physique des régions équinoxiales, et servant d'introduction à l'Ouvrage*. Chez Fr. Schœll, Paris.

von Humboldt, A. and Bonpland, A. (1814–1829). *Personal Narrative of Travels to the Equinoctial Regions of the New Continent, During The Years 1799–1804.* Longman, Rees, Orme, Brown, and Green, London.

von Humboldt, A. and Bonpland, A. (2009). *Essay on the Geography of Plants.* University of Chicago Press, Chicago, IL.

Illiger, J.K.W. (1815). Überblick der Säugthiere nach ihrer Vertheilung über die Welttheile. *Abhandlungen der physikalische Klasse der Koeniglich-Preussischen Akademie der Wissenschaften, 1804–1811*, 39–159.

Jordan, H. (1883). Zur Biogeographie der nördlich gemäßigten und arktischen Länder. *Biologisches Centralblatt*, 3(174–180), 207–217.

Kirby, W. and Spence, W. (1828). *An Introduction to Entomology or the Elements of the Natural History of Insects*, Volume 4. Longman, Rees, Orme, Brown and Green, London.

Kreft, H. and Jetz, W. (2010). A framework for delineating biogeographical regions based on species distributions. *Journal of Biogeography*, 37, 2029–2053.

de Lamarck, J-B.P.A. and de Candolle, A.P. (eds) (1805). *Flore française, ou descriptions succinctes de toutes les plantes qui croissent naturellement en France : disposées selon une nouvelle méthode d'analyse, et précédées par un exposé des principes élémentaires de la botanique*, 3rd edition. Desray, Paris.

Latreille, P.A. (1815). Introduction à la géographie générale des arachnides et des insectes : ou des climats propres à ces animaux. *Mémoire lu à l'Académie des Sciences*, 3, 37–67.

Lorenz, J.R. (1863). Physicalische Verhältnisse und Vertheilung der Organismen im Quarnerischen Golfe. *Die kaiserlich-königliche Hof- und Staatsdruckerei.* Vienna.

Mayr, E. (1944). Wallace's line in the light of recent zoogeographic studies. *Quarterly Review of Biology*, 19, 1–14.

Mayr, E. (1946). History of the North American bird fauna. *Wilson Bulletin*, 58, 3–41.

Medawar, P. (1968). *The Art of the Soluble.* Metheun and Co. Ltd., London.

Merriam, C.H. (1892). The geographical distribution of life in North America with special reference to the Mammalia. *Proceedings of the Biological Society of Washington*, 7, 1–64.

Meyen, F.J.F. (1846). *Outlines of the Geography of Plants*. Ray Society, London.

Morrone, J.J. (2015). Biogeographical regionalisation of the world: A reappraisal. *Australian Systematic Botany*, 28, 81–90.

Moss, C.E. (1910). The fundamental unites of vegetation. *New Phytologist*, 9, 18–53.

Nelson, G. (1978). From Candolle to Croizat: Comments on the history of biogeography. *Journal of the History of Biology*, 11, 269–305.

Nordenskiöld, E. (1936). *The History of Biology: A Survey*, Leonard Bucknall Eyre (translator). Tudor, New York, NY.

OED (2020). *Biogeography*. Oxford University Press, Oxford [Online]. Available at: https://www.oed.com [Accessed February 11, 2021].

Ortmann, A.E. (1902). The geographical distribution of freshwater decapods and its bearing upon ancient geography. *Proceedings of the American Philosophical Society*, 41, 267–400.

Prichard, J.C. (1826). *Researches Into the Physical History of Mankind*, 2nd edition. Houlfton and Stoneman, London.

Proches, S. and Ramdhani, S. (2012). The world's zoogeographical regions confirmed by cross-taxon analyses. *BioScience*, 62, 260–270.

Schmarda, K.L. (1853). *Die geographische Verbreitung der thiere*. Carl Gerold and Son, Vienna.

Schouw, J.F. (1823). *Grundzüge einer allgemeinen Pflanzengeographie*. Reimer, Berlin.

Sclater, P.L. (1858). On the general geographical distribution of the members of the class Aves. *Journal of the Proceedings of the Linnean Society: Zoology*, 2, 130–145.

Sendtner, O. (1854). *Die Vegetationsverhältnisse Südbayerns nach den Grundsätzen der Pflanzengeographie und mit Bezugnahme auf Landeskultur*. Geschenk des Verfassers, München.

Stromeyer, F. (1800). *Commentatio inauguralis sistens historiae vegetabilium geographicae specimen*. Heinrich Dieterich, Göttingen.

Swainson, W. (1835). *A Treatise on the Geography and Classification of Animals*. Longman, Brown, Green, and Longmans, London.

Takhtajan, A., Crovello, T.J., Cronquist, A. (1986). *Floristic Regions of the World*. University of California Press, Berkeley, CA.

Von Marilaun, A.K. (1863). *Das Pflanzenleben der Donaulander*. Wagner, Innsbruck.

Wagner, A. (1844–1846). Die geographische Verbreitung der Saugethiere. Abhandlungen der Königlich Bayerischen Akademie der Wissenschaften. *Mathematisch-Physikalische Klasse*, 4(1), 1–146; (2), 37–108; (3), 1–114.

Wallace, A.R. (1876). *The Geographical Distribution of Animals: With a Study of the Relations of Living and Extinct Faunas as Elucidating the Past Changes of the Earth's Surface*. Macmillan, London.

Zimmermann E.A.W. (1777). *Specimen zoologiae geographicae, Quadrupedum domicilia et migrationes sistens*. Theodorum Haak, Leiden.

Zimmermann E.A.W. (1778–1783). *Geographische geschichte des menschen, und der allgemein verbreiteten vierfüssigen thiere*, 3 Volumes. Weygandschen Buchhandlung, Leipzig.

2

Analytical Approaches in Biogeography: Advances and Challenges

Isabel SANMARTÍN
Real Jardín Botánico, CSIC, Madrid, Spain

2.1. Introduction

The last decades have seen an explosion of analytical approaches in biogeography. Amid the plethora of new methods competing for attention, a researcher is expected to get lost. From parsimony-based *cladistic* and *event-based* biogeography, we have moved into the expanding world of *parametric* model-based methods. This chapter mainly focuses on the latter, which are less than a decade old, but I also review previous approaches, as they provide a background on the shifting focus from phylogenetic relationships and Earth history to the integration of other disciplines (ecology, paleontology and population genetics), to understand historical processes that shaped Earth's biodiversity.

2.2. From narrative dispersal accounts to event-based methods (EBM)

As seen in Chapter 1, the introduction of the idea of evolution (Darwin 1859) provided early biogeographers with an explanation of why geographic regions

Biogeography,
coordinated by Eric GUILBERT.

sharing similar environments harbor different biotas. Since continents did not move their positions over geological time, it must be organisms that moved over Earth's surface to achieve their present distribution. These early biogeographic reconstructions were mostly narrative dispersal accounts. A paradigm changed in the mid-twentieth century with the introduction of the concept of "plate tectonics" (the 1960s). If a global process is responsible for the current distribution of biodiversity on Earth, we should see its effects in the form of congruence in biogeographic relationships across species. This general process was termed vicariance. The first analytical school, cladistic biogeography, aimed to find general patterns of relationships among areas of endemism, indicating shared biogeographic history ("area cladograms", Nelson and Platnick 1981). Dispersal and extinction were considered processes that depend on biological characteristics intrinsic to the species and, therefore, cannot generate shared distribution patterns (Humphries and Parenti 1999). Cladistic biogeographic methods are allegedly process-free: inference of the area cladogram is done with no consideration to the biogeographical events that may have generated the pattern. If any, these are inferred *a posteriori* through comparison of the area cladogram with the individual species patterns (Brooks 2005). Uncoupling the inference of biogeographic patterns from the underlying evolutionary processes made it difficult to compare alternative biogeographic scenarios (Sanmartín 2012).

The next biogeographic school was "event-based biogeography" (EBM, Ronquist 1997, 2003). Biogeographic processes or events are tied to weights or "costs", and the analysis consists of finding the pattern of area relationships with the minimum cost in terms of these processes. Four biogeographic events are considered in EBMs (Figure 2.1): vicariance, duplication, dispersal and extinction. The last two are tied to a speciation event and have also been termed "partial dispersal", or "sorting, extirpation, and range contraction" for partial extinction. Within dispersal, we may distinguish "jump dispersal", where a lineage migrates from one area to another (A to B) followed by speciation, and "range expansion", where a lineage expands its range, leading to a temporally widespread distribution (A to AB); the latter is termed "geodispersal" when it affects multiple lineages (Lieberman 2003). Two biogeographic events are not considered in EBMs because they leave no observable traces in the phylogeny, that is, no descendants survive in the ancestral range (Sanmartín 2012): "full dispersal", colonization of an area that is not followed by speciation, and "full extinction", when the lineage entirely disappears from its ancestral range, that is, lineage extinction (Figure 2.1).

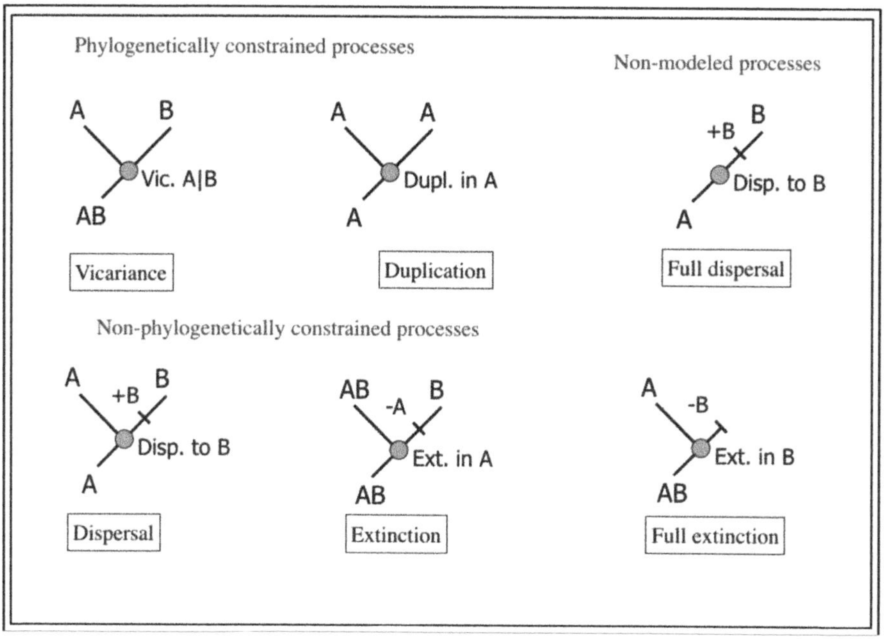

Figure 2.1. *Biogeographic processes. Four types of biogeographic processes are considered in event-based biogeography: vicariance (allopatric speciation in response to a general dispersal barrier affecting multiple lineages); duplication (speciation within the area, i.e. sympatry, or allopatric speciation in response to a temporary dispersal barrier); extinction (the disappearance of the lineage from part of its ancestral range); dispersal (colonization of a new area by crossing a pre-existent barrier). Two processes: full dispersal and full extinction (right) cannot be modeled by EBMs because they leave no observable traces in the phylogeny*

2.2.1. *Parsimony-based tree fitting*

The two most popular EBMs are parsimony-based tree fitting and dispersal-vicariance analysis (Ronquist and Sanmartín 2011). Parsimony-based tree fitting, implemented in the software TreeFitter, was born from methods used in host–parasite tree reconciliation, with which it shares many similarities (Ronquist 2003). In tree fitting, a taxon cladogram is fitted onto the area cladogram by searching for the sequence of events that explain the tip distributions according to the area cladogram and with the minimum cost. The area cladogram may be a hypothesis of relationships based on geological history, in which case we measure how much the observed distributions depart from geologically predicted vicariance (Sanmartín and Ronquist 2004). Alternatively, it may be an unknown parameter, in which case, tree fitting consists of finding the area relationships and the sequence of biogeographic events that explain the tip distributions in the phylogeny with the minimum cost. Searching for the optimal (minimum-cost) area-cladogram implies enumerating and fitting all possible hierarchical combinations of areas. If the number of areas is large, it is possible to use heuristic search tools such as those employed in parsimony phylogenetics (nearest-neighbor interchange, branch and bound, etc.). In Figure 2.2(a), the distribution of species 1– 3 in area A, which is the sister in the area cladogram to the clade BCD (Figure 2.2(b)), is explained by a sequence of events involving duplication of a widespread ancestor in ABCD, extinction of the ancestor of species 1 in part of this range (BCD), successive vicariance events, and dispersal of the ancestor of species 3 and 4 to A, followed by allopatric speciation (Figure 2.2(c)).

From the description above, it can be deduced that the most important problem in EBMs is to find the optimal cost assignments. The most common criterion is to select event costs that maximize the conservation of distribution ranges along the phylogeny (Ronquist 2003). Figure 2.1 shows that dispersal and extinction are not "phylogenetically conserved or constrained" processes because they interrupt the "vertical inheritance" of geographic ranges from ancestor to descendants. In dispersal, the colonized area B is not part of the ancestral range (Figure 2.1(c)); in extinction, part of the ancestral range (A) is lost in the right descendant (Figure 2.1(d)). Conversely, vicariance and duplication are phylogenetically constrained processes because either each descendant inherits the entire ancestral range (duplication) (Figure 2.1(b)) or the union of the two descendants' ranges equals the ancestral range (vicariance, Figure 2.1(a)). A consequence of this cost assignment is that the frequency of dispersal and extinction events is minimized relative to vicariance and duplication in EBM reconstructions. A similar phylogenetic conservation criterion is used in parsimony-based inference to minimize homoplasies (convergence and parallelism), as evolutionary changes that are not identical by descent, that is, losses and gains of traits in unrelated lineages. In the TreeFitter reconstruction in Figure 2.2(c), extinction (e) receives a cost of 1 and dispersal (i) a cost of 2; vicariance (v) and duplication (d) are given minimum costs (0.01); the lower cost of extinction relative to dispersal is due to extirpation preserving part of the ancestral range (Figure 2.1; Sanmartín and Ronquist 2004).

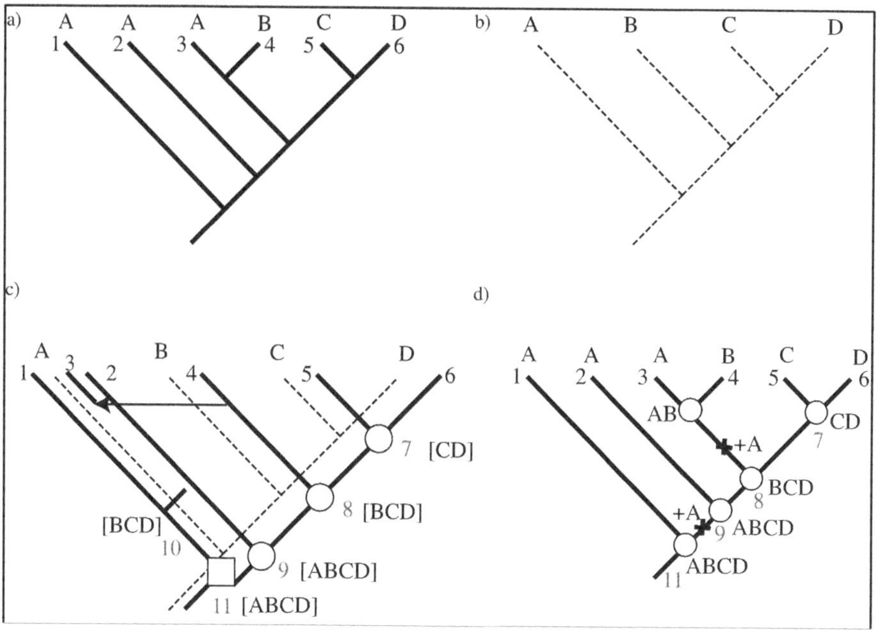

Figure 2.2. *Event-based biogeography. a) An organism phylogeny with six species (1–6) distributed in four areas (A–D). b) Area cladogram depicting relationships among areas of endemism. c) TreeFitter reconstruction: total cost = 3.04. d) DIVA reconstruction, total cost = 2.0. Notice that TreeFitter and DIVA provide a full description of ancestral ranges and explanatory events but differ in the treatment of duplication (see the text). Symbols: vicariance (circle), duplication (rhombus), dispersal (arrow or cross) and extinction (short line). For a color version of this figure, see www.iste.co.uk/guilbert/biogeography.zip*

2.2.2. *Dispersal–vicariance analysis*

Dispersal–vicariance analysis (Ronquist 1997), implemented in the software DIVA, is arguably the most popular event-based method. DIVA resembles character-state reconstruction methods employed in phylogenetic inference to map the evolution of a trait, such as the optimization algorithm of Fitch Parsimony. Given a tree and associated geographic ranges, ancestral distributions are mapped (optimized) along internal nodes by finding the sequence of biogeographic events that minimizes the total cost of the reconstruction. A major difference with Fitch Parsimony is that, while phylogenetic inference assumes single-state ancestors, DIVA allows ancestors present in two or more states, that is, widespread ancestral ranges formed by two or more discrete areas. This difference implies that DIVA optimization is more complex than Fitch's because it needs to include both anagenetic and cladogenetic events of range evolution. Changes in distribution range along the branches of the phylogeny (anagenetic change) are optimized as dispersal (range expansion: A to AB) or extinction (range contraction: AB to B) events. Range evolution at cladogenetic (speciation) nodes can involve duplication, if the ancestor occupies a single area (A/A), or vicariance, if a widespread ancestral range is divided into non-overlapping subsets (A/B). DIVA uses a slightly less complex cost-matrix than TreeFitter: dispersal and extinction are assigned a cost of 1, and duplication and vicariance, a cost of zero.

Figure 2.2(d) shows DIVA reconstruction for the same biogeographic scenario as in Figure 2.2(c). Notice that it is simpler than in TreeFitter, requiring only vicariance events interspersed with dispersal events. This example illustrates a difference between these two methods that is not always well understood (Wodcicki and Brooks 2005). As in cladistic biogeography, TreeFitter output is an area cladogram, whereas DIVA maps ancestral distributions and inferred biogeographic events onto the phylogeny. The constraint of hierarchical area relationships in TreeFitter means that the "redundant" distribution of species 1 and 2 in area A (Figure 2.2(a)) must be modeled as resulting from duplication within a widespread ancestor (ABCD), followed by extinction in part of the ancestral range (BCD). In DIVA, areas can be gained and lost along the branches of the phylogeny and their relationships do not need to follow a branching pattern. The overlapping distributions of species 1 and 2 are explained by dispersal to A along the internal branch, after the initial vicariance event that divided widespread ancestor 11 in ABCD (Figure 2.2(d)), and followed by a second vicariance event in widely distributed ancestor 9. DIVA is thus suited for reconstructing "reticulate" scenarios, in which area relationships are not dichotomous but resemble a network, with repeated cycles of dispersal and vicariance events. One example of such scenario is the Northern Hemisphere geological history, where the paleocontinents now forming Eurasia and North America recurrently merged and split during the last 150 Mya (Sanmartín *et al.* 2001). Yet, DIVA loses power and can give improbable biogeographic events when used in predominantly vicariant scenarios.

Conversely, TreeFitter is statistically more powerful to reconstruct area relationships that fit the vicariance pattern, such as the hierarchical fragmentation of the Gondwanan supercontinent (Sanmartín and Ronquist 2004). Another difference is the treatment of duplication events. In TreeFitter, duplication of ranges involving more than one area is allowed, but only if the widespread range forms an ancestral area in the area cladogram (e.g. ABCD or BCD in Figure 2.2(c)); alternative ancestral ranges such as ACD will not be accepted. DIVA accepts any combination of areas as ancestral ranges; however, as in Fitch Parsimony, duplication can only affect single areas. As a result, and unless geological constraints are used, DIVA reconstructions do not include extinction events (Kodandaramaiah 2010).

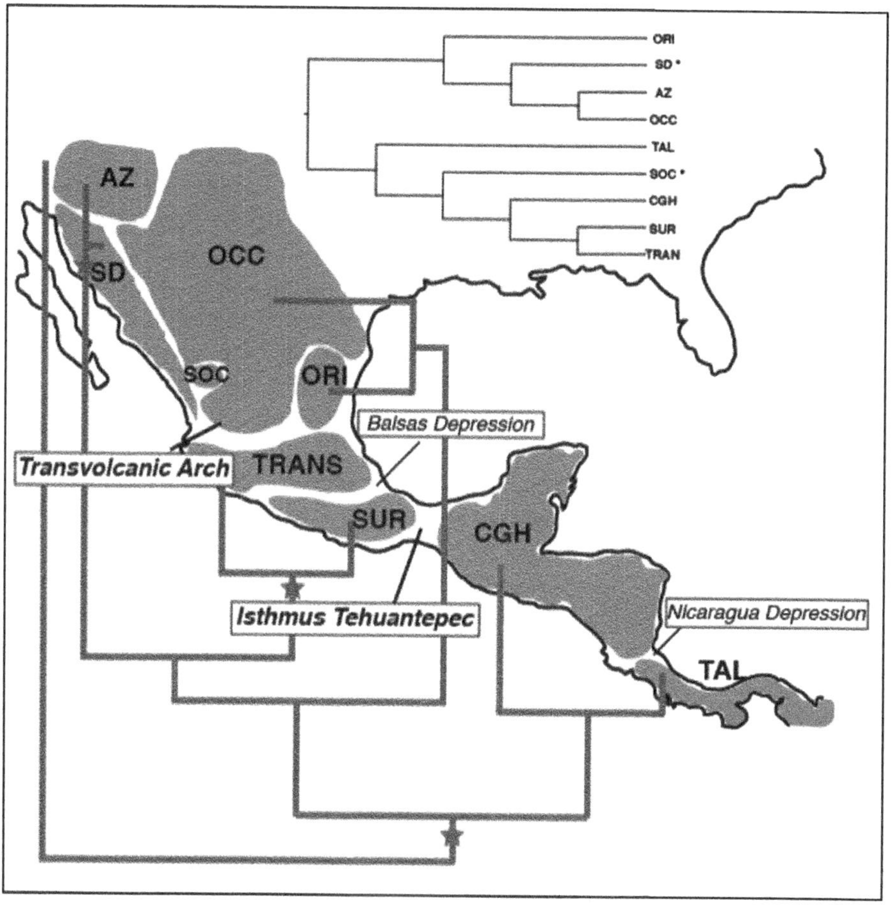

Figure 2.3. *TreeFitter reconstruction among areas of endemism in Mexico based on Marshall and Liebherr (2000)'s large insect dataset. Inset top: area cladogram obtained with Brooks Parsimony Analysis, a cladistic method. Main: area cladogram obtained with TreeFitter; "stars" represent conflicting nodes. The first division in the area cladogram involves the Transvolcanic Arch in BF and the Isthmus of Teuantepec in TF. Geological evidence indicates that disjunctions across the Isthmus are older than those involving the Transvolcanic Arch (Mastretta-Yanes* et al. *2015). Also, biogeographic relationships across the Transvolcanic Arch are supported only by widespread species, which could be explained by dispersal. Conversely, the Isthmus of Teuantepec relationship in the TF cladogram is supported only by ancestral nodes, that is, inherited ranges, suggesting shared history*

A main contribution of EBMs to historical biogeography was to demonstrate that dispersal, if constrained by abiotic factors such as wind strength or the direction of ocean currents, could be a pattern-generating process, in the same manner as vicariance is. "Concerted, directional" dispersal driven by the eastward moving West Wind Drift explains shared biogeographic patterns across southern hemisphere plant lineages (Sanmartín *et al.* 2007). Statistical testing of competing hypotheses is another contribution of EBMs. Frequencies of event types can be compared between the empirical phylogeny and a distribution of simulated phylogenies generated by randomizing geographic ranges among terminals. Significant differences indicate that biogeographic events are phylogenetically conserved (Sanmartín 2012). These randomization tests are common in parsimony-based optimization, which lacks an underlying probabilistic model (Faith and Cranston 1991); the discriminatory power of such tests, however, is questioned (Peres-Neto and Marques 2000).

While EBMs have now been superseded by parametric probabilistic approaches that integrate the time dimension (see below), they remain popular in fields where molecular data is not available, such as paleontology (Prieto-Marquez 2010; Upchurch *et al.* 2015). Treefitter is also used in host–parasite coevolution studies (Quinn *et al.* 2013). Using large datasets of phylogenetic and distributional data, event-based methods have been used to test broad-scale dispersal and vicariance hypotheses (Sanmartín *et al.* 2001; Donoghue and Smith 2004; Sanmartín and Ronquist 2004; Bremer and Jansen 2006). Figure 2.3 shows an example of this kind of analysis.

2.3. From parsimony-based to semiparametric approaches

Event-based methods were based on explicit process models and therefore represented a considerable advance over cladistic methods in terms of statistical power. However, they still have several limitations. The cost of events, for example, cannot be estimated from the data but must be fixed *a priori* using *ad hoc* criteria such as the phylogenetic conservation of ancestral ranges. As explained above, through dispersal and extinction, the descendants come to occupy a different range than the ancestor, so these two processes are underestimated ("minimized") in event-based reconstructions (Sanmartín 2012). Another constraint imposed by the use of parsimony is that biogeographic reconstructions are inferred on cladograms with no branch lengths or time information. This makes it difficult to discriminate between shared distribution history and biogeographic "pseudocongruence", that is, shared patterns that are spatially but not temporally congruent (Page 1993; Donoghue and Moore 2003) and between competing dispersal and vicariance explanations: in vicariance, the appearance of the barrier within the ancestral range is the trigger for allopatric speciation, while in dispersal, the barrier predates geographic change (Chapter 1). Finally, parsimony-based inference ignores two sources of error associated with reconstructing trait evolution: "phylogenetic uncertainty" – ancestral areas are reconstructed over a single topology, the most parsimonious tree, assuming there is no uncertainty in phylogenetic relationships – and "reconstruction uncertainty", where only the minimum-cost (most parsimonious) reconstructions are evaluated, even though alternative reconstructions could be almost as likely, or even more likely if additional information is considered (Ronquist 2004).

Incorporating phylogenetic uncertainty in EBMs is relatively straightforward: run the analysis over a distribution of trees that represent the level of clade support in the phylogeny; this distribution can be obtained from bootstrap pseudoreplication or a Bayesian posterior probability distribution. Non-bifurcating nodes (polytomies) and nodes with low clade support can then be associated with low support for ancestral ranges. In Nylander *et al.*'s (2008) Bayes-DIVA method, DIVA parsimony-based reconstructions are averaged over a distribution of trees representing the posterior probability obtained from a Bayesian phylogenetic analysis. Figure 2.4 gives an example. Node "X" is a highly supported clade including three species distributed in areas C, B and A. Phylogenetic relationships in the rest of the tree are uncertain and vary over the Bayesian sample of trees, including the potential sister-group. For example, the stem or parent node of X ("Y" in Figure 2.4) has as the other descendant: "D" in tree 1, "E" in tree 2 and "F(ED)" in tree 3. Each of these tree topologies has a different posterior probability (PP) value. Because in Bayesian inference (BI), the frequency with which a tree is sampled in the analysis is proportional to its posterior probability (Ronquist 2004), the nodal ancestral area reconstructions in Bayes-DIVA can be interpreted as "marginal probabilities" (i.e. the different wedges in the pie chart in Figure 2.4). In other words, averaging DIVA reconstructions over a Bayesian sample of trees gives us estimates of ancestral ranges at nodes that are marginalized over the variation in the remainder tree topology. Notice that DIVA does not integrate branch length information, so the only parameter that is marginalized is the tree topology; in this sense, Bayes-DIVA can be considered an empirical Bayesian method (Nylander *et al.* 2008). It is also a semiparametric model since it contains a parametric (Bayesian phylogenetic inference) and a nonparametric (parsimony biogeographic inference) component. Another important distinction is given by tree 4: Pagel *et al.*'s (2004) definition of a "floating node". Trees containing different definitions (bipartitions) of node X (tree 4, PP = 0.10) are excluded from the marginal estimations for that node in Bayes-DIVA: that is, the wedges in the pie chart sum to 0.90 (Figure 2.4). In other words, Bayes-DIVA uses a node-to-node approach in accounting for phylogenetic uncertainty: only those trees containing the node of interest (X) will be used in the estimation of ancestral-range marginal probabilities.

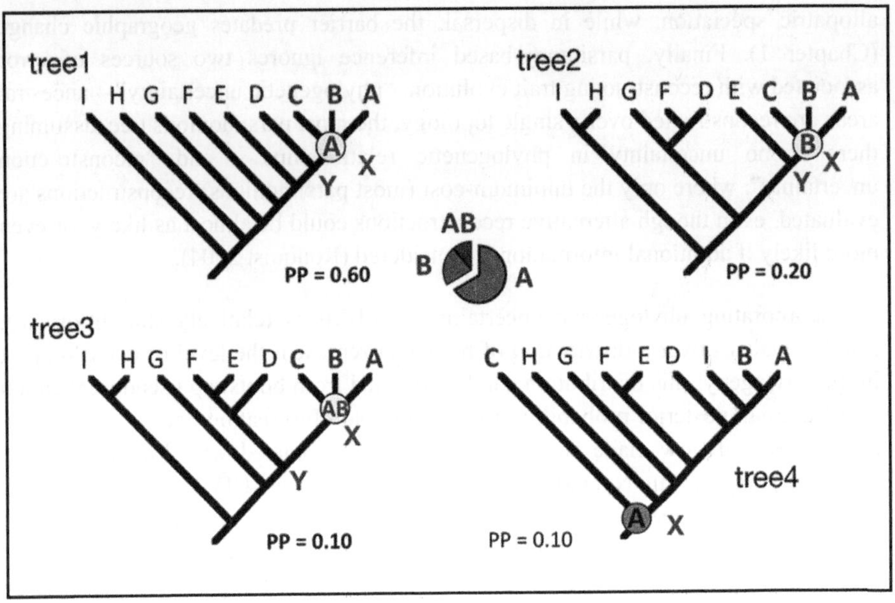

Figure 2.4. *Accounting for phylogenetic uncertainty: the Bayes-DIVA approach. Trees 1–3 represent different phylogenetic relationships among nine species distributed in areas A–I. These trees were obtained by Bayesian inference, so each of them is associated with a posterior probability (PP) value, that is, the frequency with which the tree appears in the Bayesian posterior distribution. In all trees (1–3), the species present in areas A–C form a well-supported clade, node "X"; however, the remainder of the tree topology, including the identity of the sister-group and the definition of the "parent node Y", are uncertain. In Bayes-DIVA, a DIVA analysis is run on each of these alternative trees. Because each tree is "weighted" by its posterior probability in the Bayesian sample, the end result is marginal probabilities of ancestral ranges, that is, DIVA reconstructions are marginalized or integrated over the uncertainty in the tree topology; this is represented in the central pie chart, where range A receives the highest marginal probability, followed by B and AB. Notice that only those trees containing node X are included in the computation of marginal probabilities. Tree 4, in which C, B and A do not form a clade, will not be included in the computation of marginal probabilities. For a color version of this figure, see www.iste.co.uk/guilbert/biogeography.zip*

Nylander *et al.* (2008) demonstrated that accounting for phylogenetic uncertainty may also reduce biogeographic uncertainty: that is, for a given node, some ancestral ranges that were equally optimal in DIVA will be associated with higher marginal probabilities in the Bayes-DIVA analysis; in Figure 2.4, the ancestral range for node X that receives the highest marginal probability is A.

Harris and Xiang (2009) subsequently developed S-DIVA, implemented in the software RASP (Yu *et al.* 2010), to introduce "reconstruction uncertainty" in a Bayes-DIVA analysis. In a DIVA reconstruction, there could be several equally parsimonious ancestral ranges optimized at a given node (e.g. A, B or AB in Figure 2.4), as well as for each of these equally parsimonious ancestral ranges, there could be multiple different pathways (combinations of dispersal and cladogenetic events) by which a given ancestral range (e.g. A) is optimized along the tree. In Nylander *et al.* (2008)'s Bayes-DIVA approach, biogeographic uncertainty was defined as the first option: equally parsimonious ancestral ranges were assigned an equal weight (1/N), where N is the number of most parsimonious (MP) alternative ranges for a given node. In Harris and Xiang's S-DIVA approach, biogeographic uncertainty is defined as the frequency with which a particular ancestral range appears within the pool of most parsimonious biogeographic pathways: $F(r) = i/Rt$, where i is the number of times a range (r) occurs in the total number of MP scenarios (Rt) over the tree.

Nylander *et al.*'s interpretation of ancestral range uncertainty as marginal probabilities is consistent with Bayesian probabilistic inference – nodal ancestral ranges are marginalized over a parameter (tree topology), whose values are sampled according to a probability distribution (Huelsenbeck *et al.* 2000). However, the probabilistic interpretation of $F(r)$ in S-DIVA is not that clear. This has to do with the distinction between joint and marginal reconstructions. In a likelihood context, a joint reconstruction is the single best reconstruction across all nodes in a tree, and the marginal reconstructions are the single best reconstruction for each node considered along and after integrating for all possible reconstructions in the rest of the nodes. MP pathways in DIVA are estimated jointly over the tree, whereas ancestral ranges are estimated node-by-node. Equating the output of Bayes-DIVA to marginal probabilities is straightforward: the probability for ancestral range A on node X is approximately equal to the product of the probability that node X exists (PP value) and the probability of state A on node X – assuming that the MP solution is the ML solution (Huelsenbeck *et al.* 2000). However, the output frequency scenarios from S-DIVA cannot be equated directly to the joint probabilities obtained from a model-based method (Pagel *et al.* 2004). This is because discrimination between DIVA reconstructions is based on the MP score (the number of dispersal events): optimal nodal ranges are "equally parsimonious" if they carry the same cost in terms of explanatory dispersal events. That they appear in a different frequency among the biogeographic reconstruction pathways (e.g. 0.3, 0.7) does not imply that the first has lower joint probability than the second. A simile can be found in parsimony phylogenetics, given three equally most parsimonious trees (with the same number of steps), if two of them include clade X and one includes clade Y, this does not imply that clade X has some "probability" proportional to 2/3 and clade Y, a "probability" proportional to 1/3. Therefore, I argue that the original Bayes-DIVA method, which is also implemented in RASP, is more appropriate than Harris and Xiang's (2009) extension to integrate phylogenetic uncertainty in EBMs: the reported values for ancestral ranges in Nylander *et al.*'s (2008) approach have a probabilistic interpretation within an empirical Bayesian context, which S-DIVA lacks.

2.4. A new revolution: parametric approaches in biogeography

The last decade has witnessed the introduction of parametric approaches in biogeography without the biases inherent in the parsimony framework (Ronquist and Sanmartín 2011). A common feature of these methods is the use of statistical probabilistic models, whose variables or parameters are quantifiable biogeographic processes that are dependent on time. Thus, in addition to the tree topology and tip distributions, parametric models incorporate a third source of information: branch lengths – measured in numbers or units of time – provide direct evidence on the rate or probability of geographic evolution. Longer branches imply a higher probability of change in the geographic range than shorter branches do. As more time elapses since the divergence of the species from its ancestor, there is more opportunity for biogeographic change (by dispersal, extinction or range expansion) along the branch. Branch lengths also inform on the certainty or degree of error in biogeographic inference: a species subtended by a long branch would be associated with a higher uncertainty about its ancestral range than one subtended by a short branch (Sanmartín 2020).

Two parametric models with stochastic variables that are time-dependent are typically used to model range evolution: the Brownian Motion (BM) model and the continuous-time Markov Chain (CTMC) model. BM is used to model the random dispersal of individuals within a population in a continuous landscape. We will return to this model later. CTMC models are typically used to describe range evolution at the species level. A Markov chain is a stochastic, memory-less process that models transitions between discrete states over continuous or discrete-time; the probability of each transition event depends only on the state attained in the previous event. The difference between a discrete-time MC (DTMC) and a continuous-time MC (CTMC) is that in DTMC, the chain moves (transitions or changes state) in discrete intervals (Δt), whereas in continuous-time, the moves are in infinitesimal amounts of time (dt) so that it can effectively measure instantaneous change. CTMC models are used in historical inference disciplines, such as phylogenetics or biogeography, where the states observed in the present are the result of a stochastic process that evolves over time. Stochastic means that, unlike in a deterministic process, the outcome of the process cannot be predicted with certainty, that is, we cannot predict the result of evolution. At the same time, evolutionary processes are not entirely random or unpredictable: they can be described by probability distributions with parameters, whose values we infer from the data.

Figure 2.5 provides an example of the parametric approach. In biogeography, the states of the stochastic CTMC process that governs range evolution are the set of discrete geographic areas that form the distribution range of a taxon (e.g. A, AB, B). The rates of transition or change between states in the CTMC process (e.g. A to B), within an infinitesimal amount of time (*dt*), are governed by the so-called instantaneous rate matrix Q, which has as parameters biogeographic processes that determine the probability of range evolution as a function of time, for example, dispersal, extinction, speciation. Given a phylogeny with time-calibrated branch lengths, tip distributions coded as discrete entities (A, B) and a stochastic CTMC model of range evolution (Q matrix, Figure 2.5), we can estimate the probabilities of ancestral ranges (A, B, AB) and the rate of parameters describing the transition between geographical ranges (p, q, D_{AB}, E_B, in Figure 2.5), using statistical inference approaches such as maximum likelihood (ML) or BI.

Besides the possibility of integrating time into biogeographic inference, parametric methods offer several advantages over parsimony-based approaches (Ree and Sanmartin 2009). Rather than inferring only the most parsimonious reconstruction, one can integrate over all possible biogeographical scenarios in the estimation of rates of parameters and species ancestral ranges; that is, parametric methods account for the "reconstruction uncertainty" (Ronquist 2004; Ree and Smith 2008). If BI is used, parameter estimates are not conditioned on a given phylogeny but marginalized over the tree topology and branch lengths by simultaneously estimating the parameters governing phylogenetic and biogeographic evolution; that is, BI parametric methods account for the "phylogenetic uncertainty" (Ronquist 2004; Sanmartín *et al.* 2008). Sources of evidence other than the phylogeny and tip distributions (e.g. the fossil record, geological or paleoclimate information or the species ecology) can be integrated into parametric models, either in the form of new parameters in the Q matrix or through scaling parameters that modify the baseline rate of a different parameter (Buerki *et al.* 2011; Meseguer *et al.* 2015; Quintero and Landis 2019; Landis *et al.* 2021). Finally, model selection, that is, the statistical testing of competing biogeographic scenarios, is straightforward with parametric models because, as in EBMs, assumptions about processes governing range evolution are made explicit and integral to the inference framework. Moreover, because the underlying stochastic models are based on well-known probability distributions, parametric models can make use of statistical tests employed in phylogenetics for model choice, such as likelihood ratio tests (LRT), the Akaike information criterion (AIC) or Bayes factor comparisons.

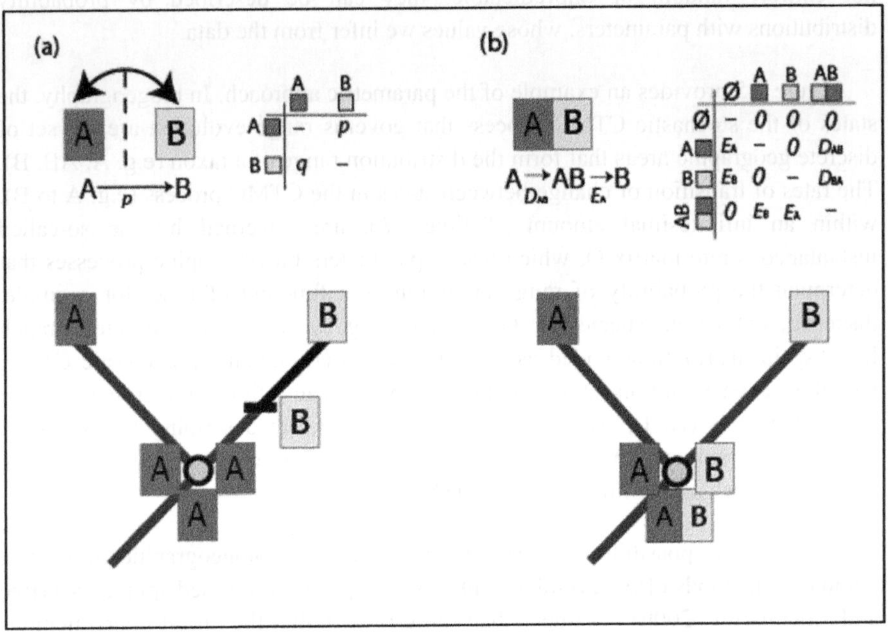

Figure 2.5. *Parametric models in biogeographic inference. Top: a continuous-time Markov chain (CTMC) process is used to describe the probability of range evolution; the states of the CTMC are discrete geographic ranges (A, B, AB), and transitions between states (A to B) are governed by an instantaneous Q matrix with rate parameters (q) that are estimated from the data. Bottom: given a two-species phylogeny with associated distributions and branch lengths measured as time since divergence, we can use maximum likelihood or Bayesian inference to estimate the ancestral geographic range and the sequence of biogeographic events that gave rise to the current distributions. a) The BIB model implements a CTMC process with only single-area ranges as discrete states (A or B); anagenetic evolution along branches is governed by parameters p and q, describing the instantaneous movement between states; ancestral states are identically inherited by the two descendants through speciation (A/A). b) The DEC model implements a CTMC process with widespread distributions formed by two or more discrete areas (AB), and thus requires an additional cladogenetic component to describe the different ways by which a widespread ancestral range is divided between the two descendants (A/B); anagenetic evolution is governed by a CTMC process with two parameters: range expansion (D_{AB}) and range contraction (E_B); direct transitions between single areas (A to B) are not allowed in the Q matrix. For a color version of this figure, see www.iste.co.uk/guilbert/biogeography.zip*

2.4.1. *Ancestral range versus single state models: DEC and BIB*

The first two parametric methods developed in biogeography (Figure 2.5) were the Bayesian island biogeography (BIB) model (Sanmartín *et al.* 2008) and the dispersal-extinction-cladogenesis (DEC) model (Ree *et al.* 2005; Ree and Smith 2008). The BIB model uses Markov-Chain Monte Carlo (MCMC) simulations and BI to estimate ancestral ranges and rates of biogeographic parameters alongside phylogenetic parameters, such as the tree topology, rates of molecular evolution and branch lengths; the input data are DNA sequences and tip distributions of the study species (Sanmartín *et al.* 2008). The DEC model was originally implemented in an ML framework (Ree and Smith 2008) and later extended to BI (Landis *et al.* 2013), but it is typically applied to a phylogeny with fixed topology and branch lengths. Though they both use CTMC processes to model range evolution, BIB and DEC models are slightly different (Figure 2.5). BIB (Figure 2.5(a)) implements a simpler character evolutionary model, in which ancestors can only occupy single areas (A or B) and range evolution along the branches if governed by a CTMC process with only one type of parameter equivalent to range switching or instantaneous dispersal; the Q matrix describes the instantaneous transition from one area as a jump dispersal event (p = A to B). At the speciation events in the phylogeny, the single-area ancestral range is inherited entirely and identically by the two descendants (A/A); in other words, there is no need to include a cladogenetic component in the BIB model because the ancestral range is not altered through speciation (Figure 2.5(a)).

DEC implements a more complex ancestral-state reconstruction model (Figure 2.5(b)), allowing for ranges in the Q matrix to comprise two or more discrete areas (i.e. widespread states, AB). Therefore, the model requires an additional cladogenetic component describing the different ways (range inheritance scenarios) by which a widespread ancestral range is divided between the two descendants. In Figure 2.5(b), this involves vicariance, range division into non-overlapping subsets (A/B), but other possible range inheritance scenarios are "peripheral isolate speciation" (Ree *et al.* 2005), in which one descendant inherits the whole ancestral range and the other descendant inherits only one area (AB/B), or "widespread sympatry", in which the two descendants inherit the widespread ancestral range (AB/AB; Landis *et al.* 2013). In the original ML implementation of DEC (Ree and Smith 2008), all these range inheritance scenarios are assigned equal relative likelihoods or "weights". In the BI implementation (Landis 2017), different prior probabilities can be assigned to the speciation modes. There are other DEC-derived models (DIVALIKE, BAYAREALIKE, Maztke 2014) which differ on the type of range inheritance scenarios allowed for widespread ranges. Like BIB, DEC implements anagenetic evolution as a CTMC process. Instantaneous transitions between geographic states are governed by two parameters (Figure 2.5(b)): range expansion, where an additional area is added to the current range (D_{AB} = A to AB), and range contraction, with the removal of an area from the ancestral range (E_B = AB to B). Notice that instantaneous transitions between single-area states are not allowed in DEC. Moving between single areas requires going through a widespread state in which the ancestor is present in both areas (Figure 2.5(b)). This can be observed in the Q matrix, with zero rate values for direct transitions between A and B (Figure 2.5(b)). A transition from A to B requires two consecutive instantaneous events: range expansion from A to AB and range contraction from AB to B.

Figure 2.6 illustrates, with a real biogeographic problem, some of the differences between BIB and DEC. In sum, the BIB model allows only anagenetic changes along branches in the phylogeny and constrains ancestors to occur in single areas, whereas DEC includes both anagenetic and cladogenetic range evolution; in fact, DEC is the parametric counterpart of DIVA (Ree *et al.* 2005). However, this additional level of complexity brings some statistical limitations discussed below. Both models implement different sources of uncertainty (phylogenetic and reconstruction in BIB, and reconstruction in DEC).

The DEC model is undoubtedly more realistic than BIB. In biogeography, widespread terminals and ancestors are biologically plausible: an extant or extinct taxon could have occupied more than one area, especially if these areas were connected and there was no dispersal barrier between them. However, such complexity comes with a cost: there are 2^N possible ancestral ranges for N areas, so the Q instantaneous rate matrix cannot be analytically calculated with more than 10 areas (1,024 states). This can be reduced by removing certain transitions from the Q matrix, for example, disallowing ancestral ranges that involve discrete areas that are non-adjacent in the physical space (Buerki *et al.* 2011), or using alternative estimation methods, such as data augmentation (Landis *et al.* 2013). Dispersal in DEC is equivalent to range expansion – the ancestor moves into a new area but keeps the original distribution for some time; in other words, moving between single areas requires going through a widespread state in which the ancestor is present in both areas (Figure 2.5(b)). This type of dispersal may be appropriate for continental settings in which areas are adjacent, that is, share a physical edge, and we expect gene flow to be maintained for some period of time between the allopatric populations (Ree and Sanmartín 2009). Yet, it comes with the necessity of modeling cladogenetic events or range inheritance scenarios (Figure 2.5(b)).

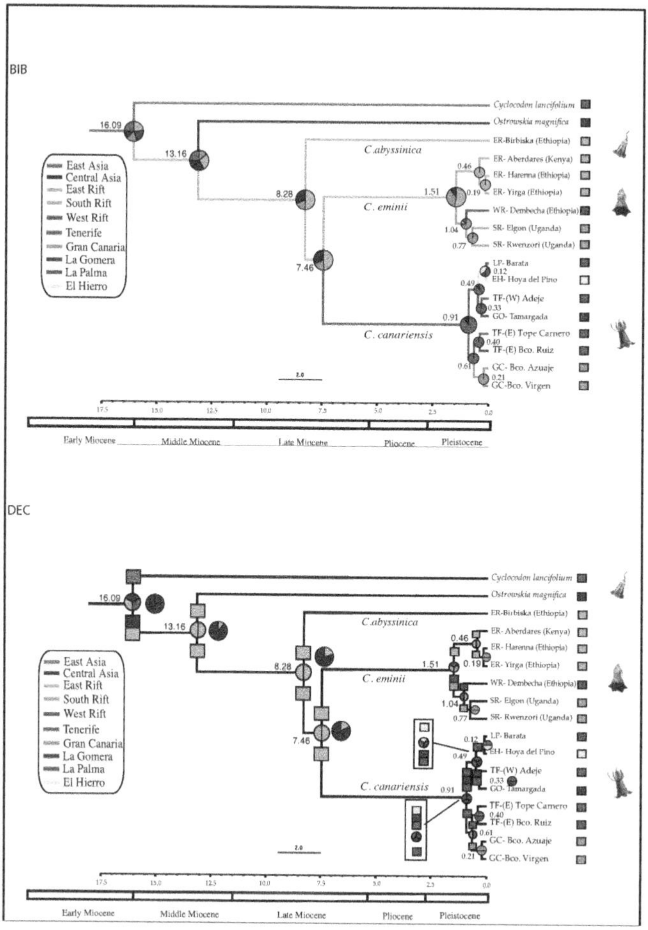

Figure 2.6. *Parametric biogeographic reconstruction of the spatio-temporal evolution of genus Canarina. Canarina is a three-species genus in the angiosperm family Campanulaceae, with a disjunct distribution between the Canary Islands in the west and the Eastern African mountains and Horn of Africa plateaus in the east (Mairal et al. 2015). Both BIB and DEC explain this geographic disjunction as a sequence of migration events from Asia, where sister-genera occur, to Eastern Africa, and to Macaronesia; the latter along the 7 million-year branch separating* C. eminii *and* C. canariensis. *DEC infers a similar scenario, but vicariance in a widespread distribution is inferred at some nodes, mostly involving short internal branches and descendants with non-overlapping distributions. Pie charts represent the uncertainty in the estimation of ancestral ranges. For a color version of this figure, see www.iste.co.uk/guilbert/biogeography.zip*

The original DEC model (Ree and Smith 2008) includes a "null" state in the Q matrix (∅), equivalent to global extinction. A species can become extinct across its entire range but only if this comprises a single area (A to ∅; Figure 2.5(b)); global extinction across a widespread range is assigned a zero rate in the DEC model (AB to ∅; Figure 2.5(b)). Also, global extinction behaves as an "absorbing state" in the Q matrix because the rate of abandoning this range is zero (e.g. ∅ to A; Figure 2.5(b)). One of the consequences of including a null range in the Q matrix is that extinction rates are typically underestimated in the DEC model, even several orders of magnitude compared with dispersal rates. Because global extinction can only occur within single-area ranges, it is often inferred at terminal branches, where dispersal cannot be countered off by a loss of areas via cladogenesis (vicariance or peripatry). Removing the null range from the Q matrix has the effect that extinction events are forced to occur within widespread ranges in order to counteract dispersal or range expansion along internal branches. In other words, loss of areas would be achieved via extinction in ancestral branches rather than via cladogenesis, and thus extinction rates are increased relative to dispersal if global extinction is disallowed (Massana *et al.* 2014). Bayesian extensions of DEC (Landis 2017) correct for this bias by conditioning the DEC inference to the survival of all lineages in the extant phylogeny, that is, never entering the null state.

The issues above do not affect the BIB model because it uses a simpler stochastic CTMC model similar to those employed in molecular character evolution. Widespread states (equivalent to "polymorphism" in nucleotide models) are not allowed in the CTMC matrix, and range evolution is limited to the anagenetic component. This is modeled as instantaneous transitions between single-area states, equivalent to "jump dispersal", but which can vary across area pairs and may also be asymmetric, that is, the rate of moving from A to B, p, is not the same as from B to A, q (Figure 2.5(a)). Modeling dispersal as an instantaneous process, without going through a widespread state, may seem unrealistic but allows the BIB model to "borrow" the sophisticated machinery and statistical algorithms used in molecular models of nucleotide substitution; in fact, initial implementations of BIB used software routinely employed in molecular phylogenetics (Sanmartín *et al.* 2008; Lemey *et al.* 2009). In standard molecular models, nucleotide substitutions within a species DNA sequence are considered as instantaneous. In-between demographic-level processes, involving increased allele polymorphism within gene trees, competition among mutations in terms of fitness, and rates of fixation differing between alleles (De Maio *et al.* 2015), are typically ignored. Similarly, in the BIB model, the species is assumed to instantaneously change the area relative to its current range, ignoring the intermediate population-level processes, such as changes in effective population size due to migration, introgression, etc. The BIB model can thus be appropriate to model scenarios in which areas are discrete entities isolated by dispersal barriers, so that migration to a new area effectively leads to speciation; in other words, the ancestor is not expected to maintain the widespread distribution for long, as in the case of founder effects in oceanic islands isolated by geographic barriers (Sanmartín *et al.* 2008), or in continental islands isolated by ecological barriers (Sanmartín *et al.* 2010). However, the assumption of single-state ancestral ranges means that BIB is most useful to explore and test general patterns of geographic movement or dispersal; if the interest lies on inferring speciation modes or possible ways in which ancestral ranges are divided, BIB is not well suited. Notice that constraining ancestors to single areas in the Q matrix does not imply that phylogenies with extant widespread species cannot be analyzed with BIB. As in molecular evolutionary models, these widespread terminals will be treated as sources of "ambiguity" in the BIB analysis: 50% of the time the MCMC chain will sample from one of the discrete states, and 50% from the other. Another solution is to use an expanded, constrained Q matrix in which transitions between widespread states are allowed only between spatially adjacent ranges, as in an ordered "character step matrix" in parsimony-based approaches (Bribiesca-Contreras *et al.* 2019).

One advantage of the direct analogy between the BIB model and nucleotide substitution models is the possibility to infer the stationary frequencies of the states in the CTMC process (Sanmartín *et al.* 2008, 2020). The standard CTMC models used in parametric biogeography and molecular evolution are "time-homogeneous" or "stationary" Markov models. They have the property that the rates of transition between states are constant and, over time, tend to reach a stationary equilibrium state. They are also often time-reversible, that is, independent from the flow of time (i.e. this is not the same as symmetric). Over time, the frequencies of the states of a time-homogeneous Markov process converge to the stationary values regardless of the starting point. In a time-reversible stationary CTMC, the state equilibrium frequencies are built into the Q rate matrix, so that the transition rates can be decomposed into two parameters: the relative exchangeability rates and the state stationary frequencies. Similarly, the rate of moving from A to B in the Q matrix of the BIB model (p) can be broken down into two parameters: the relative dispersal rate per migrating lineage (r_{AB}) and the area "carrying capacities" (π_A, π_B). The latter are the model "stationary" frequencies: the number of lineages at equilibrium conditions, or in other words, the number of lineages expected in each area if the CTMC dispersal process is let to run for a very long time without external disturbances (Sanmartín *et al.* 2008). Disentangling transition rates into two parameters allows the root states, that is, the states at the start of the process, to be drawn from the stationary frequencies of the CTMC. Also, the two parameters account for different aspects of the dispersal process. Relative dispersal rates can be informed (scaled) by the geographic distance between areas or the strength of wind and ocean currents, while carrying capacities can be partitioned by area size or the degree of environmental heterogeneity versus a species ecology; this allows researchers to measure the role played by abiotic factors and biotic factors in shaping area colonization patterns (Sanmartín *et al.* 2008; Sanmartín 2020). Finally, though there is no speciation parameter in the BIB model, carrying capacities can be used as a proxy for rates of "within-area diversification". Stationary frequencies represent the time the CTMC process spends without transitioning between states, or, in a biogeographic context, without migrating between areas. Sanmartín *et al.* (2010) used this equivalence in a continental-island context, to demonstrate that the southern African component of the Rand Flora was formed through within-area diversification, whereas the Macaronesian component was shaped by immigration events from nearby regions.

The partitioning of CTMC transition rates into stationary frequencies and relative exchange rates is not possible in DEC. The reason was pointed by Ronquist and Sanmartín (2011) and discussed extensively in Ree and Sanmartín (2018). DEC and DEC-derived models are not complete parametric models like BIB because one key component of the biogeographic model, cladogenetic scenarios of range evolution, is not part of the stochastic CTMC process that governs the evolution of geographic ranges as a function of time. In other words, there is no speciation parameter in the Q matrix, even though speciation has an effect on range evolution in the DEC model (Figure 2.5(b)). As a result, root states in DEC cannot be drawn from the stationary frequencies of the CTMC process, as can be done in BIB. In Ree and Smith's (2008) ML implementation of DEC, root states are inferred by first estimating the likelihood of alternative ancestral ranges and then selecting the one that maximizes the global likelihood. Another consequence of DEC not being a fully parametric model is that DEC-derived models that differ in the type of implemented cladogenetic scenarios cannot be compared statistically. DEC and DEC-derived models such as DIVALIKE or BAYAREALIKE contain the same number of parameters in the CTMC Q matrix that governs range evolution (i.e. the rates of dispersal and extinction), so it is erroneous to use penalty-based likelihood tests such as AIC (Matzke 2014) to statistically distinguish or identify them. Instead, we can choose between these models, which imply different speciation modes of widespread range division, using biological knowledge about the study group (Sanmartín 2020). The same issue arises when comparing time-homogeneous and time-stratified DEC models (below) because these models do not differ in the number of CTMC parameters. On the other hand, within a Bayesian framework, we can statistically compare any two models using the Bayes factor. The latter computes the ratio of the marginal likelihood of two competing models, or, in other words, the posterior against the prior odds for any of the models as the one generating the data (Goodman 1999). Unlike AIC or LRT, Bayes factor comparisons do not depend on any single set of parameters, as they integrate over all parameters in each model, while at the same time applying a penalty to overfitting, that is, a low ratio of data to parameters (Kass and Raftery 1995).

2.4.2. *Extending the DEC and BIB models*

Over time, the DEC and BIB models have been expanded to include more complexity and increasing realism. The original DEC model (Ree and Smith 2008) included dispersal or range expansion only as an anagenetic event, which was modeled as a time-dependent rate within the Q instantaneous rate matrix (Figure 2.5(b)). Matzke (2014) extended this model to include "cladogenetic dispersal" or "founder-event speciation", as an event of dispersal that is coincident with speciation, with one daughter lineage instantaneously "jumping" into a new area that was not part of the ancestral range, for example, from A to A and C in Figure 2.5(b). This new cladogenetic scenario is modeled in the DEC+J model by a separate parameter *j* (Matzke 2014), which is not part of the CTMC process that governs range evolution along branches. Therefore, this *j* parameter is not equivalent to the rate of jump dispersal *p* and *q* in the BIB model (Figure 2.5(a)), and it is also not dependent on time, unlike the D_{AB} or E_A parameters in DEC. Ree and Sanmartín (2018) showed that by decoupling "jump dispersal" from time, the DEC+J model can result in highly counterintuitive scenarios and degenerate likelihood inferences, in particular if founder speciation is assigned a higher likelihood ("weight") relative to other cladogenetic scenarios such as allopatry or peripheral isolate speciation. Moreover, when estimated as its maximum value, the inclusion of *j* can lead to underestimation of the rates of the anagenetic, time-dependent parameters: range expansion and range contraction. As a result, the DEC+J model can generate reconstructions with rates of anagenetic dispersal and (especially) of extinction close to zero, and distribution patterns that are explained almost exclusively by cladogenetic events. The end result is a diminishing of the relevance of time (branch lengths) in biogeographic inference, considered as the key advance of parametric over parsimony-based approaches (Ree and Sanmartín 2018). Figure 2.7 shows an example of this potential bias. As pointed out by Ronquist and Sanmartín (2011) and Ree and Sanmartín (2018), the proper modeling of cladogenetic events in parametric range evolution requires the use of trait-dependent speciation-extinction models (Maddison *et al.* 2007), discussed in more detail below. A different solution is adopted by another DEC-derived model, BayArea (Landis *et al.* 2013). It uses a Bayesian data augmentation approach in which parameters in the Q matrix are estimated by simulating outcomes (geographic range evolutionary histories) along branches in the phylogeny. This allows for a larger number of areas and geographic ranges in the model, including widespread states. Unlike DEC, there is no modeling of speciation scenarios: ranges are identically inherited by the two descendants, which also helps simplifying the model.

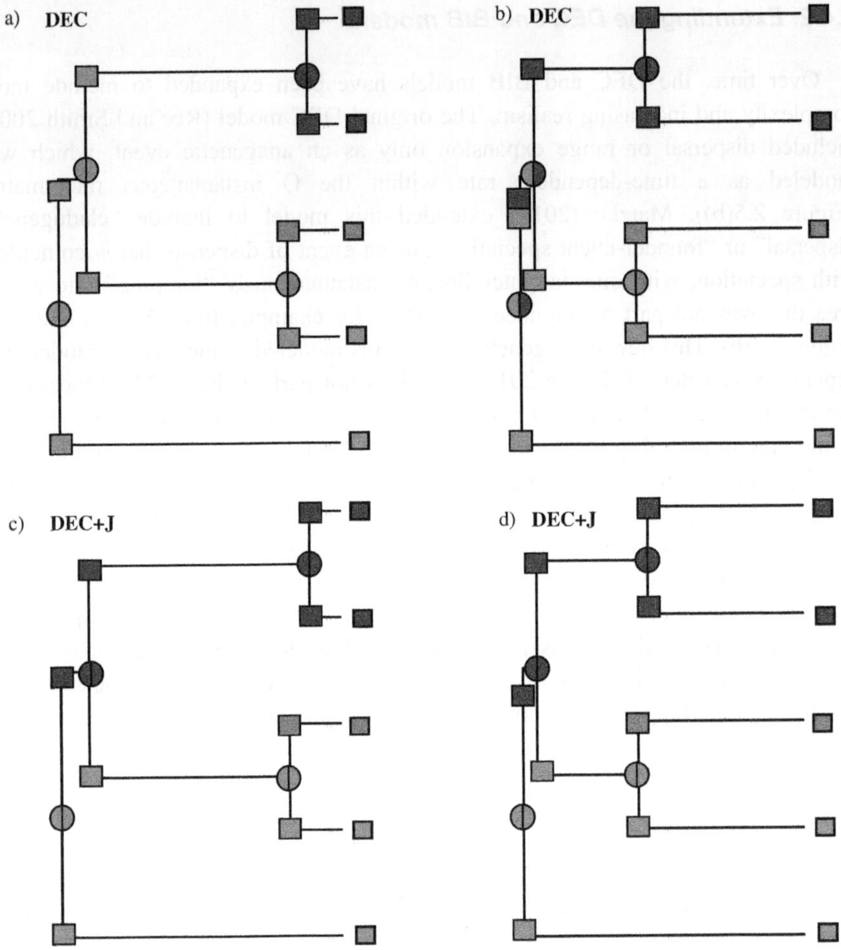

Figure 2.7. *Effect of decoupling biogeographic inference from time in DEC+J. Two phylogenies with identical topology and tip distributions, but internal branches elongated or shortened by half. The software BiogeoBears was used to infer ancestral ranges and rates of dispersal (d) and extinction (e) under the DEC and DEC+J models; the latter includes a parameter, j, for cladogenetic jump dispersal ("founder speciation"). a) DEC (short branches): d=0.0542, e=0.0436, j=0; LnL=−7.75. b) DEC (long branches): d = 0.0289, e=0.0375, j=0; LnL=−8.94. c) DEC+J (short): d=0; e=0; j=0.4265; LnL=−3.99. d) DEC+J (long): d=0; e=0; j=0.4265; LnL=−3.99. Notice that the DEC reconstruction for the most basal nodes changes with the branch lengths, but the DEC+J reconstruction does not. Only the range with the highest relative likelihood is shown; the maximum number of areas in widespread ranges was constrained to two. LnL: model log-likelihood. For a color version of this figure, see www.iste.co.uk/guilbert/biogeography.zip*

Regarding the BIB model, extensions have gone in the direction of introducing species-specific rates of geographic movement or implementing procedures for reducing the size of the Q matrix. The original BIB model was used to infer patterns of colonization in oceanic (Sanmartín *et al.* 2008) or continental (Sanmartín *et al.* 2010) islands. It implemented a hierarchical Bayesian approach in which relative dispersal rates between islands and island carrying capacities were estimated from phylogenetic and distribution data from multiple, co-distributed island lineages. Phylogenetic and biogeographic parameters were simultaneously estimated from species DNA sequence data and associated geographic distributions, but allowing each species to have their own rates of molecular and biogeographic (dispersal) evolution. This hierarchical, species-partitioned approach allows researchers to infer general, broad-scale patterns of island colonization while accounting for (marginalizing) organism-specific differences in rates of molecular evolution, age of origin or dispersal ability. The BIB model was subsequently implemented in a epidemiology context to study patterns of viral spread (Lemey *et al.* 2009). These authors also extended the BIB model to include a stepwise regression approach, Bayesian stochastic variable selection, to identify those transition rates or dispersal pathways in the CTMC Q matrix that are better supported by the data (Lemey *et al.* 2009). This BIB extension has also been used to infer migration patterns at the population, within-species level (Mairal *et al.* 2015). Other extensions of BIB have gone in the direction of making dispersal rates dependent on external factors or predictors (Faria *et al.* 2013), or allowing the inferred dispersal pathways to differ across taxa (Cybis *et al.* 2013).

The applications of BIB in epidemiology and phylogeography are probably some of the most popular uses of the model in the present. BIB in these fields is termed discrete trait analysis, DTA, or the "mugration" model because it equates migration to mutation events (De Maio *et al.* 2015). Though treating migration events as instantaneous mutations in a sequence might be acceptable at geological time scales and species levels, as was done in the original BIB (Sanmartín *et al.* 2008), it can be more problematic under the coalescent process; this is a model used at short-time scales and population-levels for building phylogenetic relationships (De Maio *et al.* 2015). Subsequent authors have extended the BIB-DTA model to allow for geographically structured populations' conditioning under the coalescent process (De Maio *et al.* 2015; Muller *et al.* 2017).

2.5. Expanding parametric models

2.5.1. *Time-heterogeneous models*

BIB and DEC models assume constant rates of dispersal and extinction as part of the CTMC process governing range evolution. As with molecular evolutionary models, relaxation of these assumptions has gone in the direction of allowing for rates to vary over time and across lineages, the so-called time-heterogeneous CTMC models. In the case of BIB, Bjelec *et al.* (2014) extended the DTA model to allow for the overall dispersal rate to vary across time slices in a stratified phylogeny; they used a piecewise-constant stochastic process in which rates of migration are constant within a given time slice but change between time slices. The temporal boundary (breakpoint) between two time slices may be estimated from the phylogenetic and distribution data alongside the biogeographic parameters.

A similar approach was implemented in the time-stratified, "epoch" DEC model (Ree and Smith 2008; Landis 2017): the phylogeny is divided into time intervals, and each interval is assigned a different set of values that scale the baseline dispersal rate according to paleogeographic information; for example, the availability of temporal land bridges facilitating migration between continents (Buerki *et al.* 2011). Time-stratified DEC models can also be used in biogeographic dating (Landis 2017). Rather than assuming a single CTMC process over time, DEC is allowed to shift between different Q matrices at discrete time points, based on paleogeographic evidence. Phylogeny, molecular dating and biogeographic parameters are jointly estimated using hierarchical BI. Paleogeographic data, that is, the formation of dispersal corridors and barriers over time, is used to inform the rates of a piecewise-constant epoch DEC model, and these time-dependent CTMC probabilities are used in turn to inform estimates of species divergence times in the phylogeny; for example, species can only diverge in allopatry if a paleogeographic barrier is present (Landis 2017).

Another exciting approach is the modeling of non-stationary CTMC models, where the equilibrium frequencies are allowed to change at discrete time points between time slices (Sanmartín 2020). Changes in area carrying capacities could result from a global extinction event that wipes out the biota of an island, decreasing its standing carrying capacity, and thus changing the stationary properties of the CTMC dispersal process. The point in time when there is a change in equilibrium frequencies and also the intensity of the extinction event (which might vary between areas) can be estimated by BI (Sanmartín 2020). Alternatively, the CTMC process may never attain equilibrium, or start with different values at root, such as in a directional CTMC process (Klofstein *et al.* 2015).

2.5.2. *Diversification-dependent models*

The latest exciting developments in parametric biogeography have been in the direction of implementing "state-dependent speciation and extinction (SSE) models", in which there is a causal relationship between range evolution and lineage diversification (Maddison *et al.* 2007, Goldberg *et al.* 2011; FitzJohn 2012). As explained above, BIB and DEC do not include a speciation parameter in the stochastic CTMC process that governs geographic evolution. This is unrealistic since diversification and range evolution clearly interact: for example, the dispersal of a species into a new region may result in increased speciation rates due to lower competition or access to novel environmental resources (Moore and Donoghue 2007). Moreover, unlike the DEC model, SSE models provide a complete parametric description of biogeographic evolution, since speciation is a rate parameter in the CTMC process. In the geological state-dependent speciation and extinction model (GeoSSE; Goldberg *et al.* 2011), the Q matrix includes parameters for anagenetic range expansion and range contraction or extinction, as well as parameters for lineage speciation within single areas (S_A, S_B) or within a widespread range (S_{AB}). There is also a parameter for lineage extinction within single areas (E_A, E_B): for widespread ranges, this is modeled as the sum of extinction events in single areas. All these parameters are time-dependent. The SSE counterpart of DEC+J is the ClaSSE model (Goldberg and Igic 2012), which allows changes in states to occur not only along branches (anagenetic) but also at speciation nodes (cladogenetic): this "founder-speciation" event is governed by its own time-dependent rate parameter in the Q matrix.

Coupling diversification with range evolution, as in GeoSSE and ClaSSE, allows statistical testing of classical hypotheses, such as whether widespread ranges lead to higher speciation rates (Goldberg *et al.* 2011) or whether extinction rates are dependent on area size or environmental heterogeneity (Meseguer *et al.* 2015). A shortcoming of SSE models is their computational complexity. The stationary distributions and parameter probabilities in SSE models are estimated through numerical integration, rather than analytically by matrix exponentiation as in DEC. One attractive avenue forward to tackle these computationally intractable models is the probabilistic programming language (PPL) framework (Ronquist *et al.* 2020).

2.5.3. *Ecology-integrative models*

Another exciting development in recent years is the long-sought-after integration of the ecological and historical (phylogenetic) sides of biogeography. Ecological biogeography is defined as dealing with environmental factors and evolutionary processes that act at short time scales and individual or population levels, such as biotic interaction (facilitation, competition), environmental filtering or random genetic drift. Historical biogeography is concerned with deep-time geological events and species-level evolutionary processes, such as dispersal, extinction or speciation (Sanmartín 2012). The distinction between these two approaches has become blurred. For example, a vicariance barrier can be geological (e.g. a mountain) or environmental (e.g. climatically inhospitable land where a species cannot maintain viable populations).

Similarly, for overland dispersal, both the physical bridge and the right environmental conditions along the corridor are a requisite (Donoghue 2008). Ecological niche models can be used to find areas that are within the environmental tolerances of a species, and this information can be used in a biogeographic analysis for modeling the probability of dispersal along corridors or across barriers (Smith and Donoghue 2010). The ecological preferences of ancestors can also be incorporated through the inclusion of fossil, extinct taxa in the analysis; this offers great potential for reconstructing species distributions over the distant past (Meseguer *et al.* 2015). Ecological processes such as competition and environmental filtering can be modeled in Quintero and Landis's (2019) composite biogeographic-trait evolutionary model: the rates of range expansion and range contraction depend on the trait values of other co-distributed species (effect of competition on biogeography), while the rate of divergence and convergence of trait values in a species depends on its sympatry with other species, gained or lost via colonization and extinction rates (effect of biogeography on traits).

2.6. Population-level and individual-based models

All models described above were designed to deal with phylogenies in which the terminal tips represent individual species (though BIB-DTA has been used in a phylogeographic context). CTMC processes are less appropriate to model the geographic evolution of individuals within a population, or between closely related populations, because they require the *a priori* definition of discrete geographic ranges and assume that movement between states is rare, that is, the chain remains in the same state and rarely jumps among states. When dealing with within-species biogeography or phylogeography, it is often difficult to define geographic ranges because boundaries are blurred by the frequent movement of individuals within populations and by gene flow. A Brownian Motion (BM) process, also termed "random walk" or diffusion model, is typically used for modeling the geographic evolution of populations and individuals (Lemey *et al.* 2011). This is a stochastic process with one parameter governing range evolution: there is a central value from where individuals move away with speed equal to this parameter. Unlike in biogeographic Markov models, tips in the phylogeny are individuals with associated geographical coordinates. Finally, models based on electric circuit-resistance theory (McRae *et al.* 2008) have been used in phylogeography to model the rate and path of movement or gene flow on heterogeneous landscapes. A special attraction of this model is the possibility to define connectivity maps based on 2D landscapes with barriers: low resistances are assigned to landscape feature types that are most permeable to movement or best promote gene flow, and high resistances assigned to movement barriers. The field of parametric phylogeography is in rapid expansion (Bloomquist *et al.* 2010), especially coalescent-based methods using approximate Bayesian computation, a likelihood-free Bayesian approach in which parameters in the model are estimated via simulation, and models are compared via summary statistics (Hickerson *et al.* 2007).

One class of expanding simulation models is forward-time, individual-based models, also termed *in silico* or automat models (Gotelli *et al.* 2009; Overcast *et al.* 2019). These models set up a series of rules by which speciation, extinction and dispersal of lineages can occur within an environmentally heterogeneous, two-dimensional gridded landscape; they are therefore spatially explicit models (Gotelli *et al.* 2009). These models have been used for testing macroecological hypotheses on species richness and distribution patterns, but some incorporate evolutionary predictions (Rangel *et al.* 2018). Recently, simulation modeling has experienced a spur forward, especially within the realm of phylogeography (Overcast *et al.* 2019), with the introduction of machine learning approaches and the integration of genetic data. Both *in silico* and machine learning approaches use simulations under pre-specified scenarios, as well as statistical comparison of observations against the distribution of simulated values to discriminate among alternative biogeographic scenarios. These models are less efficient for parameter inference than parametric approaches such as DEC or BIB, because a large range of values needs to be explored via simulation. Conversely, simulation models are more powerful in modeling complex phylogeographic scenarios involving multiple interacting parameters, since there is no need to derive the likelihood function and parameter dependencies. In particular, machine-learning methods are extremely flexible, with no cap on the number of parameters, and have been used for merging ecological and evolutionary processes (Overcast *et al.* 2019), trait-based biogeography (Sukumaran *et al.* 2016) or the integration of the spatial landscape (Tagliocollo *et al.* 2015). Some ML approaches do not rely on summary statistics and can be more efficient than ABC methods for phylogeographic inference (Fonseca *et al.* 2020).

2.7. References

Beaumont, M.A. (2010). Approximate Bayesian computation in evolution and ecology. *Annu. Rev. Ecol. Evol. Syst.*, 41, 379–406.

Bielejec, F., Lemey, P., Baele, G., Rambaut A., Suchard, M.A. (2014). Inferring heterogeneous evolutionary processes through time: From sequence substitution to phylogeography. *Syst. Biol.*, 63, 493–504.

Bloomquist, E.W., Lemey, P., Suchard, M.A. (2010). Three roads diverged? Routes to phylogeographic inference. *Trends Ecol. Evol.*, 25, 626–632.

Bremer, K. and Janssen, T. (2006). Gondwanan origin of major monocot groups inferred from dispersal-vicariance analysis. *Aliso*, 22, 22–27.

Bribiesca-Contreras, G., Verbruggen, H., Hugall, A.F., O'Hara, T.D. (2019). Global biogeographic structuring of tropical shallow-water brittle stars. *J. Biogeogr.*, 46, 1287–1299.

Brooks, D.R. (2005). Historical biogeography in the age of complexity: Expansion and integration. *Rev. Mex. Biodivers.*, 76, 79–94.

Buerki, S., Forest, F., Alvarez, N., Nylander, J.A.A., Arrigo, N., Sanmartin, I. (2011). An evaluation of new parsimony-based versus parametric inference methods in biogeography: A case study using the globally distributed plant family Sapindaceae. *J. Biogeogr.*, 38, 531–550.

Cybis, G.B., Sinsheimer, J.S., Lemey, P., Suchard, M.A. (2013). Graph hierarchies for phylogeography. *Phil. Trans. R. Soc. B*, 368, 20120206.

Darwin, C. (1859). *On the Origin of Species by Means of Natural Selection or the Preservation of Favored Races in the Struggle for Life*. John Murray, London.

De Maio, N., Wu, C.-H., O'Reilly, K.M., Wilson, D. (2015). New routes to phylogeography: A Bayesian structured coalescent approximation. *PLoS Genet.*, 11(8), e1005421.

Donoghue, M.J. (2008). A phylogenetic perspective on the distribution of plant diversity. *Proc. Natl. Acad. Sci., USA*, 105, 11549–11555.

Donoghue, M.J. and Moore, B.R. (2003). Toward an integrative historical biogeography. *Integr. Comp. Biol.*, 43, 261–270.

Donoghue, M.J. and Smith, S.A. (2004). Patterns in the assembly of temperate forests around the Northern Hemisphere. *Philos. Trans. R. Soc. London Biol.*, 359, 1633–1644.

Faith, D.P. and Cranston, P.S. (1991). Could a cladogram this short have arisen by chance alone? On permutation tests for cladisticand permutation structure. *Cladistics*, 7, 1–28.

Faria, N.R., Suchard, M.A., Rambaut, A., Streicker, D.G., Lemey, P. (2013). Simultaneously reconstructing viral cross-species transmission history and identifying the underlying constraints. *Phil. Trans. R. Soc. B.*, 368, 20120196.

Fitzjohn, R.G. (2012). Diversitree: Comparative phylogenetic analyses of diversification in R. *Methods Eco. Evol.*, 3(6), 1084–1092.

Fonseca, E.M., Colli, G.R., Werneck, F.P., Carstens, B.C. (2020). Phylogeographic model selection using convolutional neural networks. bioRxiv. doi: https://doi.org/10.1101/2020.09.11.291856.

Goldberg, E. and Igic, B. (2012). Tempo and mode in plant breeding system evolution. *Evolution*, 66, 3701–3709.

Goldberg, E.E., Lancaster, L.T., Ree, R.H. (2011). Phylogenetic inference of reciprocal effects between geographic range evolution and diversification. *Syst. Biol.*, 60, 451–465.

Goodman, S. (1999). Toward evidence-based medical statistics: The Bayes factor. *Ann. Intern. Med.*, 130, 1005–1013.

Gotelli, N.J., Anderson, M.J., Arita, H.T., Chao, A., Colwell, R.K., Connolly, S.R., Currie, D.J., Dunn, R.R., Graves, G.R., Green, J.L., Grytnes, J.-A., Jiang, Y.-H., Jetz, W., Lyons, S.K., McCain, C.M., Magurran, A.E., Rahbek, C., Rangel, T.F.L.V.B., Soberón, J., Webb, C.O., Willig, M.R. (2009). Patterns and causes of species richness: A general simulation model for macroecology. *Ecol. Lett*, 12, 873–886.

Harris, A.J. and Xiang, Q.Y. (2009). Estimating ancestral distributions of lineages with uncertain sister groups: A statistical approach to dispersal vicariance analysis and a case using Aesculus L. (Sapindaceae) including fossils. *J. Syst. Evol.*, 47, 349–368.

Hickerson, M.J., Stahl, E., Takebayashi, N. (2007). msBayes: Pipeline for testing comparative phylogeographic histories using hierarchical approximate Bayesian computation. *BMC Bioinform.*, 8, 268.

Huelsenbeck, J.P., Rannala, B., Masly, J.P. (2000). Accommodating phylogenetic uncertainty in evolutionary studies. *Science*, 288, 2349–2350.

Humphries, C.J. and Parenti, L. (1999). *Cladistic Biogeography*, 2nd edition. Oxford University Press, New York.

Klopfstein, S., Vilhelmsen, L., Ronquist, F. (2015). Nonstationary Markov model detects directional evolution in hymenopteran morphology. *Syst. Biol.*, 64, 1089–1103.

Kodandaramaiah, U. (2010). Use of dispersal-vicariance analysis in biogeography – A critique. *J. Biogeogr.*, 37, 3–11.

Landis, M.J. (2017). Biogeographic dating of speciation times using paleogeographically informed processes. *Syst. Biol.*, 66, 128–144.

Landis, M.J., Matzke, N.J., Moore, B.R., Huelsenbeck, J.P. (2013). Bayesian analysis of biogeography when the number of areas is large. *Syst. Biol.*, 62, 789–804.

Lemey, P., Rambaut, A., Drummond, A.J., Suchard, M.A. (2009). Bayesian phylogeography finds its roots. *PLoS Comput. Biol.*, 5.

Lemey, P., Rambaut, A., Welch, J.J., Suchard, M.A. (2010). Phylogeography takes a relaxed random walk in continuous space and time. *Mol. Biol. Evol.*, 27, 1877–1885.

Lieberman, B.S. (2003). Paleobiogeography: The relevance of fossils to biogeography. *Annu. Rev. Ecol. Evol. Syst.*, 34, 51–69.

Maddison, W.P., Midford, P.E., Otto, S.P. (2007). Estimating a binary character's effect on speciation and extinction. *Syst. Biol.*, 56, 701–710.

Mairal, M., Pokorny, L., Aldasoro, J.J., Alarcón, M., Sanmartín, I. (2015). Ancient vicariance and climate-driven extinction explain continental-wide disjunctions in Africa: The case of the rand flora genus *Canarina* (Campanulaceae). *Mol. Ecol.*, 24, 1335–1354.

Marshall, C.J. and Liebherr, J.K. (2000). Cladistic biogeography of the Mexican transition zone. *J. Biogeogr.*, 27, 203–216.

Massana, K.A., Beaulieu, J.M., Matzke, N.J., O'Meara, B.C. (2015). Non-null effects of the null range in biogeographic models: Exploring parameter estimation in the DEC model. bioRxiv, 026914.

Mastretta-Yanes, A., Moreno-Letelier, A., Pinñero, D., Jorgensen, T.H., Emerson, B.C. (2015). Biodiversity in the Mexican highlands and the interaction of geology, geography and climate within the Trans-Mexican Volcanic Belt. *J. Biogeogr.*, 42, 1586–1600.

Matzke, N.J. (2014). Model selection in historical biogeography reveals that founder-event speciation is a crucial process in island clades. *Syst Biol.*, 63, 951–970.

McRae, B.H., Dickson, B.G., Keitt, T.H., Shah, V.B. (2008). Using circuit theory to model connectivity in ecology, evolution, and conservation. *Ecology*, 89, 2712–2724.

Meseguer, A.S., Lobo, J.M., Ree, R., Beerling, D.J., Sanmartín, I. (2015). Integrating fossils, phylogenies, and niche models into biogeography to reveal ancient evolutionary history: The case of *Hypericum* (Hypericaceae). *Syst. Biol.*, 64, 215–232.

Moore, B.R. and Donoghue, M.J. (2007). Correlates of diversification in the plant clade Dipsacales: Geographic movement and evolutionary innovations. *Am. Nat.*, 170, 28–55.

Muller, N.T., Rasmussen, D.A., Stadler, T. (2017). The structured coalescent and its approximations. *Mol. Biol. Evol.*, 34, 2970–2981.

Nelson, G. and Platnick, N.I. (1981). *Systematics and Biogeography: Cladistics and Vicariance*. Columbia University Press, New York, NY.

Nylander, J.A.A., Olsson, U., Alstrom, P., Sanmartín, I. (2008). Accounting for phylogenetic uncertainty in biogeography: A Bayesian approach to dispersal-vicariance analysis of the thrushes (Aves: *Turdus*). *Syst Biol.*, 57, 257–68.

Overcast, I., Emerson, B.C., Hickerson, M.J. (2019). An integrated model of population genetics and community ecology. *J. Biogeogr.*, 46, 816–829.

Page, R.D.M. (1993). Genes, organisms, and areas: The problem of multiple lineages. *Syst. Biol.*, 42, 77–84.

Pagel, M., Meade, A., Barker, D. (2004). Bayesian estimation of ancestral character states on phylogenies. *Syst. Biol.*, 53, 673–684.

Prieto-Márquez, A. (2010). Global historical biogeography of hadrosaurid dinosaurs. *Zool. J. Linn. Soc.*, 159, 503–525.

Quinn, S., McFrederick, D., Taylor, R. (2013). Evolutionary history of nematodes associated with sweat bees. *Mol. Phylogen. Evol.*, 66, 847–856.

Quintero, I. and Landis, M.J. (2020). Interdependent phenotypic and biogeographic evolution driven by biotic interactions. *Syst. Biol.*, 69, 739–755.

Rangel, T.F., Edwards, N.E., Holden, P.B., Diniz-Filho, J.A.F., Gosling, W.D., Coelho, M.T.P., Cassemiro, F.A.S., Rahbek, C., Colwell, R.K. (2018). Modeling the ecology and evolution of biodiversity: Biogeographical cradles, museums, and graves. *Science*, 361, eaar5452.

Ree, R.H. and Sanmartín, I. (2009). Prospects and challenges for parametric models in historical biogeographical inference. *J. Biogeogr.*, 36, 1211–1220.

Ree, R.H. and Sanmartín, I. (2018). Conceptual and statistical problems with the DEC+J model of founder-event speciation and its comparison with DEC via model selection. *J. Biogeogr.*, 45, 741–749.

Ree, R.H. and Smith S.A. (2008). Maximum likelihood inference of geographic range evolution by dispersal, local extinction, and cladogenesis. *Syst. Biol.*, 57, 4–14.

Ree, R.H., Moore, B.R., Webb, C.O., Donoghue, M.J. (2005). A likelihood framework for inferring the evolution of geographic range on phylogenetic trees. *Evolution*, 59, 2299–2311.

Robert, E., Adrian, K., Raftery, E. (1995). Bayes factors. *J. Am. Stat. Assoc.*, 90, 773–795.

Ronquist, F. (1997). Dispersal-vicariance analysis: A new approach to the quantification of historical biogeography. *Syst. Biol.*, 46, 195–203.

Ronquist, F. (2003). Parsimony analysis of coevolving associations. In *Tangled Trees: Phylogeny, Cospeciation, and Coevolution*, RDM, 22–64. University of Chicago Press, Chicago, IL.

Ronquist, F. (2004). Bayesian inference of character evolution. *Trends Ecol. Evol.*, 19, 475–481.

Ronquist, F. and Sanmartín, I. (2011). Phylogenetic methods in biogeography. *Annu. Rev. Ecol., Evol., and Syst.*, 42, 441–464.

Ronquist, F., Kudlicka, J., Senderov, V., Borgström, J., Lartillot, N., Lundén, D., Murray, L., Schön, T.B., Broman, D. (2020). Universal probabilistic programming: A powerful new approach to statistical phylogenetics [Online]. Available at: bioRxiv preprint doi: https://doi.org/10.1101/2020.06.16.154443, June 18, 2020.

Sanmartín, I. (2012). Historical biogeography: Evolution in time and space. *Evo. Edu. Outreach*, 5, 555–568.

Sanmartín, I. (2020). Breaking the chains of parsimony: The development of parametric methods in historical biogeography. In *Biogeography: An Ecological and Evolutionary Approach*, 10th edition, Cox, B.C., Moore, P.D., Ladle, R. (eds). Wiley, Hoboken, NJ.

Sanmartín, I. and Ronquist, F. (2004). Southern hemisphere biogeography inferred by event-based models: Plant versus animal patterns. *Syst. Biol.*, 53, 216–243.

Sanmartín, I., Enghoff, H., Ronquist, F. (2001). Patterns of animal dispersal, vicariance and diversification in the Holarctic. *Biol. J. Linn. Soc.*, 73, 345–390.

Sanmartín, I., Wanntorp, L., Winkworth, R. (2007). West wind drift revisited: Testing for directional dispersal in the southern hemisphere using event-based tree fitting. *J. Biogeogr.*, 34, 398–416.

Sanmartín, I., Van Der Mark, P., Ronquist, F. (2008). Inferring dispersal: A Bayesian approach to phylogeny-based island biogeography, with special reference to the Canary Islands. *J. Biogeogr.*, 35, 428–449.

Sanmartín, I., Anderson, C.L., Alarcon, M., Ronquist, F., Aldasoro, J.A. (2010). Bayesian island biogeography in a continental setting: The Rand Flora case. *Biol. Lett.*, 6, 703–707.

Smith, S.A. and Donoghue, M.J. (2010). Combining historical biogeography with niche modeling in the *Caprifolium* clade of *Lonicera* (Caprifoliaceae, Dipsacales). *Syst. Biol.*, 59, 322–341.

Sukumaran, J., Economo, E.P., Knowles, L. (2016). Machine learning biogeographic processes from biotic patterns: A new trait-dependent dispersal and diversification model with model choice by simulation-trained discriminant analysis. *Syst. Biol.*, 65, 525–545.

Tagliacollo, V.A., Duke-Sylvester, S.M., Matamoros, W.A., Chakrabarty, P., Albert, J.S. (2015). Coordinated dispersal and pre-Isthmian assembly of the Central American ichthyofauna. *Syst. Biol.*, 66, 183–196.

Wojciki, M. and Brooks, D.R. (2005). PACT: An efficient and powerful algorithm for generating area cladograms. *J. Biogeogr.*, 32, 755–774.

Yu, Y., Harris, A.J., Blair C., He, X. (2010). RASP (Reconstruct Ancestral State in Phylogenies): A tool for historical biogeography. *Mol. Phylogen. Evol.*, 87, 46–49.

3

Phylogeography

Inessa VOET and Violaine NICOLAS
Institut de Systématique, Evolution, Biodiversité (ISYEB), National Museum of Natural History, Paris, France

3.1. Introduction

Sometimes, a novel word encapsulates a concept and becomes part of the working lexicon of a discipline. Such was the case for "phylogeography". This word was introduced by Avise in 1987 in his work "Intraspecific Phylogeography: The Mitochondrial DNA Bridge Between Population Genetics and Systematics" (Avise *et al.* 1987). Phylogeography endeavors to understand the processes that underlie the geographical distribution of genetic variation within and among closely related species. Phylogeography takes a population genetics and phylogenetic perspective on biogeography. In the early days of phylogeography, scientists used mitochondrial (mt) DNA to precisely address how conspecific individuals were genealogically linked through shared ancestors. In genetic surveys of various species, striking patterns were being uncovered in the spatial arrangements of mtDNA lineages. In other words, genealogy and geography seemed to be connected. After 1987, "phylogeography" became the convenient shorthand for referring to explicit genealogical inquiries (including those for nuclear genes) into the spatial and temporal dimensions of microevolution. The unwavering popularity of phylogeography is indisputable (Figure 3.1).

Biogeography,
coordinated by Eric GUILBERT.

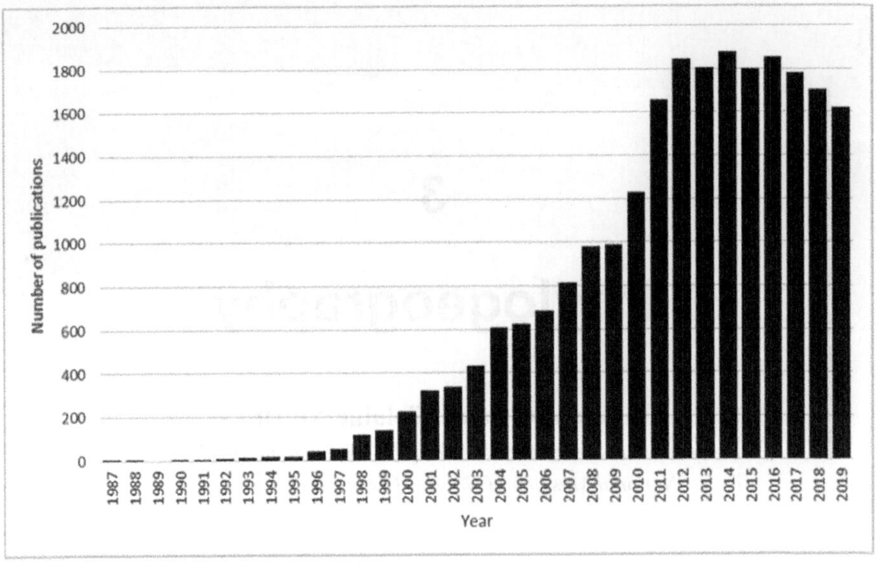

Figure 3.1. *Number of publications with the word "phylogeography" in the "topic" per year, according to Web of Science*

Phylogeography deals with historical, phylogenetic components of spatial distributions of gene lineages. It thus provides a powerful empirical and conceptual bridge between microevolutionary (population genetics) and macroevolutionary (phylogeny, historical biogeography) disciplines. Phylogeography describes how genetic variation, introduced by mutation, is geographically structured within and between populations by population genetic processes, such as drift, gene flow, selection and recombination. For populations that have been historically separated and have experienced little or no gene flow, genetic differences can accumulate by these evolutionary processes, potentially resulting in speciation.

Phylogeography is sometimes considered as a sub-discipline of biogeography. Phylogeographic studies provide detailed species-specific information on how geologic events, environmental influences and geographic factors interact with aspects of a species ecology and natural history in shaping its evolution. Phylogeographers seek to interpret the extent and mode by which historical processes in population demographics and landscape alterations may have left evolutionary footprints on the contemporary geographic distributions and phylogenetic signature of gene-based organismal traits. Thus, phylogeography serves to expand and to balance the perspectives of ecogeography (Avise *et al.* 1987; Avise 2000).

The aim of this chapter is to explain what phylogeography is and to present methodological and conceptual developments in this field.

3.2. The early days of phylogeography: cytoplasmic genomes and qualitative *post hoc* explanations of historical processes

During the first decades of phylogeography, most studies involved analyses of exclusively animal mtDNA. The wide use of this cytoplasmic genome in phylogeographic studies is explained by the fact that it has several phylogenetically favorable properties (Avise *et al.* 1987; Avise 2000, 2009): 1) it occurs in all animals, and homologous comparisons can thus be made among a wide variety or organisms; 2) it is easy to isolate (most somatic cells contain hundreds or thousands of mtDNA molecules, and heteroplasmy is uncommon); 3) it has a simple genetic structure lacking complicated features such as repetitive DNA and introns, and with many genes being free of indels; 4) due to its maternal inheritance (mtDNA inheritance is both haploid and asexual), it exhibits a straightforward mode of genetic transmission, without recombination or other genetic rearrangements suffered by the nuclear DNA; and 5) although some coding segments are highly conserved, many mitochondrial genes evolve at a rapid pace. Therefore, genetic disparity between diverging lineages accumulates rapidly, and can be detected on a relatively short temporal scale. MtDNA genotypes are referred to as haplotypes, which differ from one another by particular mutations accumulated since female ancestors were last shared. Because of the rapid pace of mtDNA evolution, many different haplotypes typically coexist within a species. These haplotype sequences can be used to estimate the matrilineal histories of individuals and populations.

Plants have two cytoplasmic genomes: mtDNA and chloroplastic DNA (cpDNA). Contrary to animals, plant mtDNA is poorly suited for phylogeographic studies (Avise 2009) within species because of its slow pace of sequence evolution (about 100 times slower than in animals) and several other technical hurdles (e.g. inheritance is not always maternal, gene order rearrangements are common and plant mtDNA is highly variable in size). In the early days of plant phylogeography, most studies were conducted on cpDNA because, like mtDNA in animals, it has several phylogenetically favorable properties (Avise 2000): 1) cpDNA typically occurs in many copies per cell, and most individuals display a single cpDNA haplotype sequence; 2) several chloroplastic genes evolve at a relatively rapid pace and various individuals within a species can show detectable differences in cpDNA sequences, 3) it is haploid and non-recombining, even if the transmission can be matrilineal, patrilineal or a mixture of the two depending on the mode of genetic transmission of cpDNA in the species being investigated.

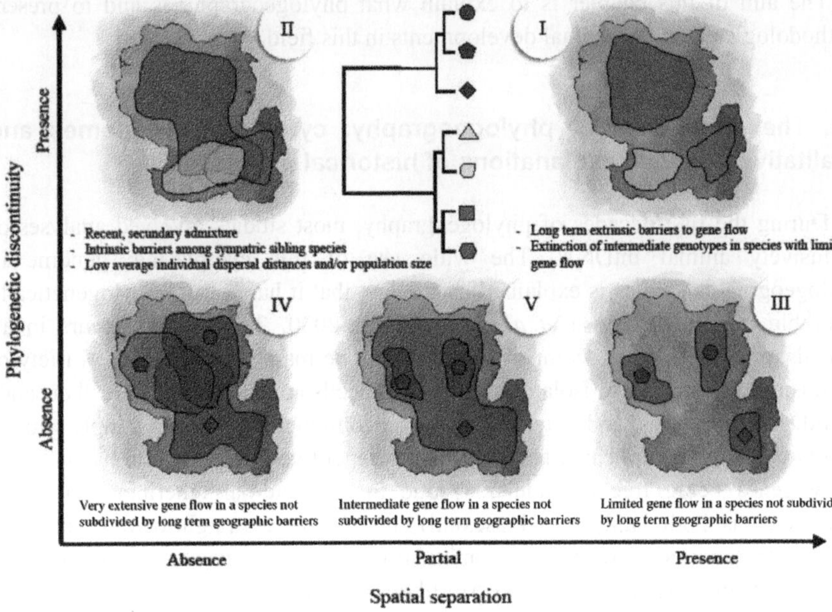

Figure 3.2. *Schematic representation of the five categories of phylogeographic patterns theoretically observable in mtDNA surveys. For a color version of this figure, see www.iste.co.uk/guilbert/biogeography.zip*

Using mtDNA, Avise (Avise *et al.* 1987; Avise 2000) showed that intraspecific phylogenies overlaid on geographic maps can lead to five major categories (Figure 3.2) of possibilities, depending on the combination of phylogenetic discontinuity (or genetic "breaks": arrays of related genotypes differing from other such arrays by many mutational steps) and spatial separation levels. The authors provided interpretations for these five categories:

– I – Phylogenetic discontinuities with spatial separation: genetically divergent populations occupy separate geographic regions within the range of the species. This situation can be the result of long-term extrinsic (i.e. zoogeographic) barriers to gene flow, or of the extinction of intermediate genotypes in widely distributed species with limited dispersal and gene flow capabilities.

– II – Phylogenetic discontinuities with the lack of spatial separation: genetically divergent populations co-occur geographically. This can be the result of recent secondary admixture zones or intrinsic (e.g. reproductive isolation) barriers acting among sympatric sibling species. Phylogeographic breaks can also arise in continuously distributed species if the average individual dispersal distances and/or population size of the species are low (Irwin 2002).

– III – Phylogenetic continuity with spatial separation, which involves historically limited gene flow between populations in a species not subdivided by firm long-term zoogeographic barriers to dispersal.

– IV – Phylogenetic continuity with a lack of spatial separation which arises when there is very extensive gene flow in a species not subdivided by long-term zoogeographic barriers.

– V – Phylogenetic continuity with partial spatial separation: in this category, some mtDNA haplotypes (the ancestral ones) are geographically widespread, while allied haplotypes (newly arisen mutations) are localized. This pattern can appear in taxa with historically intermediate levels of gene flow between geographic populations.

3.3. Statistical phylogeography

In the early days of phylogeography, mtDNA or cpDNA gene trees were used for demographic inference. However, there is an inherent contradiction in using gene trees as a literal interpretation of a species past, because gene trees contain historical signal, as well as stochasticity: there is a random loss of gene lineages by genetic drift because of the different number of descendants individuals can or will produce. Phylogeography has undergone both marked methodological and philosophical transformations in how genetic data are used for demographic inference. Because many events may have occurred in the past (e.g. bottlenecks, population expansion, migration and vicariance), and gene lineages may be lost by chance, a species history might not be easily inferred from a gene genealogy. In the early 2000s, several researchers criticized the narrative nature and lack of statistical rigor of phylogeography. They emphasized that it was important to consider the stochasticity of population genetic processes and to explicitly assess the confidence of various phylogeographic conclusions. There are several reasons why historical inferences directly derived from a gene genealogy may be inaccurate or misleading, which are not mutually exclusive (reviewed by Knowles 2004): 1) the stochastic process of lineage sorting can produce discordance between the population's and the gene's history; 2) the actual history can be obscured by a deep coalescence of gene lineages; 3) the resolution of the genetic marker can be insufficient for recovering the population history from a gene tree estimate; 4) the gene genealogy can reflect the action of selection rather than the population's history; and 5) alternative hypotheses can be indistinguishable because of the high stochastic variance of independent gene trees. A transition occurred from describing to testing the processes underlying patterns of genetic variation. This was concomitant with a shift in the way historical inferences were made, namely the explicit consideration of stochastic variance and reference to predictions based on models that are defined *a priori*. This was the rise of what is called "statistical phylogeography" (Knowles and Maddison 2002; Knowles 2004, 2009). Inferences about a species' history are now based on statistical tests of historical hypotheses and estimates of demographical parameters. A wide range of historical scenarios can be modeled, from very simple to very complex, including different geographical scenarios, timing of population divergence and estimates of effective population sizes, migration, population admixture and growth.

Statistical phylogeographical methods generally fall into one of two categories: those that take a summary-statistic approach versus likelihood or Bayesian analyses. In likelihood analyses, the probability of observing the data under alternative models is calculated, while in the summary-statistic approach, a simple statistic summarizes the data. Summary-statistic approaches are easy to implement, but their interpretations are sensitive to the defined models and different population histories can sometimes produce the same summary statistic, because they do not necessarily take advantage of all the information in the data. For example, a low Fst-value could indicate an older population divergence with substantial gene flow, or a recent split with no gene flow where the shared lineages between populations reflect their common ancestry. Knowledge of the biological system under study is essential for deciding which approach to use for phylogeographic analyses. What might be an appropriate method for one study might not be for another, depending on the specifics of each species history and the type of data collected.

Each gene has its own history, and gene trees can be different from the species tree. The analysis of multiple unlinked loci is critical for accommodating coalescent stochasticity and improving the accuracy of inferences about demographic history and estimates of divergence times (Knowles 2004). Single-copy nuclear genes can provide voluminous sequence data for phylogeographic assessments at the intraspecific level. However, in sexually reproducing organisms, meiosis and syngamy continually re-assort unlinked genes into new multi-locus combinations. These recombinational processes can also manifest within a locus whenever successful meiotic crossover events occur between different alleles. Each recombinant allele that emerges is then an amalgamated stretch of DNA, the subsets of which might have quite different evolutionary pasts. Thus, intragenic (or inter-allelic) recombination tends to complicate and garble what would otherwise be cleaner genealogical signatures from the nuclear genome. This contrasts with the historical clarity characteristic of the nonrecombining cytoplasmic genomes. Three technical and biological hurdles have for a long time impeded progress (Avise

2009): 1) the relatively slow pace of sequence evolution at many nuclear loci; 2) the difficulty in isolating nuclear haplotypes, one at a time, from diploid organisms; and 3) the phenomenon of intragenic recombination. The first complication was first partly overcome by sequencing single-copy nuclear sequences that evolve more rapidly than average (e.g. intron sequences instead of exon sequences). The second complication was overcome by sequencing genes that are present in haploid conditions in each member of the heterogametic sex (e.g. X- and Y-chromosomes of mammals, Z- and W-chromosomes of birds) or by resolving allele ambiguities in diploid conditions through cloning or via computation using software like PHASE (Stephens *et al.* 2001; Stephens and Donnelly 2003). Generating such multilocus data was time expensive: most practitioners of biogeography have spent a significant portion of their time developing and screening molecular markers suitable to their study system and appropriate to their evolutionary timescale of interest (i.e. with enough variability), amplifying and sequencing DNA for each sample at each locus, and phasing nuclear data.

The development of next-generation sequencing (NGS) was very appealing in the field of phylogeography as it could have the ability to condense the many steps of multilocus data generation for non-model organisms into a more time-efficient and cost-effective process. However, NGS has been slow to take root in phylogeography compared to other fields like metagenomics and disease genetics. According to McCormack *et al.* (2013), this lag has been caused by the predominant focus on non-model organisms and the need for sequencing large numbers of samples per species. Obtaining whole-genome data for hundreds of individuals is still too costly; thus, a method of genome size reduction is required. Phylogeography requires homologous genomic regions from multiple individuals to infer gene genealogies. In traditional Sanger sequencing, this is straightforward through the design of PCR primers specific to a particular locus. With NGS, DNA is not necessarily enriched for a single locus via PCR-based amplification, although this is one possible application, but for many loci through a variety of methods involving reduction of the size of the genome. One of the main drawbacks of NGS methods is that there is little control over exactly what regions of the genome are sequenced. On the other hand, the number of targets (and the number of reads associated with each target) is increased by orders of magnitude. Using NGS for phylogeography is only cost-effective if many individuals can be combined (multiplexed) in the same sequencing run and the costs subdivided among many samples. To this aim, short identifying DNA sequences (called "indexes", "barcodes" or "tags") need to be incorporated into the DNA fragments either by PCR or ligation. These tags identify an individual prior to pooling with other tagged samples. Sequences are later sorted using bioinformatics. Several reviews of the different NGS sequencing platforms and chemistries already exist, and we invite interested readers to read these papers

(Shendure and Ji 2008; Glenn 2011). Methods that reduce the genome by restriction digest and manual size selection are particularly appropriate for studies at the intraspecific level, whereas methods that target specific genomic regions (i.e. target enrichment or sequence capture) have wider applicability from the population level to deep-level phylogenomic (McCormack *et al.* 2013). An overview of how to analyze NGS data to arrive at data sets applicable to the standard toolkit of phylogeography, including initial data processing to alignment and genotype calling, can be found in McCormack *et al.* (2013). Most existing NGS technologies produce short reads and therefore are best suitable for generating single nucleotide polymorphisms (SNPs), not whole loci featuring many linked SNPs. This has for a long time limited the application of NGS data with respect to the standard analytical toolkit of phylogeography that rely mostly on gene trees. Individual SNPs have low information content on a per-locus basis and have been predominately used with classification methods such as structure (Pritchard *et al.* 2000), principal component analysis (e.g. Dufresnes *et al.* 2020) or K-means clustering (Myers *et al.* 2020). However, the use of SNPs for phylogenetic inference is growing rapidly (reviewed by Leaché and Oaks 2017). A fundamental difference between genome-wide-SNP phylogenetics and traditional approaches that use gene sequences is that DNA sequences are linear strings of nucleic acids composed of constant and variable characters, while SNP data are composed exclusively of variable characters, either as a direct result of the data collection strategy, or after bioinformatic steps removing the constant characters. This can lead to extreme overestimation of evolutionary rates and inaccurate topologies, branching times and support values (Leaché *et al.* 2015). Another important problem with the use of SNPs is that the amount of missing data increases with the number of loci and sample size, due to allelic dropout. More work is needed to evaluate and select missing data thresholds that avoid negative consequences for phylogenetic analyses (Leaché and Oaks 2017). Several methods for estimating SNP phylogenies can be used. Data concatenation, in which SNPs are put together end-to-end to construct one exceptionally long super locus, is problematic because it ignores incomplete lineage sorting, and in doing so assumes that all SNPs share the same coalescent history. If concatenation is used, studies suggest it is better to apply this method to the original sequence alignments from which the SNPs were extracted: removing constant sites results in acquisition bias that can inflate branch length estimates and in extreme cases produce an inaccurate phylogeny (Leaché *et al.* 2015). Methods for estimating species trees in the presence of incomplete lineage sorting have been developed. The most popular of those methods are summary methods like ASTRAL (Mirarab *et al.* 2014) and ASTRAL-2 (Mirarab and Warnow 2015). Several other methods were recently developed, such as SNAPP (Bryant *et al.* 2012), PoMo (De Maio *et al.* 2015) and SVDquartets (Chifman and Kubatko 2014).

To conclude, phylogeography relied heavily on non-recombining and rapidly evolving mtDNA or cpDNA to match gene genealogies with geography, but the advent of genomic data in combination with the development of coalescent theory has revolutionized the field. In general, the application of genome-wide approaches uncovers detailed population structure that is often missed by traditional markers such as mtDNA. In addition, applying phylogeographic methods to genome-wide data sets has shown to what a great extent different genes often result in different gene trees. This phylogenetic incongruence can provide a more detailed picture of population history because different gene trees capture particular historical events and population genetic processes that have shaped the present patterns of genetic diversity. However, recent work has also uncovered high levels of reticulation due to recombination and gene flow (Ottenburghs *et al.* 2019). New statistical methods are being developed to deal with such reticulated scenarios.

3.4. Comparative phylogeography

Comparative phylogeography emerged three decades ago as an integrative approach to historical biogeography (Arbogast and Kenagy 2001). Comparative phylogeography can be defined as the study of the effects of evolutionary history and biogeography on the distribution of genetic variation in co-distributed species. If spatially consistent genealogical splits appear across multiple co-distributed species, the responsible evolutionary forces must have had widespread effects at the level of biotic communities or ecosystems. Comparative phylogeographic studies can reveal ways in which entire communities or assemblages are structured by shared responses to a common past and unique species-specific historical event. Thus, they can contribute to broader studies of ecology and evolution in a number of ways (Bermingham and Moritz 1998): 1) Phylogeographic analysis can historically and evolutionarily identify independent geographic regions that can be considered as natural replicates among which generalizations about specific processes can be tested statistically. For example, the evolutionary response to selection gradients can be compared across different historical isolates; 2) Phylogeography can provide an evolutionary and geographical context for the species comprising ecological communities, thus permitting determination of historical and spatial influences on patterns of species richness; and 3) Understanding of historical responses to changes in the landscape and the identification of evolutionarily isolated areas can inform conservation strategies.

In general, a comparative phylogeographic study has two phases (Gutiérrez-García and Vázquez-Domínguez 2011). It begins with a descriptive survey that includes the collection of genetic data from the species and phylogenetic analyses of that data. The intraspecific data obtained at this phase provide insight into phylogenetic relationships of populations, genetic diversity and genetic distance between lineages, effective population size, gene flow and the timing of diversification events. The second phase is the comparative component and includes diverse analyses to test for congruence between the evolutionary and distributional histories of each species; it also includes evaluation of the geographical, ecological and biological hypotheses that could explain those histories. The aim at this phase is to evaluate whether the evolutionary histories of these species show a shared response to the same historical events.

Since its beginning, comparative phylogeography uses concordance in patterns of genetic variation as a criterion for evaluating hypotheses. Phylogeographic congruence among several co-distributed taxa has provided key evidence of the impact of geological events, biogeographic barriers or past environmental change on today's distribution of genetic variation (Avise 2000; Hewitt 2004a, 2004b). Comparative phylogeographic analyses have been used to address a broad array of ecological and evolutionary questions, from inferring Quaternary refugia and assigning conservation priorities (Marske *et al.* 2013; Dauby *et al.* 2014; Médail and Baumel 2018) to evaluating the stability of ecological associations between interacting organisms (Stone *et al.* 2012). Commonly applied tests include the evaluation of spatially concordant phylogeographic breaks, concerted demographic expansion/contraction among members of an ecological community or temporally synchronous population divergences (reviewed by Papadopoulou and Knowles 2016). Methodological advances of the last decade, especially coalescent-based tools for hypothesis testing and parameter estimation that include statistical assessment of concordance across taxa, such as the widely used hierarchical approximate Bayesian computation (Hickerson *et al.* 2006; Huang *et al.* 2011), also promoted the utility of the concordance criterion.

Rigorously testing the congruence of phylogeographical patterns among taxa (as a result of a common evolutionary history) is no easy task, considering that even sister species can differ in their ecological characteristics and life-history features. Species-specific demographic and ecological characteristics can result in different outcomes of the same historical and biogeographical events and, consequently, in the loss of concordant phylogeographical patterns between species. Therefore, in the planning of a comparative phylogeographic study, the information obtained through intraspecific phylogeographic evaluations is of fundamental value. It is also important to consider, among other things, the stochastic variation of loci, the degree of taxonomic resolution in the species of interest, their phenotypic characteristics (ecology, traits that limit or promote movement, traits having an impact on demography) and fossil evidence and biogeographical, geological and historical information relating to the geographical distributions being studied.

The term "comparative phylogeography" was also recently used not only for comparing co-distributed species (i.e. in a spatial sense), but also in at least three other senses: (i) genomic – sharing phylogeographic patterns across multiple loci within a species, (ii) taxonomic – emphasizing how phylogeographic findings have impacted ecological and evolutionary thought in a diversity of organismal groups and (iii) conceptual – addressing the place of phylogeography in relation to various allied disciplines in the biodiversity sciences (Avise *et al.* 2016).

3.5. Integrative studies

The need to incorporate other sources of data corroborating the genetic data used in phylogeographic tests was acknowledged from the beginning of the development of statistical phylogeography (Knowles 2004, 2009). Fossil data are commonly used as an independent source of data (e.g. Gavin *et al.* 2014; Lalis *et al.* 2016; Lalis *et al.* 2019). However, such data are not available for all species.

3.5.1. *Integration of ecological niche modeling in phylogeographic studies*

Species distribution modeling (SDMs; also known as ecological niche models or ENMs) techniques generate information about abiotic preferences and tolerances of species, and hence estimates of the current, past and potential future distribution of species. ENMs employ independent data sources that can be used to evaluate or develop phylogeographic hypotheses about the processes generating patterns of genetic variation in disparate taxa (Carstens and Richards 2007; Richards *et al.* 2007; Alvarado-Serrano and Knowles 2014). For generating alternative biogeographical hypotheses, the necessary data consist of 1) a set of georeferenced localities that describe where the species has been documented to occur, 2) a set of GIS layers containing information about the pertinent aspects of the current environment for the geographic area and species of interest and 3) for the case of paleodistributions, a second set of GIS layers describing an estimate of the environment at a particular time period of interest in the past (Figure 3.3). Using these inputs and an SDM algorithm, both the current and past distributions of the focal species can be estimated. These estimates of a species' past distributions, or paleodistribution models, can then be directly used to explain the obtained genetic structure in a correlative (or corroborative) manner, or to guide the generation of alternative biogeographical hypotheses (or predictions) that can be tested with genetic data (Figure 3.3).

Increased availability of species distribution data on the Internet (e.g. GBIF, VERTNET), as well as increased availability of climatic or environmental variables for both current (e.g. Worldclim, SRTM 90m DEMs, HYDRO1k) and past data (e.g. WorldClim data for the last glacial maximum or last interglacial), and development of easy-use algorithms (e.g. Maxent) explain the rapid development of ecological niche modeling approaches and their use in phylogeographic studies. A rapid search in the Web of Science database using the words "phylogeography" and "ecological niche modeling" shows that the first papers combining these two approaches were published in the early 2000s, and increased considerably after 2010. During the last five years, this topic combination has been represented by almost 100 papers per year.

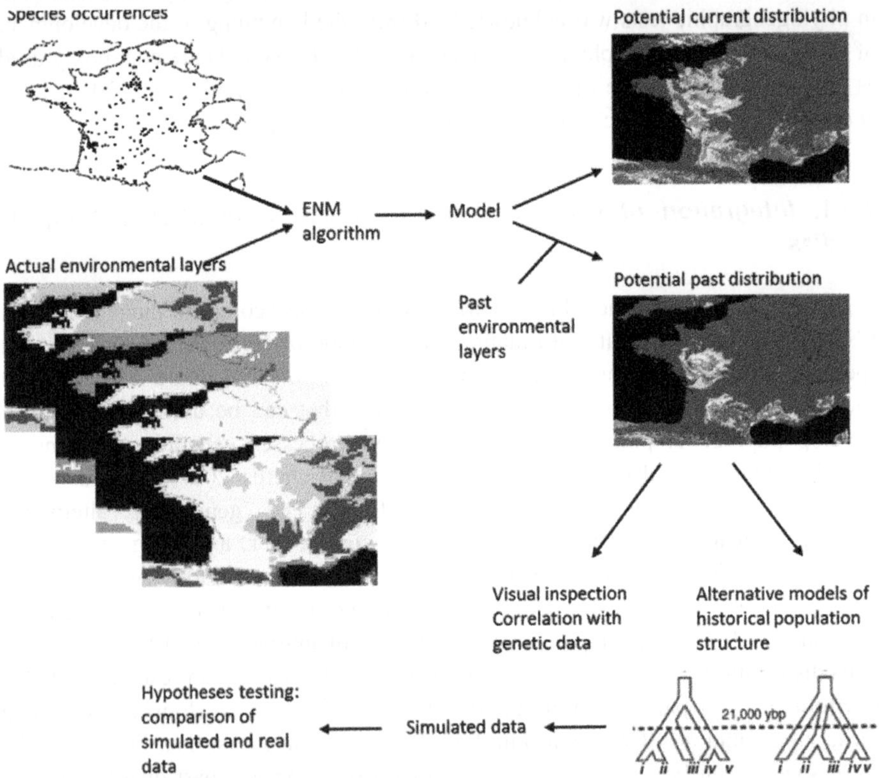

Figure 3.3. *Schematic drawing explaining how ecological niche modeling (ENM) can be used in an integrative framework with phylogeographic data. For a color version of this figure, see www.iste.co.uk/guilbert/biogeography.zip*

One of the most common applications of ENMs in phylogeography is to interpret patterns of genetic variation based on the *post hoc* concordance between patterns of genetic divergence and projections of the species distribution. For example, a deep phylogenetic split between geographically close populations may be suggestive of long-term isolation. Support for this hypothesis might be obtained by visual inspection of projections of past species distributions (i.e. from ENMs based on paleoclimatic variables), where the distributional disjunctions correspond to the genetic differentiation observed between regions (e.g. Allal *et al.* 2011). Alternatively, actually allopatric populations can be genetically similar due to past connectivity, and projections of past species distributions will show spatial continuity between populations in the past (e.g. Nicolas *et al.* 2018). Because these applications are correlative, *ad hoc* interpretations might be misleading given that other processes might have generated the observed genetic patterns (Alvarado-Serrano and Knowles 2014).

Habitat suitability scores derived from ENMs can be used in a correlative manner to test the impact of the landscape on population genetic connectivity (e.g. Row *et al.* 2010). However, it should be emphasized that once again, the correlation between the landscape, or habitat suitability and genetic data, may not be causal because genetic patterns are not exclusively linked to the present landscape configuration but can reflect past configuration of habitat (Alvarado-Serrano and Knowles 2014).

Rather than seeking visual corroboration or correlations between information from ENMs and patterns of genetic variation, ENMs can instead be used to generate hypotheses that are subsequently tested with genetic data. Statistical phylogeographic inferences rely on explicit models of historical scenarios. For any set of multiple populations, there is an extremely large set of possible histories with different combinations of connections or routes of colonization among contemporary populations, as well as varying numbers or spatial configurations of populations in the past. The pertinent question is how to decide what model would be an appropriate representation. ENMs, coupled with paleoclimate estimates, can provide the information necessary for generating alternative models in cases for which no external information on past distributions has previously been available (Carstens and Richards 2007; Richards *et al.* 2007). Estimates of a species' past distributions, or paleodistribution models, can be used to guide the generation of alternative biogeographical hypotheses. The alternative population structures suggested by paleodistribution models can be evaluated by constructing null distributions for expected patterns of genetic variation (or a summary statistic that is used to characterize the data) from data simulated by a neutral coalescent process under a specific population model. This approach provides a statistical framework for determining whether the empirical data differed significantly from theoretical expectations under a given model, while taking into account the inherent stochasticity of genetic processes. For example, Portik *et al.* (2017), by combining genome-wide variation and ENM in a model-testing framework, showed that although forest refugia played a prominent role in the intraspecific diversification of the amphibian *Scotobleps gabonicus,* there was also evidence for potential interactions between landscape features and historical refugia, including major rivers and elevational barriers.

Predictions for species distributions at different time periods from ENMs can be used to identify regions of environmental stability where a species may (in principle) have persisted over time (i.e. refugia), in contrast to unstable areas (i.e. areas where climatic changes would have made the region uninhabitable during particular periods). For example, Faye *et al.* (2016) show that, in the palm genus *Podococcus*, high and unique genetic diversity is strongly correlated with inferred areas of climatic stability since the last glacial maxima in Central Africa. Carnaval *et al.* (2009), using frogs as indicator species, ecological niche models under paleoclimates and simultaneous Bayesian analyses of multispecies molecular data, compared alternative hypotheses of assemblage-scale response to late Quaternary climate change. They show that the southern Atlantic forest of South America was climatically unstable relative to the central region, which served as a large climatic refugium for neotropical species in the late Pleistocene.

An essential aspect of phylogeography is to determine the relationship between the diversification of lineages and variation in an ecological niche. By using climatic data to estimate potential geographic distributions, ENMs make it possible to explicitly test if discrete lineages have diverged into different climatic environments or alternatively have retained a particular niche despite population divergence (niche conservatism, Wiens *et al.* 2010; Peterson 2011). For example, ecological divergence would show that phylogeographic estimates supporting divergence with gene flow and ENM comparisons would suggest that sister lineages occur in ecologically distinct environments with no evidence of a biogeographic barrier. Alternatively, if biogeographic barriers are promoting divergence, phylogeographic models should show reduced gene flow while ENM demonstrates little or no ecological divergence. Wan *et al.* (2018) showed that in Asian shrew-like moles, much genetic diversification has occurred without evident niche divergence, and that topographical diversity has provided strong geographical isolation.

Recently, the temporal focus of phylogeographic and niche studies has broadened, exploring both historical and current geographic patterns to produce ecological projections of past and future distributions. For instance,

Martínez-Méndez *et al.* (2015) integrated analyses of evolutionary history, contemporary geographic distribution and future distribution projections for the lizard *Sceloporus serrifer*, and concluded that populations inhabiting the lowest lands will become extinct before the year 2070.

To conclude, the rapid methodological advances in the construction of ENMs, together with the increasing availability of geospatial environmental and species distribution data, will no doubt expand the ways in which ENMs might be used to address phylogeographic questions, continuing the impressive trajectory of their applications in phylogeography to date. However, understanding the relationship between the distribution of genetic variation of species and the environmental conditions around them remains challenging, especially given the errors and biases associated with the data, the relative uncertainty regarding the use of algorithms, the discrepancies in method application and interpretation, and the current gaps in the theory regarding the integration of different information sources. Thus, there are significant challenges that limit progress in the application of ENM to phylogeographic studies, as well as various approaches currently under development that seek to solve them (reviewed by Luna-Aranguré and Vázquez-Domínguez 2020).

3.5.2. *Integration of life-history traits in phylogeographic studies*

Genetic structure of neutral genes primarily reflects demographic processes (e.g. drift, expansion, changes in effective population size) that are a consequence of historical biotic and abiotic conditions during a species' evolutionary history. Phenotypes are targets of selection and affect the performance of organisms in variable environments. Furthermore, different classes of phenotypes vary in how they impact processes, such as dispersal, colonization and persistence, thereby providing a window into the importance of various evolutionary processes in current and historical selective environments. For example, phenotypes related to locomotor efficiency, physiological tolerance or body size influence migration and gene flow among subdivided populations. Recruitment rate, life span and time to maturity affect population size and turnover, and thus the amount of genetic variation in subdivided populations. Given that species-specific phenotypes can dictate spatial variation in population responses to environmental change, phylogeography would benefit from a more integrative and inclusive framework, one that incorporates predictions based on those phenotypes, an approach that has been termed "trait-based phylogeography" (Paz *et al.* 2015). This is important because variation in life-history traits may explain variation in phylogeographic structure among species, which is important to take into account in comparative phylogeography. According to Zamudio *et al.* (2016), a new conceptual framework that explicitly integrates quantitative analyses of phenotypic variation within a phylogeographic framework can greatly enhance our knowledge of how genetic and phenotypic divergence arise, how they are linked and how they respond to changing ecological and evolutionary contexts.

Lack of concordance in temporal and spatial clustering in co-distributed taxa may not mean that taxa are not responding to a common landscape or climatic barrier; rather, discrepancies may reflect variation in ecological traits and dispersal capabilities of taxa sampled across the presumed barrier (Papadopoulou and Knowles 2016). These efforts refine expectations for clustered divergences by explicitly including geography and trait-based responses for each species (Massatti and Knowles 2014).

All lineages harbor phenotypic and genetic variation among individuals, and this variation can be geographically partitioned in different ways, as shown in Figure 3.4 (reviewed by Zamudio *et al.* 2016). Integrative approaches that combine high-throughput sequencing, experimental manipulations and high-quality phenotypic data sets should allow different biological mechanisms underlying phenotype–genotype concordance to be distinguished. Quantifying functional genetic variation within the context of the phylogeographic history of a species and across the range of environments it inhabits can reveal how regional variation in selective regimes and demographic processes drives the evolution of adaptive phenotypes. An analytical framework that advocates genealogical and spatially explicit analyses of intraspecific functional genetic and phenotypic variations will bridge microevolutionary processes acting on individual populations and macroevolutionary patterns at larger spatial and temporal scales. These approaches have been possible only in relatively few systems to date, such as model organisms with extensive genomic resources like stickleback (Deagle *et al.* 2013), in well-known genetic pathways causing shifts in coloration (Hoekstra *et al.* 2004) or in physiological adaptation to high altitude (Bulgarella *et al.* 2012). However, identifying the underlying genetic basis of phenotypic variation within species is becoming increasingly tractable. Examining functional phenotypic variation in a phylogeographic framework holds great promise for exploring links between genotypic and phenotypic diversities and adaptation across variable environments.

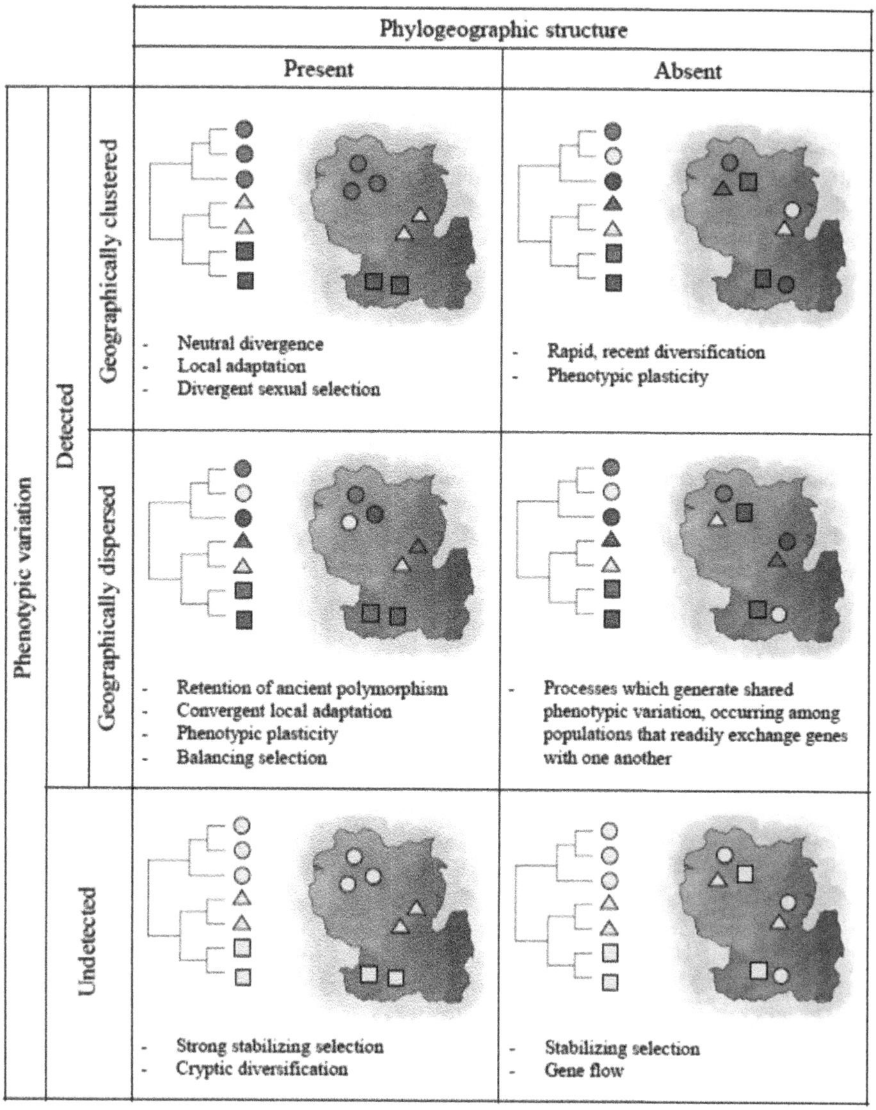

Figure 3.4. *Six potential patterns of phylogeographic structure and phenotypic diversity among populations. The three colors represent different phenotypes. Although discrete morphs are depicted here, similar patterns can arise for phenotypes with continuous variation. For a color version of this figure, see www.iste.co.uk/guilbert/biogeography.zip*

3.6. Conclusion

To conclude, phylogeographic studies have yielded hundreds of individual and comparative case studies which, together, provide invaluable insights into the roles of isolation, niche conservatism and environmental stability in generating patterns of alpha- and beta-diversities, and help identify key processes leading to and resulting from extinction events, including the population dynamics of species range reduction and its effects on the strength and temporal flexibility of networks of species interactions. At the same time, studies uniting evolutionary patterns of diversification with ecological processes have shed new light on classic questions such as the evolutionary factors controlling the composition of local species assemblages, large-scale patterns of species richness or the susceptibility of whole clades to extinction under environmental change (Marske *et al.* 2013). Other applications of phylogeographic analyses include defining species boundaries and assigning and assessing conservation priorities (Richards *et al.* 2007).

Owing to the development of statistical phylogeography and dense genomic sampling, it is now possible to reconstruct complex demographic histories that involve multiple processes, including population growth and decline, lineage splitting and post-divergence gene flow. It also allows a stronger focus on the relative importance of neutral versus selective forces in driving microevolutionary change over time. Phylogeography is clearly at a turning point (Garrick *et al.* 2015): data set size and information content are improving dramatically due to the large numbers of independent autosomal loci being assayed, and this increase in genomic sampling seems not to have come at the expense of geographic sampling.

3.7. References

Allal, F., Sanou, H., Millet, L., Vaillant, A., Camus-Kulandaivelu, L., Logossa, Z.A., Lefevre, F., Bouvet, J.M. (2011). Past climate changes explain the phylogeography of *Vitellaria paradoxa* over Africa. *Heredity*, 107(2), 174–186.

Alvarado-Serrano, D.F. and Knowles, L.L. (2014). Ecological niche models in phylogeographic studies: Applications, advances and precautions. *Mol. Ecol. Resour.*, 14(2), 233–248.

Arbogast, B.S. and Kenagy, G.J. (2001). Comparative phylogeography as an integrative approach to historical biogeography. *J. Biogeogr.*, 28(7), 819–825.

Avise, J.C. (2000). *Phylogeography: The History and Formation of Species*. Harvard University Press, Cambridge, MA.

Avise, J.C. (2009). Phylogeography: Retrospect and prospect. *J. Biogeogr.*, 36, 3–15.

Avise, J.C., Arnold, J., Ball, R.M., Bermingham, E., Lamb, T., Neigel, J.E., Reeb, C.A., Saunders, N.C. (1987). Intraspecific phylogeography: The mitochondrial DNA bridge between population genetics and systematics. *Ann. Rev. Ecol. Syst.*, 18, 489–522.

Avise, J.C., Bowen, B.W., Ayala, F.J. (2016). In the light of evolution X: Comparative phylogeography. *Proc. Natl. Acad. Sci. USA*, 113(29), 7957–7961.

Bermingham, E. and Moritz, C. (1998). Comparative phylogeography: Concepts and applications. *Mol. Ecol.*, 7(4), 367–369.

Bryant, D., Bouckaert, R., Felsenstein, J., Rosenberg, N.A., Roy Choudhury, A. (2012). Inferring species trees directly from biallelic genetic markers: Bypassing gene trees in a full coalescent analysis. *Mol. Biol. Evol.*, 29(8), 1917–1932.

Bulgarella, M., Peters, J.L., Kopuchian, C., Valqui, T., Wilson, R.E., McCracken, K.G. (2012). Multilocus coalescent analysis of haemoglobin differentiation between low- and high-altitude populations of crested ducks (*Lophonetta specularioides*). *Mol. Ecol.*, 21(2), 350–368.

Carnaval, A.C., Hickerson, M.J., Haddad, C.F., Rodrigues, M.T., Moritz, C. (2009). Stability predicts genetic diversity in the Brazilian Atlantic forest hotspot. *Science*, 323(5915), 785–789.

Carstens, B.C. and Richards, C.L. (2007). Integrating coalescent and ecological niche modeling in comparative phylogeography. *Evolution*, 61(6), 1439–1454.

Chifman, J. and Kubatko, L. (2014). Quartet inference from SNP data under the coalescent model. *Bioinformatics*, 30(23), 3317–3324.

Dauby, G., Duminil, J., Heuertz, M., Koffi, G.K., Stevart, T., Hardy, O.J. (2014). Congruent phylogeographical patterns of eight tree species in Atlantic Central Africa provide insights into the past dynamics of forest cover. *Mol. Ecol.*, 23(9), 2299–2312.

De Maio, N., Schrempf, D., Kosiol, C. (2015). PoMo: An allele frequency-based approach for species tree estimation. *Syst. Biol.*, 64(6), 1018–1031.

Deagle, B.E., Jones, F.C., Absher, D.M., Kingsley, D.M., Reimchen, T.E. (2013). Phylogeography and adaptation genetics of stickleback from the Haida Gwaii archipelago revealed using genome-wide single nucleotide polymorphism genotyping. *Mol. Ecol.*, 22(7), 1917–1932.

Dufresnes, C., Nicieza, A.G., Litvinchuk, S.N., Rodrigues, N., Jeffries, D.L., Vences, M., Perrin, N., Martinez-Solano, I. (2020). Are glacial refugia hotspots of speciation and cytonuclear discordances? Answers from the genomic phylogeography of Spanish common frogs. *Mol. Ecol.*, 29(5), 986–1000.

Faye, A., Deblauwe, V., Mariac, C., Richard, D., Sonke, B., Vigouroux, Y., Couvreur, T.L.P. (2016). Phylogeography of the genus *Podococcus* (Palmae/Arecaceae) in Central African rain forests: Climate stability predicts unique genetic diversity. *Mol. Phylogenet. Evol.*, 105, 126–138.

Garrick, R.C., Bonatelli, I.A., Hyseni, C., Morales, A., Pelletier, T.A., Perez, M.F., Rice, E., Satler, J.D., Symula, R.E., Thome, M.T., Carstens, B.C. (2015). The evolution of phylogeographic data sets. *Mol. Ecol.*, 24(6), 1164–1171.

Gavin, D.G., Fitzpatrick, M.C., Gugger, P.F., Heath, K.D., Rodriguez-Sanchez, F., Dobrowski, S.Z., Hampe, A., Hu, F.S., Ashcroft, M.B., Bartlein, P.J., Blois, J.L., Carstens, B.C., Davis, E.B., de Lafontaine, G., Edwards, M.E., Fernandez, M., Henne, P.D., Herring, E.M., Holden, Z.A., Kong, W.S., Liu, J., Magri, D., Matzke, N.J., McGlone, M.S., Saltre, F., Stigall, A.L., Tsai, Y.H., Williams, J.W. (2014). Climate refugia: Joint inference from fossil records, species distribution models and phylogeography. *New Phytol.*, 204(1), 37–54.

Glenn, T.C. (2011). Field guide to next-generation DNA sequencers. *Mol. Ecol. Resour.*, 11(5), 759–769.

Gutiérrez-García, T.A. and Vázquez-Domínguez, E. (2011). Comparative phylogeography: Designing studies while surviving the process. *BioScience*, 61(11), 857–868.

Hewitt, G.M. (2004a). Genetic consequences of climatic oscillations in the Quaternary. *Philos. Trans. R. Soc. B*, 359(1442), 183–195.

Hewitt, G.M. (2004b). The structure of biodiversity – Insights from molecular phylogeography. *Front Zool.*, 1(1), 4.

Hickerson, M.J., Stahl, E.A., Lessios, H.A. (2006). Test for simultaneous divergence using approximate Bayesian computation. *Evolution*, 60(12), 2435–2453.

Hoekstra, H.E., Drumm, K.E., Nachman, M.W. (2004). Ecological genetics of adaptive color polymorphism in pocket mice: Geographic variation in selected and neutral genes. *Evolution*, 58(6), 1329–1341.

Huang, W., Takebayashi, N., Qi, Y., Hickerson, M.J. (2011). MTML-msBayes: Approximate Bayesian comparative phylogeographic inference from multiple taxa and multiple loci with rate heterogeneity. *BMC Bioinform.*, 12, 1.

Irwin, D.E. (2002). Phylogeographic breaks without geographic barriers to gene flow. *Evolution*, 56(12), 2383–2394.

Knowles, L.L. (2004). The burgeoning field of statistical phylogeography. *J. Evol. Biol.*, 17, 1–10.

Knowles, L.L. (2009). Statistical phylogeography. *Annu. Rev. Ecol. Evol. Syst.*, 40, 593–612.

Knowles, L.L. and Maddison, W.P. (2002). Statistical phylogeography. *Mol. Ecol.*, 11, 2623–2635.

Lalis, A., Leblois, R., Liefried, S., Ouarour, A., Beeravolu, C., Michaux, J., Hamani, A., Denys, C.V.N. (2016). New molecular data favour an anthropogenic introduction of the wood mouse (*Apodemus sylvaticus*) in North Africa. *J. Zool. Syst. Evol. Res.*, 54(1), 1–15.

Lalis, A., Mona, S., Stoetzel, E., Bonhomme, F., Souttou, K., Ouarour, A., Aulagnier, S., Denys, C., Nicolas, V. (2019). Out of Africa: Demographic and colonization history of the Algerian mouse (*Mus spretus* Lataste). *Heredity*, 122(2), 150–171.

Leaché, A.D. and Oaks, J.R. (2017). The utility of single nucleotide polymorphism (SNP) data in phylogenetics. *Annu. Rev. Ecol. Evol. Systematics*, 48(1), 69–84.

Leaché, A.D., Banbury, B.L., Felsenstein, J., de Oca, A.N., Stamatakis, A. (2015). Short tree, long tree, right tree, wrong tree: New acquisition bias corrections for inferring SNP phylogenies. *Syst. Biol.*, 64(6), 1032–1047.

Luna-Aranguré, C. and Vázquez-Domínguez, E. (2020). Analysis of the application of ecological niche modeling in phylogeographic studies: Contributions, challenges, and future. *Therya*, 11(1), 47–55.

Marske, K.A., Rahbek, C., Nogués-Bravo, D. (2013). Phylogeography: Spanning the ecology-evolution continuum. *Ecography*, 36, 1169–1181.

Martínez-Méndez, N., Mejía, O., Méndez de la Cruz, F.R. (2015). The past, present and future of a lizard: The phylogeography and extinction risk of *Sceloporus serrifer* (Squamata: Phrynosomatidae) under a global warming scenario. *Zool. Anzeiger*, 254, 86–98.

Massatti, R. and Knowles, L.L. (2014). Microhabitat differences impact phylogeographic concordance of codistributed species: Genomic evidence in montane sedges (*Carex* L.) from the Rocky Mountains. *Evolution*, 68(10), 2833–2846.

McCormack, J.E., Hird, S.M., Zellmer, A.J., Carstens, B.C., Brumfield, R.T. (2013). Applications of next-generation sequencing to phylogeography and phylogenetics. *Mol. Phylogenet. Evol.*, 66(2), 526–538.

Médail, F. and Baumel, A. (2018). Using phylogeography to define conservation priorities: The case of narrow endemic plants in the Mediterranean Basin hotspot. *Biol. Conserv.*, 224, 258–266.

Mirarab, S. and Warnow, T. (2015). ASTRAL-II: Coalescent-based species tree estimation with many hundreds of taxa and thousands of genes. *Bioinformatics*, 31(12), 44–52.

Mirarab, S., Reaz, R., Bayzid, M.S., Zimmermann, T., Swenson, M.S., Warnow, T. (2014). ASTRAL: Genome-scale coalescent-based species tree estimation. *Bioinformatics*, 30(17), 541–548.

Myers, E.A., McKelvy, A.D., Burbrink, F.T. (2020). Biogeographic barriers, Pleistocene refugia, and climatic gradients in the southeastern Nearctic drive diversification in cornsnakes (*Pantherophis guttatus* complex). *Mol. Ecol.*, 29(4), 797–811.

Nicolas, V., Mataame, A., Crochet, P.A., Geniez, P., Fahd, S., Ohler, A. (2018). Phylogeography and ecological niche modelling unravel the evolutionary history of the African green toad, *Bufotes boulengeri boulengeri* (Amphibia: Bufonidae), through the Quaternary. *J. Zool. Syst. Evol. Res.*, 56(1), 102–116.

Ottenburghs, J., Lavretsky, P., Peters, J.L., Kawakami, T., Kraus, R.H.S. (2019). Population genomics and phylogeography. In *Avian Genomics in Ecology and Evolution*, Kraus, R. (ed.). Springer, Cham.

Papadopoulou, A. and Knowles, L.L. (2016). Toward a paradigm shift in comparative phylogeography driven by trait-based hypotheses. *Proc. Natl. Acad. Sci. USA*, 113(29), 8018–8024.

Paz, A., Ibanez, R., Lips, K.R., Crawford, A.J. (2015). Testing the role of ecology and life history in structuring genetic variation across a landscape: A trait-based phylogeographic approach. *Mol. Ecol.*, 24(14), 3723–3737.

Peterson, A.T. (2011). Ecological niche conservatism: A time-structured review of evidence. *J. Biogeog.*, 38, 817–827.

Portik, D.M., Leache, A.D., Rivera, D., Barej, M.F., Burger, M., Hirschfeld, M., Rodel, M.O., Blackburn, D.C., Fujita, M.K. (2017). Evaluating mechanisms of diversification in a Guineo-Congolian tropical forest frog using demographic model selection. *Mol. Ecol.*, 26(19), 5245–5263.

Pritchard, J.K., Stephens, M., Donnelly, P. (2000). Inference of population structure using multilocus genotype data. *Genetics*, 155, 945–959.

Richards, C.L., Carstens, B.C., Knowles, L.L. (2007). Distribution modelling and statistical phylogeography: An integrative framework for generating and testing alternative biogeographical hypotheses. *J. Biogeogr.*, 34, 1833–1845.

Row, J.R., Blouin-Demers, G., Lougheed, S.C. (2010). Habitat distribution influences dispersal and fine-scale genetic population structure of eastern foxsnakes (*Mintonius gloydi*) across a fragmented landscape. *Mol. Ecol.*, 19(23), 5157–5171.

Shendure, J. and Ji, H. (2008). Next-generation DNA sequencing. *Nat. Biotechnol.*, 26(10), 1135–1145.

Stephens, M. and Donnelly, P. (2003). A comparison of Bayesian methods for haplotype reconstruction from population genotype data. *Am. J. Hum. Genet.*, 73(5), 1162–1169.

Stephens, M., Smith, N.J., Donnelly, P. (2001). A new statistical method for haplotype reconstruction from population data. *Am. J. Hum. Genet.*, 68(4), 978–989.

Stone, G.N., Lohse, K., Nicholls, J.A., Fuentes-Utrilla, P., Sinclair, F., Schonrogge, K., Csoka, G., Melika, G., Nieves-Aldrey, J.L., Pujade-Villar, J., Tavakoli, M., Askew, R.R., Hickerson, M.J. (2012). Reconstructing community assembly in time and space reveals enemy escape in a Western Palearctic insect community. *Curr. Biol.*, 22(6), 532–537.

Wan, T., He, K., Jin, W., Liu, S.-Y., Chen, Z.-Z., Zhang, B., Murphy, R.W., Jiang, X.-L. (2018). Climate niche conservatism and complex topography illuminate the cryptic diversification of Asian shrew-like moles. *J. Biogeogr.*, 45, 2400–2414.

Wiens, J.J., Ackerly, D.D., Allen, A.P., Anacker, B.L., Buckley, L.B., Cornell, H.V., Damschen, E.I., Jonathan Davies, T., Grytnes, J.A., Harrison, S.P., Hawkins, B.A., Holt, R.D., McCain, C.M., Stephens, P.R. (2010). Niche conservatism as an emerging principle in ecology and conservation biology. *Ecol. Lett.*, 13(10), 1310–1324.

Zamudio, K.R., Bell, R.C., Mason, N.A. (2016). Phenotypes in phylogeography: Species' traits, environmental variation, and vertebrate diversification. *Proc. Natl. Acad. Sci. USA*, 113(29), 8041–8048.

4

Geophysical Biogeography

Laurent HUSSON[1] and Pierre SEPULCHRE[2]
ISTerre, CNRS, Université Grenoble Alpes, France
LSCE, IPSL, CNRS, Université Versailles Saint Quentin, France

4.1. Introduction

Attempts to understand the present-day patterns of biodiversity often seek to relate them to the current environmental conditions. However, the present-day distribution of species is only a snapshot of a process that must be seen in geological time. This distribution results from the interaction of biotic and abiotic factors at time scales that range from the instantaneous to the time of the earliest appearance of life on Earth. Of course, while the current knowledge of species distribution at the surface of the Earth is already patchy, the accessibility of data over geological time even more significantly challenges the biogeographer on their quest. The fossil record gets sparser and sparser as we venture in the far past of the Earth, all the way back to Australia's 2.7 Ga Pilbara sediments, where the earliest indications of prokaryotes have been found.

To start with, a theater needs a stage, and the physical conditions need to be set. Zooming out, we are left with few physical inducements to invoke: only geology and climate can set the environmental conditions and prompt the development of a biosphere. Australia has kangaroos, South America has anteaters and Madagascar has lemurs; these charismatic endemics readily attest to the long-term environmental isolation of their native lands. Likewise, extinct species, such as mammoths and saber-toothed tigers, witnessed rapid climate changes. Time and space scales are crucial to linking physical geography and evolutionary forces. Unraveling these

Biogeography,
coordinated by Eric GUILBERT.

relationships would be facilitated if the characteristic time and space scales were unique and constant, but the challenge is to unravel their consequences on the biota, in light of their tempos. Rapid changes can overprint the slow pace of continental drift. Mass extinctions are good examples, where massive and swift changes in geophysical conditions – both climatic and geological – lead to the brutal collapse of numerous species. The current mass extinction of the biosphere has a subtler pathway: humans first modified the physical environment to an extent and pace unrivaled in geological time, and it is only as a feedback relationship that this disruption has led to the current erosion of biodiversity. Deciphering the physical causes of evolution of the biosphere as we know it today, as well as in the past, from the phylogenetic and fossil archive, within the not-so-serene geological history, is the overarching objective of *geophysical biogeography*.

4.2. Geophysical biogeography at large

4.2.1. *Present day*

Interactions between the physical environment and the biosphere are reflected in the common classifications used by biogeographers and conservationists: realms, biomes and ecoregions (Figure 4.1 shows realms, biomes and ecoregions on Earth). Realms are mostly in line with the current distribution of continents. Biomes, which are defined according to the dominant vegetation cover, clearly mirror the geological and climatic conditions, as epitomized by, for instance, high latitude environments and mountain belts.

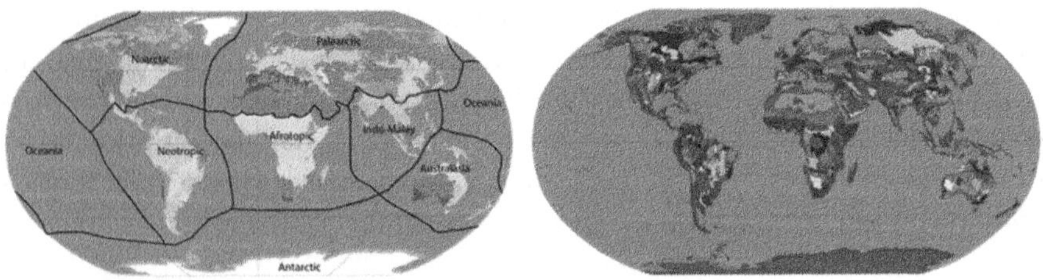

Figure 4.1. *Terrestrial realms and biomes (left), and ecoregions (right), reflect the physical environments (data from Olson* et al. *(2001), where details are also given). For a color version of this figure, see www.iste.co.uk/guilbert/biogeography.zip*

Ecoregions, according to their common explanation, are meant to give a detailed depiction of their biotas, but their definition also integrates the broad notion of environmental conditions. In fact, this shows that these divisions have likely been made using the spatial extent of common biotic and ecological structures, and perhaps noting *a posteriori* that geomorphic and climatic conditions can conveniently be used to further delineate ecozones. The busy world map of ecoregions (Figure 4.1, right) lets the physical conditions shine through. Mountain belts and coastal regions are narrow fringes, and the vast tracts of desert are covered by broad ecoregions. Overall, biogeographical partitions of the world equally mirror the distribution of the living organisms and the regions of continuous climatic and geomorphic conditions. The adequation between the two is at once an input parameter that was used to define those ranges and an outcome, by revealing how the biosphere is subordinate to the solid Earth and the climate.

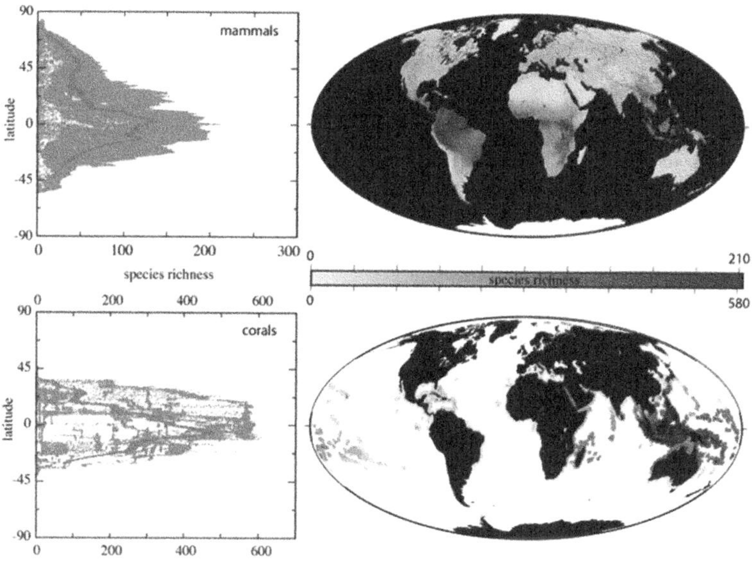

Figure 4.2. *Geographical distribution of species richness for mammals (top, data from Jenkins* et al. *2016) and corals (bottom, data from Jenkins* et al. *(2013)). Graphs show the latitudinal distribution of species richness. Brown data show the raw distribution, and magenta curves show the average distribution. For a color version of this figure, see www.iste.co.uk/guilbert/biogeography.zip*

Biomes and ecoregions thus delineate, initially, the roles of the abiotic factors of climate and geology. An enduring related observation is the latitudinal diversity gradient (LDG). This primordial poleward decrease in species richness from the equator prevails for most taxa, both terrestrial and marine (Figure 4.2). The usual suspect is climate, which has a comparable latitudinal dependency. However, the mechanisms are not as straightforward as they might initially seem. Why, after all, would colder climates at high latitudes deplete species richness? Are species going extinct at a faster rate at high latitudes than in the tropics? Or is it, conversely, that speciation is faster in the intertropical cradle? Do species converge and accumulate more in the intertropical domain? In an attempt to unravel the mechanisms at play, Mannion *et al.* (2014) and Meseguer and Condamine (2020) tracked the time dependency of the LDG by digging in the fossil archive or in the phylogenetic record. They found that the LDG is not a stationary feature, and narrowed around the equator during the transition from the greenhouse Earth towards the icehouse Earth. While it does not answer those questions, it nevertheless suggests that high latitude extinction during cold climatic periods is an efficient lever.

4.2.2. *The dynamic Earth: continental drift*

The outermost layer of the mantle, the lithosphere, forms at mid-oceanic ridges and returns to the deep mantle at subduction zones, such as, for example, in the Pacific ring of fire. Continents are part of the lithosphere and subsequently rift and drift apart from each other, and collide over tens to hundreds of millions of years. The first argument for continental drift was provided by Ortelius (1587), who noted the similarities between the coastlines on both sides of the Atlantic Ocean. The theory mostly stalled until the biogeographical argument led to the advent of the concept. Wegener (1915) realized that similar fossil terrestrial taxa, of similar ages, were found on various continents: either ancient land bridges allowed for terrestrial species to travel across large oceans (which was not an absurd idea at the time), or formerly united continents have drifted apart from each other, and common species rafted away from one another. Wegener took up the latter theory, which became corroborated only during the second half of the 20th century and today is considered common sense.

The geological archive allows the jigsaw to fall into place (Figure 4.3), although reconstructions become increasingly uncertain in older periods. During ridge spreading, the newly formed oceanic crust fossilizes the magnetic field. The relative position of continents across oceans in past times is accessible by reclosing oceans according to their magnetic record (Vine and Matthews 1963). Onshore, a variety of geological data attest for continental collisions and rifting. This change sets the stage for biodiversification processes, and that stage is transient. It is essential because it defines the physiography and the long-term evolution of the climate, both being determinants to speciation, extinction and dispersal. Continental drift constantly remodels physical barriers that hamper species range expansion. At first glance, the tectonic puzzle slowly develops, on time scales of 10–100 My. This time scale is the background noise of the slow biogeographical evolution, and was likely a determinant to the present-day distribution of species richness (e.g. Pelissier *et al.* 2018). It is also the most remarkable, for the field of biogeography grew from initial observations that are consequences of this evolution: Wallace (1863) noticed the contrast between Australasian and Indo-Malay ecozones, which results from their long-term isolation from one another. Indeed, plate reconstructions indicate that Australia has been isolated from all other plates since ~120 Ma (Figure 4.3). Although the supercontinent Pangaea broke up much longer ago, this reconstruction also shows that continents were generally partly, intermittently, connected until 100 Ma. Observing that Africa and South America were imbricated into a single continent as late as 120 Ma would have pleased Wegener. Opening the North Atlantic gradually isolated North America from Eurasia over the last 100 My. The Cretaceous (145–65 Ma) was mostly a time of continental isolation, while the Cenozoic (from 65 Ma to the present day) is marked by multiple episodes of continental collisions. The Tethys Ocean, at the southern rim of Eurasia, gradually shrunk, and only minor bathtubs like the Mediterranean Sea currently remain. India docked into Eurasia in the East, soon followed by Africa. South America was uprooted from Antarctica to become a giant island for millions of years before the formation of the Panama isthmus and its connection with North America. Each of these long-term processes has permanently reshaped the vertices and edges of the graph describing biota connections at the global scale. A classic example is given by the Indian lineage of spiny frogs *Paini*, which landed in Eurasia in the aftermath of the Indian collision (Bossuyt and Milinkovitch 2001) prior to its dispersal and diversification throughout Southeast Asia (Che *et al.* 2010).

A closer consideration of continental drift (Figure 4.3) reveals many smaller features that are determinant to species routing and niche availability: collisions and continental breakup cause mountain belts and rugged topographies. During the Cretaceous tectonic quiescence, erosion erased all remnants of topographic complexity. Conversely, from 60 Ma onwards, orogenies spring up at all continental margins and rejuvenate the topographic complexity. Figure 4.3 shows the physiography from 140 Ma to the present, revealing that fringes of highly elevated regions gradually festoon continents. The most striking event is the rise of the Himalayan–Tibetan orogeny, creating opportunities for diversification (e.g. Ding *et al.* 2020). In North America, the Sevier (160–50 Ma) and Laramide (~80–35 Ma) orogenies formed a high range in the western half of North America. These reliefs were later tectonically dismantled during a phase of orogenic collapse that led to the formation of the Basin and Range Province, which is home to unique ecoregions. In South America, the rise of the Andes had continental scale consequences on biotic evolution (Hoorn *et al.* 2010). A range cutting through East Africa, from the Red Sea to Malawi, also hastened biodiversification (Couvreur *et al.* 2020) – we could keep inventorying these ubiquitous relationships between tectonic activity and biodiversification. The worldwide emergence of tectonic activity in most continents during the Cenozoic has remodeled their physiography, and each of these events triggered diversification processes that were previously dormant.

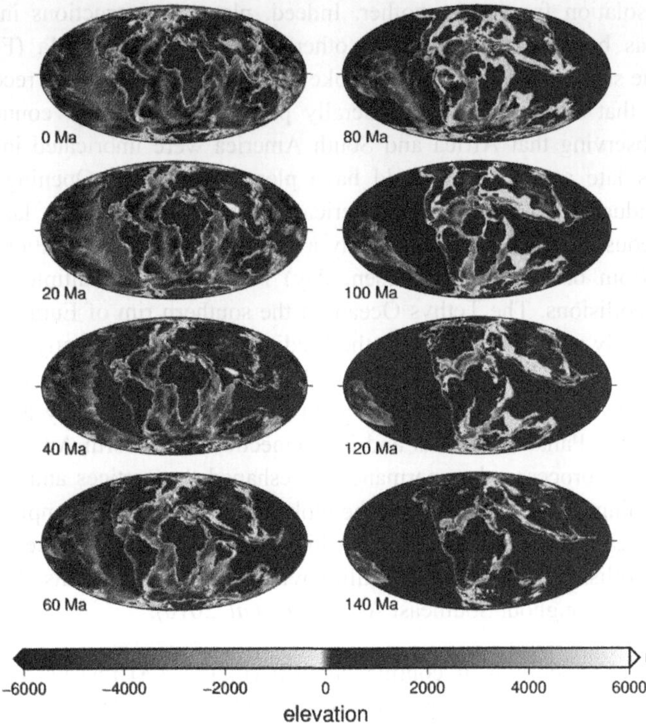

Figure 4.3. *Continental drift, oceanic rifting, mountain building over the last 140 Ma (from Scotese and Wright, 2018). Uniform oceanic regions are undocumented. For a color version of this figure, see www.iste.co.uk/guilbert/biogeography.zip*

Lastly, reconstructions also show that oceans pervade continents with different extents through time (Figure 4.3), which modifies the land–sea mask. For instance, continental flooding peaks during the Late Cretaceous (80 Ma). Again, geodynamics controls sea level, by modifying the shape of the oceanic reservoirs: smaller oceanic basins hold less oceanic water, and overflow adjacent continents (e.g. Cogné and Humler 2008; Conrad and Husson 2009). In addition, the convecting mantle warps the surface of the Earth. Sluggish movements underneath the lithosphere make the surface uplift or subside. While interacting with the sea level, this process (dynamic topography) modifies the absolute sea level (Husson and Conrad 2006) as well as the relative sea level (Gurnis *et al.* 1993). A canonical example is given by the North American seaway (Liu *et al.* 2008) that crosscuts North America during the Late Cretaceous (80 Ma, Figure 4.3). Such profound modifications of the land–sea mask are crucial to the transit and isolation of the flora and fauna at continental scales.

4.2.3. *Continental drift and climate*

Climate through geological time is only indirectly evaluated: for example, oxygen isotopes $\delta^{18}O$ are used as a proxy for global temperature (e.g. Hansen *et al.* 2008; Figure 4.4, after Zachos *et al.* 2001). Overall, the temperature of the Earth is primarily driven by the balance between the energy it receives from the Sun and the energy radiated back to space. The incoming radiation depends on orbital parameters that determine the seasonality of an unevenly distributed energy flux on the spherical Earth. The outcoming radiation is linked to the albedo (the reflection of solar radiations that depends on the cloud cover and the surface "color") and to greenhouse gases, whose concentrations are controlled by the geosphere. In other words, while extra-terrestrial interactions set solar radiations at a ky time scale, the solid Earth redistributes this energy (via atmospheric and ocean dynamics) at a My time scale. Geology and climate are commonly considered as independent abiotic drivers of evolution, but, in fact, they co-evolve.

The size, shape and latitudinal position of the continents directly influence surface energetics and thus the atmospheric and oceanic dynamics. Supercontinents, which occasionally formed when many continents were agglomerated, cause an extremely unbalanced distribution of landmasses on the surface of the Earth. For example, the supercontinent Pangaea, which, during the Triassic, was mostly located at low latitudes and embedded a large oceanic embayment on its eastern side, is thought to have triggered megamonsoons on its western coasts and strong continentality inland (Preto *et al.* 2010). Conversely, the breakup of Pangaea into smaller continents between 225 and 95 Ma enhanced the hydrological cycle, decreased subtropical aridity and led to the onset of the mid-latitude warm temperate environments in which flowering plants diversified (Chaboureau *et al.* 2014). These results infer that breakups, earlier thought of as geographical disconnectors for the biota, conversely reduce the bioclimatic barriers such as deserts, and have instead favored dispersion. At the geological timescale, the breakup and formation of supercontinents also alter the global climate through the carbon cycle. Silicate weathering balances volcanic outgassing and ultimately determines the amount of atmospheric CO_2. Because the size of continents and their latitudinal positions determine rainfall intensity and the nature of the weathered rock, continental drift lays down weathering rates, atmospheric CO_2 (pCO_2) and atmospheric temperatures (Goddéris *et al.* 2014).

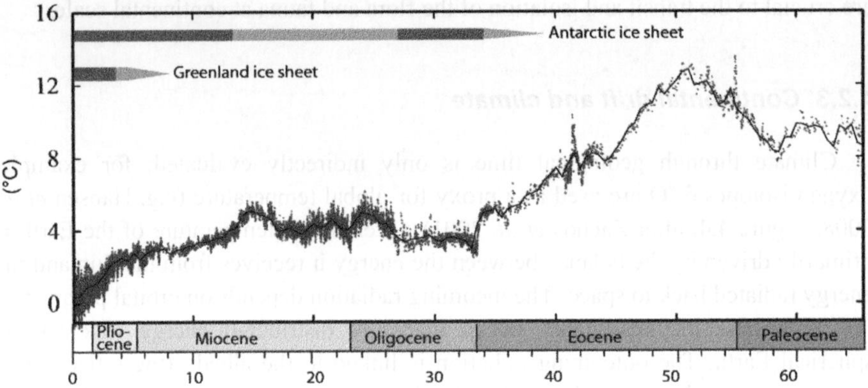

Figure 4.4. *Global temperature of the deep ocean, estimated from* ^{18}O *measured on benthic foraminifera (modified from Hansen* et al. *(2008)). Blue dots and black curve are 5-point and 500,000 year running means, respectively. Blue rectangles are the estimated permanent (dark blue) or ephemeral (light blue) polar ice sheets. For a color version of this figure, see www.iste.co.uk/guilbert/biogeography.zip*

Likewise, the slow dance of continents leads to the occasional formation of isthmus and seaways. Whereas, from the point of view of the biogeographer, land bridges are corridors that foster dispersal of terrestrial species, from the climatologist viewpoint, they are barriers that separate climate compartments from one another and modify their physical characteristics. At the Eocene–Oligocene boundary, ca. 34 Ma, the opening of the Drake and Tasman passages (between South America and Antarctica, and between Australia and Antarctica, respectively) triggered the Antarctic circumpolar current (Kennett 1977; Hill *et al.* 2013) and the reorganization of the global ocean circulation, redistributing the locations of deep water formations as well as marine productivity (Ladant *et al.* 2018; Toumoulin *et al.* 2020). Associated with a drawdown of atmospheric CO_2 concentration, it led to the initiation of the Antarctic ice sheet (Ladant *et al.* 2014). Simultaneously, the Drake and Tasman seaways also isolated the Antarctic fauna by closing a dispersal route that connected South America to Australia via Antarctica. The fossil record from Seymour Island (off the Antarctic peninsula) indeed shows that its endemic fauna, including marsupials, diverged from the one from Patagonia during the

Eocene, while the Australian fauna shows no connection with southern elements ever since (Reguero *et al.* 2002).

In the northern high latitudes, the tectonic opening of the Bering Strait, between Eurasia and North America, occurred during the Late Miocene, but, with the onset of the large Plio-Pleistocene glacial/interglacial cycles, the strait repeatedly opened and closed during sea level variations. These oscillations alternated periods of seawater exchange and isolation of the Arctic and Pacific oceanic reservoirs, which impacted the climate at a continental scale. Numerical simulations predict that the closure of the Bering Strait modified the oceanic conveyor belt by consolidating the Atlantic Meridional Overturning Circulation (AMOC); it enhanced northward heat transport and warmed the surface waters of the North Atlantic Ocean. Warmer surface ocean temperatures affected the global hydrological cycle and rainfall patterns of remote continents (Hu *et al.* 2015). Locally, opening a strait such as Bering hinders Asian and American faunal and floral dispersal opportunities: sustained connection between northeastern Asia and northwestern America is imprinted in the phylogenies of vascular plants until the Late Cenozoic intercontinental disjunction, at the time of the onset of the Bering strait (Wen *et al.* 2016).

In the tropics, the constriction of the Central American Seaway (CAS) from the Miocene to the Pliocene prevented longitudinal water flows between the Pacific and the Atlantic Oceans, which reinforced the western boundary current of the Atlantic Ocean, and strengthened the AMOC. Regionally, it isolated very distinctive marine provinces in the tropical Pacific and Atlantic Oceans, as their salinity, temperature and productivity quickly adapted to the revised circulation scheme. These changes added to the physical barrier to marine dispersal between basins. Biogeographical consequences are observable in the phylogenies of coral reef fishes, mollusks, echinoids and crustaceans, which record several events of disruption in gene flow between the East Pacific and the Atlantic that likely mirror vicariant episodes (e.g. Cowman and Bellwood 2013). While preventing the interchange between marine clades between the Atlantic and Pacific Oceans, the newly created isthmus opened a new dispersal route for terrestrial flora and fauna to connect both Americas. The two American continents long remained apart by only a stone's throw, but land bridges were rare. Their absence prevented biotic exchange, although land connectivity permitted the occasional sloths and porcupines to cross over during the Paleogene (e.g. Philippon *et al.* 2020). The Late Miocene to Pleistocene connection (O'Dea *et al.* 2016) sparked the Great American Biotic Interchange (GABI), the biggest ever biogeographical exchange between the two Americas (see Jaramillo (2018) for a review). The GABI occurred in several pulses, starting from the Late Miocene until its climax in the Pleistocene. However, the biggest pulses of GABI not only occurred because of the physical connection between South and North America through the isthmus, but also because the revised climatic conditions facilitated it. During the Pliocene and Pleistocene interglacials, tropical rainforests in Central America acted as an ecological barrier to savanna-adapted clades, which only migrated during glacial conditions, under dryer climates and less extensive forests (Woodburne 2010).

The Miocene (19–14 Ma) closure of the eastern Tethys seaway separated eastern Africa from Eurasia. This connection launched a corridor for terrestrial fauna, the so-called "Gomphotherium land bridge", as evidenced by multiple waves of proboscideans migrating from Africa to Europe (Harzhauser *et al.* 2007). Together with the CAS constriction, the closure of the eastern Tethys seaway ended the existence of the circum-equatorial surface current and reinforced the AMOC while redirecting saline waters flowing to the Indian Ocean towards the Atlantic Ocean.

The Earth does not bear so many large-scale continents, and the above case studies depict most of the large-scale events that have disrupted the long-term, slow interaction between the solid Earth and the soft Earth since the breakup of Pangaea. In all cases, these events are marked by somewhat sudden (up to a few My) connection and disruption events. Also, in all cases, the story is manifold: continental events are also oceanic events, and vice versa: land bridges are oceanic gateways. For the biosphere, it works as a traffic light at crossroads: when gates are opened for marine taxa, they are closed to terrestrial ones and, conversely, land bridges close oceanic passages. The effects are important on all spheres of the Earth's system, not just for the physical connectivity of continental and oceanic masses. Connecting or disconnecting continents modifies the oceanic circulation, as well as the atmospheric and biological circulation, and feedback interactions are found at all levels. For instance, the revised atmospheric conditions following the dismantlement of a land bridge reset the chemical equilibrium with the biosphere, which, in turn, alters atmospheric and oceanic circulation. All spheres are in dynamic equilibrium from thermodynamic or chemical standpoints, on geological time scales.

4.2.4. *The fast pace of mass extinctions*

The current pattern of species distribution (e.g. Figure 4.2) might appear as a result of the slowly evolving physical conditions, as revealed by the history of plate tectonics and climate. The dynamic Earth mostly gently evolves within the time scale of mantle convection. However, this view should not conceal some more brutal events, also acting on a global scale, which occasionally lead to mass extinctions. In the far geological past, five such ecological crises, which led to an overwhelming decrease in the number of species, are commonly reported. All were triggered by some modifications of the geophysical environment. Perhaps, the most iconic is the end-Cretaceous extinction, which sees, in particular, the sudden collapse of the dinosaurs, along with countless marine and continental species. Several explanatory theories rivaled for decades, and the few that invoked biotic interactions were quickly abandoned. Massive volcanism, asteroid impact, sea level change and combinations thereof are more plausible explanations. Sea level drop is less plausible, because the process needs to be global, and irrespectively hit oceanic and continental realms at once. Because massive volcanism can modify the chemical composition of the atmosphere and the climate, the Deccan Traps in India, coeval to the end-Cretaceous extinction, remained a serious candidate for a long time (Courtillot and Fluteau, 2010). It mostly competed with the Chicxulub asteroid impact hypothesis (in Yucatan, Mexico, Alvarez *et al.* 1985), an even more radical phenomenon to change the atmosphere and climate at global scale. The timing of the two events – mass volcanism and asteroid impact – is so tight that Renne *et al.* (2015) suggested that the asteroid might have supplied enough energy to trigger mass volcanism in the antipodal Deccan; their joint effects would have even more efficiently conduced to the extinction of countless marine and terrestrial species. Akin to the end-Cretaceous extinction event, other events were triggered by changes in the interplay between the solid Earth, the atmosphere and the hydrosphere. Geophysical agents worked in concert to achieve the massacre, and the biosphere was, in these far times, only passively enduring the mayhem in the physical environment.

More recently, during the Late Quaternary, the situation changed significantly: at the time when early humans spread across the world (from the early *H. ergaster* to the more recent *H. sapiens*, past the long-lived *H. erectus)*, an impressive amount of species went massively extinct, mostly megafauna (essentially, large vertebrates). The development of these hominins occurred at a time of rapid climate change, during Pleistocene glaciations. Both events possibly acted in concert, but what is important here is that, unlike the previous mass extinctions, elements of the biosphere – namely, hominins – have proven capable of modifying the prevailing equilibrium of their biological environment at a very rapid rate. This notion of rate is essential, as, of course, at all times biotic interactions have dynamically modified the equilibrium between living organisms; for instance, the rise of angiosperms has provoked the demise of gymnosperms (Condamine *et al.* 2020). In that case, it occurred over tens of million years. However, during the Late Quaternary, the extinction of megafauna occurred over only a few thousand years.

At the present day, the rules are changing again. The Anthropocene is not solely marked by the doom of living organisms as a direct consequence of human predation, but also by indirect action: *Homo sapiens* modifies its physical environment at an unprecedented rate, and the speed at which these modifications operate exceeds the capacity of the biota to keep pace. In that case, for the first time ever, some elements of the biosphere – namely, *H. sapiens* – modify their physical environment enough to affect the biosphere and put the ecosystem out of balance: species cannot adjust to the constantly revised physical conditions. Current climate change not only requires species to geographically shift out of habitats latitudinally, but also vertically, going higher into mountain refugias. But biome migration might not be swift enough to keep pace with climate change, for they require some consistency of the climatic conditions to adapt (Loarie *et al.* 2009). Ecotones, which mark the transition zones between different biomes, are good markers to monitor the speed at which those changes operate. The tree line is flickering in mountain belts, on average reaching higher elevations (Harsch *et al.* 2009). But this escape migration, which could be viewed as a plan B to save biomes, is massively hampered by other factors, such as wildfires (e.g. Nacarella *et al.* 2020). The modern situation worsens the adaptive capacity of ecosystems: for example, deforestation leads to extinction because habitats are destroyed. Together with the induced climate change, the consequences are dramatic. This is not novel: A. von Humboldt noticed it during his trip to the Northern Andes.

Of course, the rate of extinctions is a key factor; in far geological times, the biota have profoundly modified the geophysical environment, but at time scales on the order of 10 Myr and beyond. The Anthropocene is a matter of millennia, thousands of times faster than the ordinary changes made by the biosphere on its physical environment. It is now well understood that these changes, because of their rate, challenge the possibility of maintaining *H. sapiens*' very living conditions on the surface of the Earth, along with those of an overwhelming number of species that share the same biomes, within an even smaller time scale: centuries, if not decades.

4.3. Geophysical biogeography at regional scale

Large-scale tectonics and climate set the physical conditions on which diversification processes depend, and continental drift is the key ingredient. However, how they impact the biosphere can be envisioned at a range of space and time scales. The nanoscale interactions of life and minerals constitute an entire field of research. At the meso-scale, the critical zone (where the atmosphere, hydrosphere, geosphere and biosphere interact) is easier to perceive. Arguably, soil best exemplifies those subtle feedback interactions. However, the detailed complexity of the critical zone makes it difficult to explore these interactions from a biogeographical perspective. A more favorable scale is that of the tectonic environment. Mountain belts, active rifts, passive margins and collisional arcs structure the landscape onshore and offshore. Lowlands, too, are dynamic in terms of landscape evolution: their flatness makes them extremely sensitive to vertical land motion triggered by mantle convection, namely, dynamic topography.

Within this framework, interactions between the biosphere and its physical environment can be envisioned from a variety of perspectives to determine the processes at play. An enduring issue is caused by the difficulty in determining net speciation rates, and even immigration and extinction rates. The most robust datasets are patterns of species distribution at the present day (e.g. Figure 4.1, for mammals and corals). However, the rate of net speciation is seldom accessible, and even less so in the past. Patterns of species distribution result from the balance between immigration, speciation and extinction, but we have limited access to these quantities. Alternatively, it is useful to take up those: given the geological and climatic conditions, determining the most likely explanation to the biological observations.

For the speciation term of the balance, in particular, classes of mechanisms have been identified (Figure 4.5). They describe the response of the biota to the changes in the environmental conditions. Allopatric speciation is the most intuitive class. Sometimes referred to as vicariance, it describes the physical separation of the range of the initial species by a physical barrier, for instance, a mountain belt or a body of water, which will ultimately yield distinct species. Parapatric and peripatric speciation refer to the colonization of a neighboring niche by a land bridge or a water channel. Sympatry defines the genetic drift of a subspecies within the spatial range of the initial species. In all cases, the final diversity is increased. Observed patterns of distribution do not bear any *a priori* indication of the triggering mechanism, but it can, in turn, be inferred by unraveling how the physical environment has changed.

As mentioned above, speciation is not necessarily a prominent contributor to species richness. Over geological time, species have not endlessly accumulated, because extinction plays a essential role, and restricts or even lowers the amount of species on the surface of the Earth. Conversely, and complementarily to speciation, immigration from a populated niche to another is key to increasing the number of species. For example, parapatry, followed by long-distance migration, has been crucial to the development and migration of tropical reef diversity, which has grown and prograded from the closing gigantic Tethys Ocean since Cretaceous times towards Southeast Asia (Renema *et al.* 2008; Leprieur *et al.* 2016).

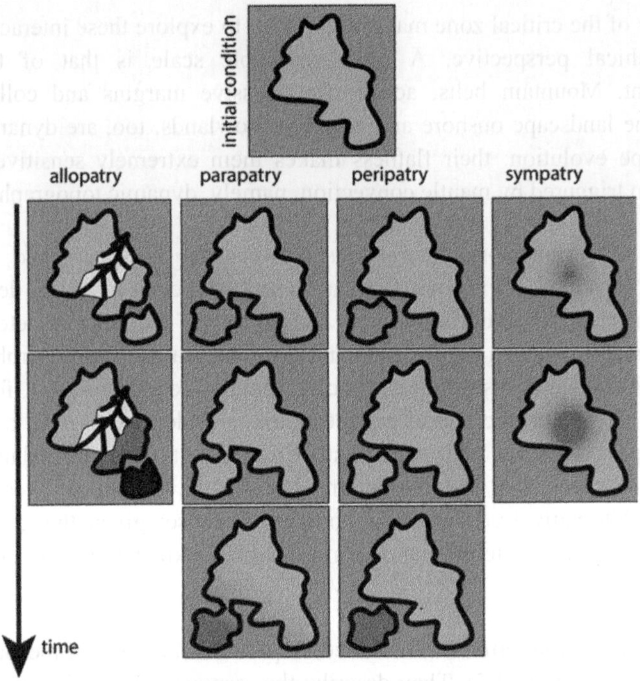

Figure 4.5. *Speciation mechanisms. All start from an identical condition with a prevailing taxa (green), leading to the development of a daughter taxa (pink and blue). For a color version of this figure, see www.iste.co.uk/guilbert/biogeography.zip*

Because the details of the individual contributions of speciation, immigration and extinction are often missing in the present day, and almost systematically in the past, it is often convenient to describe the geophysical environments at a macro-scale, only considering the net result. This is best exemplified in island biogeography (MacArthur and Wilson 1967). Similarly, the notions of museums – where species accumulate at faster rates than elsewhere, graves – where species go extinct at faster rates than elsewhere, or the cradle – where speciation occurs at faster rates than elsewhere – are convenient. This is on this basis that Hoorn *et al.* (2010) hypothesized that the Andes have supplied Amazonia with many species that then further diversified. Of course, some understanding on the details of physiographic evolution is useful to more specifically evaluate the processes at play. Habitat connectivity, for instance, is crucial to the development of species (e.g. MacArthur and Wilson (1967) or, for a more contemporaneous application, Salles *et al.* (2019)). Such macro-ecological depictions of the diversification processes constitute the basis for biogeographical studies at the scale of their physical environments.

4.3.1. *Mountain belts and rifts*

Wandering continents do not dance a peaceful waltz. Dramatic events of compression or extension disrupt their drift and scar the surface of the Earth. After some time, erosion scatters the landscape away and only the geometry of the rock units recalls these events. From a biogeographical perspective, it is crucial to consider these events as extremely dynamic. Mountain ranges and valleys wax and wane over geological time, which, accordingly, impacts diversification processes. Mountain belts are overwhelmingly found at convergent margins (Figure 4.2). Oceanic lithospheres subduct into the mantle, underneath overriding plates, which subsequently deform. Regardless of the duration of convergence, tectonic deformation occurs at variable rates. In some places, tectonic contraction is capable of building mountain belts like the Andes, but in other places, it occurs at modest rates, like in the Indonesian arc; these extreme variations mirror the effect of the deep forces on plate convergence (e.g. Husson 2012). Ultimately, continents collide, often increasing tectonics and crustal shortening. The Himalayan belt and Tibetan plateau form, in the present day, the most spectacular example, but there are many more modest examples, such as the Pyrenees or the Alps. In the past, the Sevier–Laramide orogenic plateau expanded over almost the whole western half of the North American continent, during the Late Cretaceous to the Eocene (e.g. Fan and Carrappa 2014), and collapsed to form the Basin and Range Province. Meanwhile, the Central Andean Altiplano rose to 4 km in South America (e.g. Barnes and Ehlers (2009), for a review) and the East African Rift ripped Africa in two (e.g. Couvreur *et al.* 2020). The surface of the Earth is extremely dynamic; mountains are anything but stationary features over geological time, and this is a key aspect to biogeography (Figure 4.6). Even if they only slowly develop over geological time, their fate profoundly shapes the patterns of biodiversity as we observe them today, and, of course, in the past as well.

A seminal contribution to the understanding of orogenic biogeography is that of A. von Humboldt (as in Johnston *et al.* (1848), Figure 4.7). After wandering around the Earth and visiting many mountain belts, he noted the comparable influence of elevation and latitude on the layering of the flora and ecotones. While latitude and altitude are metrics that refer to the solid Earth (the *geosphere*), the relationship is not straightforward, and Humboldt called forth an intermediate interloper, the *atmosphere*: the geosphere, which sets the climatic conditions upon which the flora appear contingent. Thus, the latitudinal and altitudinal gradients are comparable: moving upward has a similar incidence on the biosphere to moving poleward.

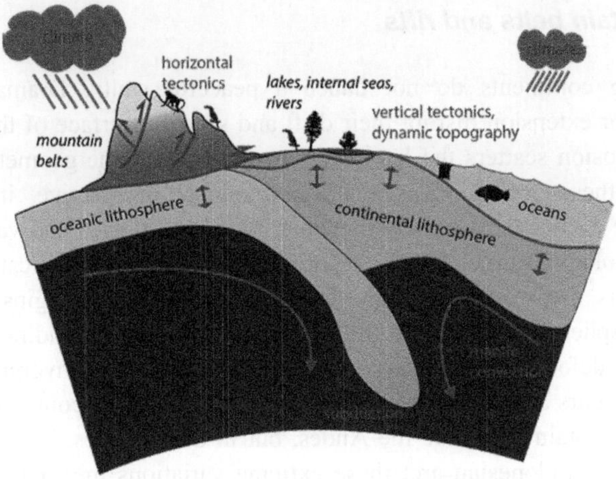

Figure 4.6. *Cartoon showing the relationships between geodynamics, climate and the biosphere. Mountain building is the response to horizontal tectonics, while dynamic topography is the response to vertical tectonics. Both interact with the hydrosphere, atmosphere and biosphere. For a color version of this figure, see www.iste.co.uk/guilbert/biogeography.zip*

How about moving through time? As emphasized above, mountain belts are transient features; the Andes, the Himalayas and Tibet, and the smaller Carpathians or Pyrenees have not always been as high and wide as they are today. The rugged and ~2000 m high Basin and Range Province in North America was a much higher and wider plateau until Miocene times before collapsing. Did nascent mountain belts display different flora and fauna than they do today? Observations convey evidence that species do adapt to the changing elevation of mountain belts (e.g. Favre *et al.* 2015). The biota during orogenic evolution mirrors the situation at any given time, as well as its past history. This is, for instance, how Ding *et al.* (2020) explained the biogeography of the entire Himalayan–Tibetan realm. They suggest that many lineages emerged from the Hengduan Mountains, to the east of the plateau, owing to its anterior geological history. Interestingly, such a relationship is now deemed robust enough to use it to define the time a mountain was built (e.g. Picard *et al.* 2008), and provide a welcome source of independent data to geologists. Most modern mountain ranges uplifted during the last 30 Ma, but more detailed chronologies are elusive, even for the Tibetan plateau (e.g. Botsyun *et al.* 2019) or the Andes (Garzione *et al.* 2008). Burgeoning phylogenetic data collections are growing at a fast rate, and offer a promising alternative in quantifying their tectonic histories.

The biota thus responds to the current physical environment, as well as to the precursor conditions: the current biota, at a given location, can be viewed as the time integral of the interactions between the biosphere and its geophysical environment. The archive is contained in the patterns of the current species, but without a proper user manual to unravel the threads that caused these patterns, only the final product is accessible. Ideally, when searching for a species of a temperate climate, we could either explore low latitudes, low elevations or search the fossil archive in modern orogens at the time they were smaller. However, this macro-ecological view is too restrictive for several reasons, due to the fact that mountain building modifies the environment conditions at scales that depart from that of the belt itself.

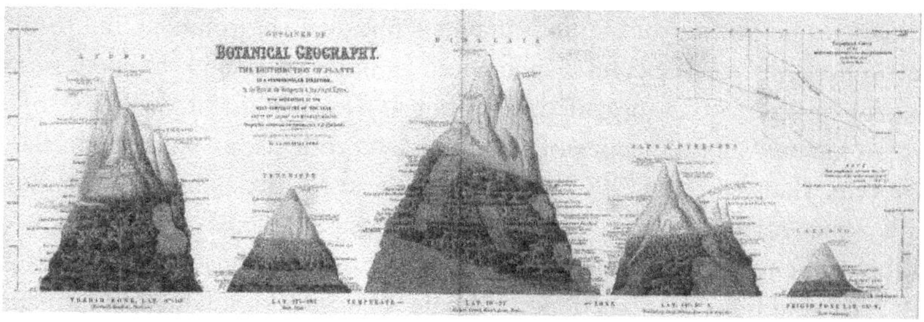

Figure 4.7. *Alexander von Humboldt's views on the relationships between latitude, altitude and flora. Subtitle indicates "The distribution of plants in a perpendicular direction, in the torrid, the temperate, and the frigid zones" to emphasize the latitudinal hierarchy of these mountain ranges, followed by "with indications of the mean temperature of the year and the coldest and warmest months" to highlight the effects of the local climate on the layering of the flora (from Johnston* et al. *1848). For a color version of this figure, see www.iste.co.uk/guilbert/biogeography.zip*

Firstly, at large geographic scales, uplifting mountains, plateaus and rift shoulders involve deformations of the land surface which alter the climate via atmospheric and ocean dynamics and geochemical cycles (Maffre *et al.* 2018a, 2018b). Orography reinforces aridity leeward and favors rainfall windward, and most of the mid-latitude dry areas of the northern hemisphere are maintained by mountains (Broccoli and Manabe 1992). At the present day, the North and South American cordilleras or the Tibetan Plateau deflect surface winds and alter moisture transports between the oceans and the land. At the continent scale, the rise of the Andes modified the global atmospheric circulation, redefined the moisture export from Amazonia to the plains of northern Argentina and shaped the Humboldt current (Garreaud *et al.* 2009; Sepulchre *et al.* 2009, 2011). Likewise, major rift systems, like the East African Rift System, alter the climate and biota. Overall, the fossil and isotopic records from East Africa reveal a Miocene transition from woodland to grassland over the eastern part of the continent (e.g. Bonnefille 2010; Uno *et al.* 2011), while climate model simulations show that the rising rift shoulders coevally hampered moisture transport from the Indian Ocean and led to aridification (Sepulchre *et al.* 2006). Of course, complexity arises at more local scales, which makes it more difficult to identify a single scenario linking tectonics, climate and biogeography (Couvreur *et al.* 2020). Like the horizontal displacement of continents, regional ups and downs of the Earth's surface not only reshape the physiography on which biota evolve, but also its continental scale decorum, that is, the climate, which may facilitate or hamper dispersal routes.

Secondly, the temporal scale plays a role: the biota has a memory and kinetics, some lineages do not mirror the instantaneous physical environment, but result from the evolution of past species: speciation, immigration and extinction are all contingent on the availability of past lineages. The congruent evolution of North American tectonics and lineages of mammals and rodents, as recorded in the fossil archive (e.g. Badgley *et al.* 2017), illustrates these relationships. Species evolution through time reveals that diversification processes closely track the tectonic evolution of the region, as lineages migrate along with the westward drift of tectonic activity. In addition to that trend, climate changes simultaneously distort the evolutionary pattern: warm periods like the Miocene climatic optimum (17–14 Ma) and the Pliocene warm period (4–2.5 Ma) overprint the initial adjustment between the migrating locus of tectonic activity and diversification (Silvestro *et al.* 2018).

Does the conformity of biotic evolution and tectonic and climatic trends at large scales suggest that the processes coevolve at the same spatial scales? From a macroecological point of view, it gives a fair description of the kinetics of diversification processes, but short scales are needed to decipher the mechanisms. In fact, the correspondence between large-scale tectonics and diversification processes points to a third limitation to the power of the latitudinal and altitudinal gradients. Landscapes are shaped by physical conditions, and these might locally dictate the behavior of diversification processes. Rahbek *et al.* (2019) inventoried a number of mountain processes and assigned to them their biogeographical functions. To list a few roles that mountains may carry, we could cite cradles or graves – where speciation is either boosted or hampered by the changing local environment, innovation hubs – where species adapt by evolving, corridors – where migration is stimulated by the connectivity of the landscape, barriers causing vicariance, refugia and museums – whenever prior conditions over a larger range shrunk, for climatic reasons, in particular. All these processes act at the landscape scale.

Landscapes change over time, and the capability of species to disperse evolves according to the connectivity of the landscape, such that the time-varying physical conditions recast the migration pathways. The landscape elevation connectivity (LEC, Bertuzzo *et al.* 2016) quantifies the landscape resistance to migration based on the connectivity between different regions of the landscape (Figure 4.8). Of course, quantifying the LEC requires some *a priori* choices: is it more difficult for a given species to cross an elevated pass or to travel long distances to bypass a mountain range? What is the breadth of a river that makes it a barrier to dispersal? Calibrated on observations from the phylogenetic or fossil archive, these numerical tools open an avenue to evaluate the relative importance of each process, such as recently listed by Rahbek *et al.* (2019), but more than 150 years after Wallace envisioned them.

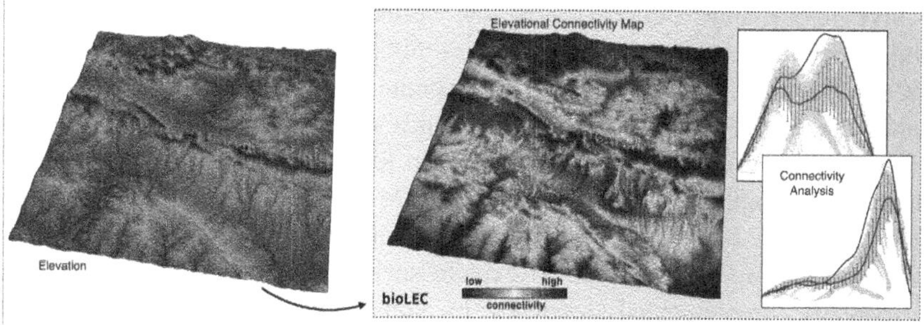

Figure 4.8. *From elevation to landscape connectivity (Salles* et al. *2019). Left and right panels show a digital elevation model and its associated landscape elevation connectivity. Graphs show the distribution of connectivity with elevation. For a color version of this figure, see www.iste.co.uk/guilbert/biogeography.zip*

4.3.2. *Epeirogenies, dynamic topography*

Mountain belts result from horizontal stresses that compress the crust of the Earth. Those stresses arise from the viscous flow of the Earth's mantle, which also induces vertical stresses, which are as important in magnitude as the plate-driving stresses. Instead of causing collisions and rifting, vertical stresses warp the surface of the Earth. This dynamic topography is conceptually simple: the mantle is heterogeneous, and overly dense units tend to sink into the mantle while lighter units tend to rise. This flow is responsible for surface motion and plate tectonics, and it also dynamically deflects the surface by possibly more than 1,000 m (e.g. Ricard *et al.* 1993). However, because it has a long wavelength – that of the underlying pattern of mantle convection, typically 1,000–10,000 km – and because it is aseismic, it is more elusive than the isostatic counterpart that we experience at a glance at the nearby landscape, yet the consequences are crucial to surface processes. At large scales, dynamic topography efficiently modifies the land–sea mask (Gurnis, 1993), and this will, of course, determine the fate of terrestrial lineages. More regionally, because it changes the timing of uplift and subsidence of mountain belts, it also tunes the physical conditions, as in the Himalayas (Husson *et al.* 2014; Webb *et al.* 2017), on which the biosphere relies (Favre *et al.* 2016; Ding *et al.* 2020).

But, perhaps more surprisingly, dynamic topography is also very efficient in the lowlands, where the dynamics of the deep Earth could paradoxically seem to be the most irrelevant: owing to their flatness and low elevations, lowlands are very sensitive to dynamic deflections of the surface of the Earth. The somewhat antiquated term of epeirogeny encompasses this concept. Such regions are prone to river avulsions and frequent drainage reorganization, and even flooding from neighboring seas (Figure 4.6). Amazonia stands as one of the richest biogeographic provinces (Figure 4.3). Discovering the reason for its richness is a long-lived quest dear to biogeographers. A key element is the Miocene rise of the adjacent Andes, which might have fed Amazonia, downstream, with new species (Hoorn *et al.* 2010). However, this might not suffice to explain such a high species richness. Bicudo *et al.* (2020) complemented this view by considering Amazonia as a dynamic landscape, as opposed to an entirely passive region. Indeed, dynamic topography in Amazonia efficiently remodels the drainage patterns; Shephard *et al.* (2010) demonstrated that mantle dynamics have modified the long-term course of the Amazon and Orinoco. In Amazonia, the dynamics of species diversification have also benefited from its own landscape dynamics. As it changes due to dynamic uplift and subsidence, the habitats and migration pathways are accordingly remodeled, which has fostered speciation. Likewise, dynamic subsidence fosters diversification processes in Sundaland (Southeast Asia). Because of its low elevation, subsided by mantle flow, this vast continental region of Southeast Asia is sensitive to sea level fluctuations. The Late Pleistocene interaction of dynamic subsidence with changing sea levels determined the diversification scheme in this iconic biogeographical region (Husson *et al.* 2019).

4.3.3. *Glacial cycles*

On the geological time scales, at odds with the anthropogenic time scale, the Earth is currently in the Quaternary "icehouse" age, but this is only a recent setting. Although the onset of a permanent ice sheet on Antarctica dates back to the Eocene–Oligocene transition (34 Ma), glacial/interglacial cycles with ice ages including very large ice sheets are a recent feature, from the Plio-Pleistocene onwards, and encompass a very small fraction of the Earth's history. The causes of such exceptional regimes are multiple. Theories alternatively invoke extra-terrestrial processes, starting with the orbital parameters of the Earth which determine the amount of energy that the Earth receives. Earth-based causes invoke its geological configuration: the distribution of continents and their impact on oceanic circulation are essential to the Earth's climate. In addition, the Cenozoic rise of mountain belts, Tibet, in particular, likely profoundly altered the chemical budget of the atmosphere (e.g. Raymo *et al.* 1988) or by modifying the atmospheric circulation (Molnar *et al.* 1993).

An important characteristic of Quaternary ice ages is their temporal cyclicity, which is superimposed on the long-term climatic trend (Figure 4.9 for the last 2.6 Ma; Figure 4.4 for longer time scales). The inception of glacial periods over the slow Late Cenozoic cooling trend is marked by some form of cyclicity. Most climatic reconstructions permit to decipher the combined effects of the orbital parameters of the Earth (Hays *et al.* 1976, in the footsteps of Milutin Milankovitch, who hypothesized in the 1920s that the orbit of the Earth should markedly impact the climate of the Earth). A period of 100,000 yr – which dominates the last Ma – marks the effect of eccentricity, while the 41,000 yr period – between 3 and 1 Ma – locks to obliquity. But why did the expression of Milankovitch cycles only become extreme during the Late Cenozoic? What are the causes of this change of periodicity? Why is the record of the latter period dominated by obliquity, and not precession (23,000 yr period), that provides much stronger insolation changes (Paillard 2006)? How is the carbon cycle, which shows a 400,000 yr oscillation pattern in the Cenozoic $\delta^{13}C$ record, articulated with the insolation forcing during icehouse periods (Paillard 2017)?

The Quaternary glaciation is a response to an atypical configuration of the solid Earth, which in turns impacts the biota. Although recent highly resolved records provide a much more dynamic story (Westerhold *et al.* 2020), the slow Cenozoic "descent into the icehouse" (Thomas *et al.* 2008, and Figure 4.4), before the onset of the Quaternary glaciations, required some slow adaptation of the biota. Plants had to adapt to frost, and grasses that now dominate cool temperate regions at high latitudes had to transit from tropical to more temperate environments (Schubert *et al.* 2019). North American mammal diversity throughout the Cenozoic also correlates with the global cooling trend inferred from the benthic record (Figueirido *et al.* 2012). In the ocean realm, the evolution of diversity of whales and diatoms together correlate with this same climate record (Marx and Uhen 2010), and also drove extinction of marine megafauna (Pimiento *et al.* 2017).

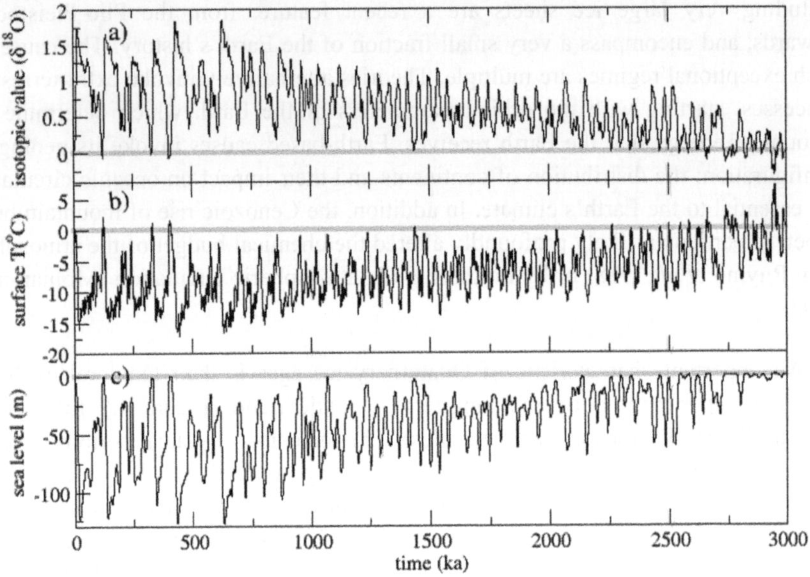

Figure 4.9. *Quaternary climate indicators relative to present day (Bintanja and Van de Wal, 2008). Marine oxygen isotopes $\delta^{18}O$ (a, relative to present day, from Lisiecki and Raymo 2005) are converted into surface temperatures (b) and equivalent sea level (c). Gray lines at zero denote the present-day reference*

On shorter time scales, since the onset of the Quaternary glaciations and maybe since the onset of permanent ice sheets earlier (Westerhold *et al.* 2020), the surface of the Earth became more dynamic: geochemical and geophysical feedback interactions with the biosphere have been boosted, and this caused a complex, yet major upheaval on its dynamics. In the oceans in particular, large quantities of seawater are drawn in and out of the oceans in the ice caps during the glacial cycle, thereby modulating sea level height by an amplitude of up to 120 m (Figure 4.9(c)). Coastlines followed accordingly, forcing the biota in this fragile environment to adapt. Consequences and feedback interactions seem infinite, either hindering or stimulating the development of species.

One such interaction is more straightforward than others: sea level oscillations during the glacial cycle constantly remodel the landscape, which determines migration pathways and isolation (see above). As sea level waxes and wanes, archipelagos in particular undergo periods of connectivity during low sea level stands, and periods of isolation during sea level highstands. This process modifies the landscapes and most ensuing diversification processes. In the highly fragmented Philippines archipelago, these alternating conditions have been conceptualized as the Pleistocene Aggregate Island Complex (PAIC) diversification model (e.g. Brown *et al.* 2013), which posits that these conditions create transient environments that are prone to diversification, inducing the so-called *species pump* (Heaney 1985). Similarly, continental shelves constitute vast expanses of land that are sensitive to sea level oscillations, alternating periods of sea flooding and aerial exposure. For instance, in the vast and flat Sundaland, changes in the drainage network during periods of exposure quickly redefine the relative distribution of land masses of Borneo, Java, Sumatra and peninsular Malaysia, and land bridges between them. Conversely, during sea level highstands, rising sea levels isolate land masses from one another and modify edaphic conditions once the sea withdraws. These intermittent and quickly evolving conditions are crucial to the development of this unique biogeographic region (Lim *et al.* 2010; Slik *et al.* 2011; Husson *et al.* 2019).

Some consequences are subtler and involve feedback relationships. For instance, by quickly vertically shifting the coastline, sea level oscillations constantly rejuvenate the accommodation space for coral reefs, which could have increased their productivity 10-fold (e.g. Husson *et al.* 2018). Because coral reefs are a major player in the oceans, this in turn modified the geochemical balance between surface runoff on continents, and the oceanic and atmospheric reservoirs. Another positive impact is, by increasing the productivity of coral reefs, the increased speciation in the entire trophic chain of associated marine species.

Last, glacial cycles are by definition related to the cryosphere. Poleward, the latitudinal expanse of ice caps obviously shrinks the biosphere towards more tropical regions, but a more subtle interaction is that of glaciers that expand and retreat in temperate mountain belts. The current anthropogenic warming permits direct monitoring of the impact of the retreat of glaciers: species richness increases in many instances at their margins (Cauvy-Fraunié and Dangles 2019). Integrated over the ice age, the role of glaciers on diversification processes is manifold, and the net result remains debated. Allopatry is recurrent, for glacial vicariance triggers divergence during glacial stages (Wallis *et al.* 2016; Wang *et al.* 2018). In the Alps, high altitudes with extreme climatic conditions are the locus of speciation, which served as cradles throughout the Pleistocene (Boucher *et al.* subm.). At the other end, interglacials permitted the export of species out of their biogeographic cradles, and also promoted diversification (Ebersbach *et al.* 2017). These drivers of diversification alternate over glacial cycles, in clusters of mountain "archipelagos" that resemble the Pleistocene Aggregate Island Complex, as described above.

4.4. Conclusions

Perhaps, the most insightful lesson of geophysical biogeography is that analyzing one compartment of the Earth system by itself should be precluded. Surely, the biosphere responds to climate and geology, but considering a linear flow from the solid Earth to the oceans, to the atmosphere, and last to the biosphere, would be erroneous. Interactions and feedback are plural, and one sphere of the Earth system at once responds and places demands on the others; this is the legacy of A. von Humboldt (1769–1859), and we now need to unravel these interactions more comprehensively. Extracting an example from the above, we could cite the role of land bridges. At first sight, continental connection is simply useful to the dispersal of terrestrial lineages, while it simultaneously becomes a red light to marine taxa. But these connections also modify the land–sea mask, and therefore remodel the coupled oceanic and atmospheric circulations, which in turn require some adaptation from the biosphere. Reciprocally, the biota itself modifies the abiotic factors; for example, primary production and continental physiography change accordingly, which ultimately modify the geochemical balance between the spheres of the Earth. We could perpetually follow the dizzying description of these retroactions. Nevertheless, conceiving a heuristic diagram of the Earth system is conversely conceivable. For that purpose, bypassing the petrifying holistic views and prioritizing the interactions at play are required. The interplay between the biosphere and its physical environment needs to be done at once, but at the cost of avoiding secondary meanders.

In an attempt to explore the relationships between the biosphere and its physical environment, time is certainly the Rosetta stone to decipher the patterns of species over the surface of the Earth. First, because transfer functions between the spheres of the Earth likely include time, in addition to space. The species patterns at present day only give a snapshot, which is the final product of the processes and tells little about those. Instead, the time derivative of the species patterns would be more indicative of these very processes of speciation, extinction and migration. In that sense, investigating the expected coevolution of biotic and abiotic processes over geological times is helpful. A second reason is most exciting: the promise to reach the temporal dimension might turn into reality, thanks to quickly mounting phylogenetic datasets, which appropriately complement the fossil archive (e.g. Silvestro *et al.* 2021). Global compilations are quickly becoming available (birds, for example, Jetz *et al.* 2012). Jointly decoding the phylogenetic and fossil archives at the light of the geophysical tempo is a promising perspective. Further, the phylogenetic archive in return constitutes an unparalleled database to quantify the timing of the changing physical world, for which the geological archive is often silent (e.g. Picard *et al.* 2008). Of course, phylogenetic data are a Pandora's box and also come with some caveats (e.g. Louca and Pennell 2020).

Time as well as space scales are crucial to decipher the complexity of these relationships. They operate from the nanoscale to that of the solar system, and time shapes the Earth system like matriochkas. The characteristic scales of landscape evolution align with those of ecoregions, and are appropriate to unravel and quantify the mechanisms of speciation, extinction and migration. At a larger macroecological scale instead, biomes better align with global patterns of climate and plate tectonics. Timewise, it is the time scales of physiographic or landscape changes that should be considered, and those time scales align with that of the geophysical forcings, from plate tectonic time scales to glacial cycles time scales. Do these reference time and space scales hold for all times? Over geological times, these scales have set the interplay between the biotic and abiotic spheres. Exceptions existed, when the slowly evolving geophysical forcings abruptly changed their paces, leading to mass extinctions of marine and terrestrial species at once. Anthropocene is another most curious and unique exception. For the first time in Earth history, one species modifies the physical conditions of all biomes, meticulously and at an unprecedented rate. For the first time, the modifications brought to the geophysical environment by a living organism are first order and are not a feedback interaction with a species adapting to a changing environment. For the first time, one species modifies the physical environment to a level that causes the extinctions of countless species on which it relies, at an unrivaled rate, and threatens its own living conditions.

4.5. References

Alvarez, L.W., Alvarez, W., Asaro, F., Michel, H.V. (1980). Extraterrestrial cause for the Cretaceous–Tertiary extinction. *Science*, 208, 1095–1108.

Badgley, C., Smiley, T.M., Terry, R., Davis, E.B., DeSantis, L.R.G., Fox, D.L., Hopkins, S.S.B., Jezkova, T., Matocq, M.D., Matzke, N., McGuire, J.L., Mulch, A., Riddle, B.R., Roth, V.L., Samuels, J.X., Strömberg, C.A.E., Yanites, B.J. (2017). Biodiversity and topographic complexity: Modern and geohistorical perspectives. *Trends in Ecology and Evolutionary Biology*, 32(3), 211–226.

Barnes, J.B. and Ehlers, T.A. (2009). End member models for Andean Plateau uplift. *Earth Science Reviews*, 97, 117–144.

Bertuzzo, E., Carrara, F., Mari, L., Altermatt, F., Rodriguez-Iturbe, I., Rinaldo, A. (2016). Geomorphic controls on species richness. *Proceedings of the National Academy of Sciences USA*, 113(7), 1737–1742.

Bicudo, T.C., Sacek, V., de Almeida, R.P., Bates, J., Ribas, C. (2019). Andean tectonics and mantle dynamics as a pervasive influence on Amazonian ecosystem. *Scientific Reports*, 9, 16879 [Online]. Available at: https://doi.org/10.1038/s41598-019-53465-y.

Bintanja, R. and van de Wal, R.S.W. (2008). North American ice sheet dynamics and the onset of 100,000-year glacial cycles. *Nature*, 454, 869–872.

Bonnefille, R. (2010). Cenozoic vegetation, climate changes and hominid evolution in tropical Africa. *Global and Planetary Change*, 72(4), 390–411.

Bossuyt, F. and Milinkovitch, M.C. (2001). Amphibians as indicators of Early Tertiary "out-of-India" dispersal of vertebrates. *Science*, 292, 93–95.

Botsyun, S., Sepulchre, P., Donnadieu, Y., Risi, C., Licht, A., Caves Rugenstein, J.K. (2019). Revised paleoaltimetry data show low Tibetan Plateau elevation during the Eocene. *Science*, 363(6430), eaaq1436.

Boucher, F.C., Dentant, C., Ibanez, S., Capblancq, T., Boleda, M., Boulangeat, L., Smyčka, J., Roquet, C., Noûs, C., Lavergne, S. (2021). Discovery of cryptic plant diversity on the rooftops of the Alps. *Scientific Reports*, 11, 11128 [Online]. Available at: https://doi.org/10.1038/s41598-021-90612-w.

Broccoli, A.J. and Manabe, J. (1992). The effects of orography on midlatitude Northern Hemisphere dry climates. *Journal of Climate*, 5(11), 1181–1201.

Brown, R.M., Siler, C.D., Oliveros, C.H., Esselstyn, J.A., Diesmos, A.C., Hosner, P.A., Linkem, C.W., Barley, A.J., Oaks, J.R., Sanquila, M.B., Welton, L.J., Blackburn, D.C., Moyle, R.G., Townsend Peterson, A., Alcala, A.C. (2013). Evolutionary processes of diversification in a model island archipelago. *Annual Review of Ecology, Evolution, and Systematics*, 44(1), 411–435.

Cauvy-Fraunié, S. and Dangles, O. (2019). A global synthesis of biodiversity responses to glacier retreat. *Nature Ecology and Evolution*, 3, 1–11.

Chaboureau, A.-C., Sepulchre, P., Donnadieu, Y., Franc, A. (2014). Tectonic-driven climate change and the diversification of angiosperms. *Proceedings of the National Academy of Sciences USA*, 111(39), 14066–14070.

Che, J., Zhou, W.-W., Hu, J.S., Yan, F., Papenfuss, T.J., Wake, D.B., Zhang, Y.P. (2010). Spiny frogs (Paini) illuminate the history of the Himalayan region and Southeast Asia. *Proceedings of the National Academy of Sciences USA*, 107(31), 13765.

Cogné, J.-P. and Humler, E. (2008). Global scale patterns of continental fragmentation: Wilson's cycles as a constraint for long-term sea-level changes. *Earth and Planetary Science Letters*, 273, 251–259.

Condamine, F.L., Silvestro, D., Koppelhus, E.B., Antonelli, A. (2020). The rise of angiosperms pushed conifers to decline during global cooling. *Proceedings of the National Academy of Sciences USA*, 46, 28867–28875.

Conrad, C.P. and Husson, L. (2009). Influence of dynamic topography on sea level and its rate of change. *Lithosphere*, 1, 110–120.

Courtillot, V. and Fluteau, F. (2010). Cretaceous extinctions: The volcanic hypothesis. *Science*, 328, 973–974.

Couvreur, T.L., Dauby, G., Blach-Overgaard, A., Deblauwe, V., Dessein, S., Droissart, V., Hardy, O.J., Harris, D.J., Janssens, S.B., Ley, A.C., Mackinder, B.A., Sonké, B., Sosef, M.S., Stévart, T., Svenning, J.-C., Wieringa, J.J., Faye, A., Missoup, A.D., Tolley, K.A., Nicolas, V., Ntie, S., Fluteau, F., Robin, C., Guillocheau, F., Barboni, D., Sepulchre, P. (2021), Tectonics, climate and the diversification of the tropical African terrestrial flora and fauna. *Biological Reviews*, 96, 16–51 [Online]. Available at: https://doi.org/10.1111/brv.12644.

Cowman, P.F. and Bellwood, D.R. (2013). Vicariance across major marine biogeographic barriers: Temporal concordance and the relative intensity of hard versus soft barriers. *Proceedings of the Royal Society B: Biological Sciences*, 280(1768), 20131541.

Ding, W.N., Ree., R.H., Spicer, R.A., Xing, Y.W. (2020). Ancient orogenic and monsoon-driven assembly of the world's richest temperate alpine flora. *Science*, 369, 578–581.

Ebersbach, J., Muellner-Riehl, A., Michalak, I., Tkach, N., Hoffmann, M., Röser, M., Sun, H., Favre, A. (2017). In and out of the Qinghai-Tibet Plateau: Divergence time estimation and historical biogeography of the large arctic-alpine genus *Saxifraga* L. *Journal of Biogeography*, 44, 900–910.

Fan, M., and Carrapa, B. (2014). Late Cretaceous–early Eocene Laramide uplift, exhumation, and basin subsidence in Wyoming: Crustal responses to flat slab subduction, *Tectonics*, 33, 509–529 [Online]. Available at: https://doi.org/10.1002/2012TC003221.

Favre, A., Päckert, M., Pauls, S., Jähnig, S., Uhl, D., Michalak, I., Muellner-Riehl, A. (2015). The role of the uplift of the Qinghai-Tibetan Plateau for the evolution of Tibetan biotas. *Biological Reviews*, 90, 236–253.

Figueirido, B., Janis, C.M., Perez-Claros, J.A., De Renzi, M., Palmqvist, P. (2012). Cenozoic climate change influences mammalian evolutionary dynamics. *Proceedings of the National Academy of Sciences USA*, 109(3), 722–727.

Garreaud, R.D., Vuille, M., Compagnucci, R., Marengo, J. (2009). Present-day South American climate. *Palaeogeography, Palaeoclimatology, Palaeoecology*, 281(3–4), 180–195.

Garzione, C.N., Hoke, G.D., Libarkin, J.C., Withers, S., MacFadden, B., Eiler, J., Ghosh, P., Mulch, A. (2008). Rise of the Andes, *Science*, 320(5881), 1304–1307.

Goddéris, Y., Donnadieu, Y., Le Hir, G., Lefebvre, V., Nardin, E. (2014). The role of palaeogeography in the phanerozoic history of atmospheric CO_2 and climate. *Earth-Science Reviews*, 128, 122–138.

Harsch, M.A., Hulme, P.E., McGlone, M.S., Duncan, R.P. (2009). Are treelines advancing? A global meta-analysis of treeline response to climate warming. *Ecology Letters*, 12, 1040–1049.

Harzhauser, M., Kroh, A., Mandic, O., Piller, W.E., Göhlich, U., Reuter, M., Berning, B. (2007). Biogeographic responses to geodynamics: A key study all around the Oligo-Miocene Tethyan seaway. *Zoologischer Anzeiger – A Journal of Comparative Zoology*, 246(4), 241–256.

Hays, J.D., Imbrie, J., Shackleton, N.J. (1976). Variation in the Earth's orbit: Pacemaker of the ice ages. *Science*, 194, 1121–1132.

Heaney, L.R. (1985). Zoogeographic evidence for Middle and Late Pleistocene land bridges to the Philippine Islands. *Modern Quaternary Research in Southeast Asia*, 9, 127–144.

Hill, D.J., Haywood, A.M., Valdes, P.J., Francis, J.E., Lunt, D.L., Wade, B.S., Bowman, V.C. (2013). Paleogeographic controls on the onset of the Antarctic circumpolar current. *Geophysical Research Letters*, 40(19), 5199–5204.

Hoorn, C., Wesselingh, F.P., ter Steege, H., Bermudez, M.A., Mora, A., Sevink, J., Sanmartin, I., Sanchez-Meseguer, A., Anderson, C.L., Figueiredo, J.P., Jaramillo, C., Riff, D., Negri, F.R., Hooghiemstra, H., Lundberg, J., Stadler, T., Sarkinen, T., Antonelli, A. (2010). Amazonia through time: Andean uplift, climate change, landscape evolution, and biodiversity. *Science*, 12(330), 927–931.

Hu, A., Meehl, G.A., Han, W., Otto-Bliestner, B., Abe-Ouchi, A., Rosenbloom, N. (2015). Effects of the Bering strait closure on AMOC and global climate under different background climates. *Progress in Oceanography, Oceanography of the Arctic and North Atlantic Basins*, 132, 174–196.

Husson, L. (2012). Trench migration and upper plate strain over a convecting mantle. *Physics of the Earth and Planetary Interiors*, 212, 32–43.

Husson, L. and Conrad, C.P. (2006). Tectonic velocities, dynamic topography, and relative sea level. *Geophysical Research Letters*, 33, L18303.

Husson, L., Bernet, M., Guillot, S., Huyghe, P., Mugnier, J.-L., Replumaz, A., Robert, X., Van der Beek, P. (2014). Dynamic ups and downs of the Himalaya. *Geology*, 42, 839–842.

Husson, L., Boucher, F.C., Sarr, A.-C., Sepulchre, P., Cahyarini, S.Y. (2020). Evidence of Sundaland's subsidence requires revisiting its biogeography. *Journal of Biogeography*, 47, 843–853.

Jaramillo, C.A. (2018). Evolution of the isthmus of Panama: Biological, paleoceanographic and paleoclimatological implications. In *Mountains, Climate and Biodiversity*, Hoorn, C., Perrigo, A., Antonelli, A. (eds). Wiley, Hoboken, NJ.

Jenkins, C.N. and Van Houtan, K. (2016). Global and regional priorities for marine biodiversity protection. *Biological Conservation*. 204(Part B), 333–339 [Online]. Available at: https://doi.org/10.1016/j.biocon.2016.10.005.

Jenkins, C.N., Pimm, S.L., Joppa, L.N. (2013). Global patterns of terrestrial vertebrate diversity and conservation. *Proceedings of the National Academy of Sciences USA*, 110(28), E2602–E2610.

Jetz, W., Thomas, G.H., Joy, J.B., Hartmann, K., Mooers, A.O. (2012). The global diversity of birds in space and time. *Nature*, 491(7424), 444–448.

Johnston, A.K., Brewster, D., Berghaus, H.K.W. (1848). *The Physical Atlas: A Series of Maps and Notes Illustrating the Geographical Distribution of Natural Phenomena*. William Blackwood, Edinburgh.

Kennett, J.P. (1977). Cenozoic evolution of Antarctic Glaciation, the Circum-Antarctic Ocean, and their impact on global paleoceanography. *Journal of Geophysical Research*, 82(27), 3843–3860.

Ladant, J.-B., Donnadieu, Y., Lefebvre, V., Dumas, C. (2014). The respective role of atmospheric carbon dioxide and orbital parameters on ice sheet evolution at the Eocene–Oligocene transition: Ice sheet evolution at the EOT. *Paleoceanography*, 29(8), 810–823.

Ladant, J.-B., Donnadieu, Y., Bopp, L., Lear, C.H., Wilson, P.A. (2018). Meridional contrasts in productivity changes driven by the opening of Drake passage. *Paleoceanography and Paleoclimatology*, 33(3), 302–317.

Leprieur, F., Descombes, P., Gaboriau, T., Cowman, P.F., Parravicini, V., Kulbicki, M., Melián, C.J., de Santana, C.N., Heine, C., Mouillot, D., Bellwood, D.R., Pellissier, L. (2016). Plate tectonics drive tropical reef biodiversity dynamics. *Nature Communications*, 6(7), 11461.

Lim, H.C., Rahman, M.A., Lim, S.L., Moyle, R.G., Sheldon, F.H. (2011). Revisiting Wallace's haunt: Coalescent simulations and comparative niche modeling reveal historical mechanisms that promoted avian population divergence in the Malay Archipelago. *Evolution*, 65, 321–334.

Liu, L., Spasojevic, S., Gurnis, L. (2008). Reconstructing Farallon plate subduction beneath North America back to the Late Cretaceous. *Science*, 322, 934–938.

Louca, S. and Pennell, M.W. (2020). Extant timetrees are consistent with a myriad of diversification histories. *Nature*, 580, 502–505.

MacArthur, R.H. and Wilson, E.O. (1967). *The Theory of Island Biogeography*. Princeton University Press, Princeton, NJ.

Maffre, P., Ladant, J.-B., Donnadieu, Y., Sepulchre, P., Goddéris, Y. (2018a). The influence of orography on modern ocean circulation. *Climate Dynamics*, 50, 1277–1289.

Maffre, P., Ladant, J.-B., Moquet, J.-S., Carretier, S., Labat, D., Goddéris, Y. (2018b). Mountain ranges, climate and weathering. Do orogens strengthen or weaken the silicate weathering carbon sink? *Earth and Planetary Science Letters*, 493, 174–185.

Mannion, P.D., Upchurch, P., Benson, R.B., Goswami, A. (2014). The latitudinal biodiversity gradient through deep time. *Trends in Ecology and Evolution*, 29, 42–50.

Marx, F.G. and Uhen, M.D. (2010). Climate, critters, and cetaceans: Cenozoic drivers of the evolution of modern whales. *Science*, 327(5968), 993–996.

Meseguer, A.S. and Condamine, F.L. (2020). Ancient tropical extinctions at high latitudes contributed to the latitudinal diversity gradient. *Evolution*, 74, 1966–1987 [Online]. Available at: https://doi.org/10.1111/evo.13967.

Molnar, P., England, P., Martinod, J. (1993). Mantle dynamics, uplift of the Tibetan Plateau, and the Indian monsoon. *Reviews of Geophysics*, 31, 357–396.

Müller, R.D., Liu, L., Gurnis, M. (2010). Miocene drainage reversal of the Amazon River driven by plate-mantle interaction. *Nature Geoscience*, 3, 870–875.

Naccarella, A., Morgan, J.W., Cutler, S.C., Venn, S.E. (2020). Alpine treeline ecotone stasis in the face of recent climate change and disturbance by fire. *PLOS One*, 15(4), e0231339.

O'Dea, A., Lessios, H.A., Coates, A.G., Eytan, R.I., Restrepo-Moreno, S.A., Cione, A.L., Collins, L.S., Queiroz, A., Farris, D.W., Norris, R.D., Stallard, R.F., Woodburne, M.O., Aguilera, O., Aubry, M.P., Berggren, W.A., Budd, A.F., Cozzuol, M.A., Coppard, S.E., Duque-Caro, H., Finnegan, S., Gasparini, G.M., Grossman, E.L., Johnson, K.G., Keigwin, L.D., Knowlton, N., Leigh, E.G., Leonard-Pingel, J.S., Marko, P.B., Pyenson, N.D., Rachello-Dolmen, P.G., Soibelzon, E., Soibelzon, L., Todd, J.A., Vermeij, G.J., Jackson, J.B.C. (2016). Formation of the isthmus of Panama. *Science Advances*, 2(8), e1600883.

Olson, D.M., Dinerstein, E., Wikramanayake, E.D., Burgess, N.D., Powell, G.V.N., Underwood, E.C., D'Amico, J.A., Itoua, I., Strand, H.E., Morrison, J.C., Loucks, C.J., Allnutt, T.F., Ricketts, T.H., Kura, Y., Lamoreux, J.F., Wettengel, W.W., Hedao, P., Kassem, K.R. (2001). Terrestrial ecoregions of the world: A new map of life on Earth, *Bioscience*, 51(11), 933–938.

Ortelius, A. (1596). *Thesaurus Geographicus*. Piantin, Antwerp.

Paillard, D. (2006). What drives the ice age cycle? *Science*, 313(5786), 455–456.

Paillard, D. (2017), The Plio-Pleistocene climatic evolution as a consequence of orbital forcing on the carbon cycle. *Climate of the Past*, 13(9), 1259–1267.

Pellissier, L., Heine, C., Rosauer, D., Albouy, C. (2018). Are global hotspots of endemic richness shaped by plate tectonics? *Biological Journal of the Linnean Society*, 123, 247–261.

Philippon, M., Cornée, J.J., Münch, P., van Hinsbergen, D.J.J., BouDagher-Fadel, M., Lydie Gailler, L., Boschman, L.M., Quillevere, F., Montheil, L., Gay, A., Lebrun, J.F., Lallemand, S., Marivaux, L., Antoine, P.O., GARANTI Team (2020). Eocene intra-plate shortening responsible for the rise of a faunal pathway in the northeastern Caribbean realm. *PLOS One*, 15(10), 20.

Picard, D., Sempere, T., Plantard, O. (2008). Direction and timing of uplift propagation in the Peruvian Andes deduced from molecular phylogenetics of highland biotaxa. *Earth and Planetary Science Letters*, 271, 326–336.

Pimiento, C., Griffin, J.N., Clements, C.F., Silvestro, D., Varela, S., Uhen M.D., Jaramillo C. (2017). The Pliocene marine megafauna extinction and its impact on functional diversity. *Nature Ecology & Evolution*, 1, 1100–1106.

Preto, N., Kustatscher, E., Wignall, P.B. (2010). Triassic climates – State of the art and perspectives. *Palaeogeography, Palaeoclimatology, Palaeoecology*, 290(1), 1–10.

Rahbek, C., Borregaard, M.K., Antonelli, A., Colwell, R.K., Holt, B.G., Nogues-Bravo, D., Rasmussen, C.M.Ø., Richardson, K., Rosing, M.T., Whittaker, R.J., Fjeldså, J. (2019), Building mountain biodiversity: Geological and evolutionary processes. *Science*, 365(6458), 1114–1119.

Raymo, M.E., Ruddiman, W.F., Froelich, P.N. (1988). Influence of late Cenozoic mountain building on ocean geochemical cycles. *Geology*, 16(7), 649–653.

Reguero, M.A., Marenssi, S.A., Santillana, S. (2002). Antarctic peninsula and South America (Patagonia) aleogene terrestrial faunas and environments: Biogeographic relationships. *Palaeogeography, Palaeoclimatology, Palaeoecology*, 179(3), 189–210.

Renema, W., Bellwood, D.R., Braga, J.C., Bromfield, K., Hall, R., Johnson, K.G., Lunt, P., Meyer, C.P., McMonagle, L.B., Morley, R.J., O'Dea, A., Todd, J.A., Wesselingh, F.P., Wilson, M.E., Pandolfi, J.M. (2008). Hopping hotspots: Global shifts in marine biodiversity. *Science*, 321(5889), 654–657.

Renne, P., Sprain, C., Richards, M., Self, S., Vanderkluysen, L., Pande, K. (2015). State shift in Deccan volcanism at the Cretaceous–Paleogene boundary, possibly induced by impact. *Science*, 350, 76–78.

Ricard, Y., Richards, M., Lithgow-Bertelloni, C., Le Stunff, Y. (1993). A geodynamic model of mantle density heterogeneity. *Journal of Geophysical Research*, 98(B12), 21, 895–21,909.

Salles, T., Rey, P., Bertuzzo, E. (2019). Mapping landscape connectivity as a driver of species richness under tectonic and climatic forcing. *Earth Surface Dynamics*, 7, 895–910.

Schubert, M., Marcussen, T., Meseguer, A.S., Fjellheim, S. (2019). The grass subfamily *Pooideae*: Cretaceous–Palaeocene origin and climate-driven Cenozoic diversification. *Global Ecology and Biogeography*, 28, 1168–1182 [Online]. Available at: https://doi.org/10.1111/geb.12923.

Scotese, C.R. and Wright, N. (2018). PALEOMAP Paleodigital Elevation Models (PaleoDEMS) for the Phanerozoic PALEOMAP Project [Online]. Available at: www.earthbyte.org/paleodem-resource-scotese-and-wright-2018/.

Sepulchre, P., Ramstein, G., Fluteau, F., Schuster, M., Tiercelin, J.-J., Brunet, M. (2006). Tectonic uplift and eastern Africa aridification. *Science*, 313(5792), 1419–1423.

Sepulchre, P., Sloan, L.C., Snyder, M., Fiechter, J. (2009). Impacts of Andean uplift on the Humboldt Current system: A climate model sensitivity study: Andes uplift and Humboldt current system. *Paleoceanography*, 24, PA4215 [Online]. Available at: https://doi.org/10.1029/2008PA001668.

Sepulchre, P., Sloan, L., Fluteau, F. (2010). Modelling the response of Amazonian climate to the uplift of the Andean mountain range. *Amazonia: Landscape and Species Evolution*, Hoorn, C. and Wesselingh, F.P. (eds). Wiley, Hoboken, NJ.

Silvestro, D. and Schnitzler, J. (2018). Inferring macroevolutionary dynamics in mountain systems from fossils. In *Mountains, Climate and Biodiversity*, Hoorn, C., Perrigo, A., Antonelli, A. (eds). Wiley, Hoboken, NJ.

Silvestro, D., Bacon, C.D., Ding, W., Zhang, Q., Donoghue, P.C.J., Antonelli, A., Xing, Y. (2021). Fossil data support a pre-Cretaceous origin of flowering plants. *Nature Ecology and Evolution*, 5, 449–457 [Online]. Available at: https://doi.org/10.1038/s41559-020-01387-8.

Slik, J.W.F., Aiba, S.-I., Bastian, M., Brearley, F.Q., Cannon, C.H., Eichhorn, K.A.O., Fredriksson, G., Kartawinata, K., Laumonier, Y., Mansor, A., Marjokorpi, A., Meijaard, E., Morley, R.J., Nagamasu, H., Nilus, R., Nurtjahya, E., Payne, J., Permana, A., Poulsen, A.D., Raes, N., Riswan, S., van Schaik, C.P., Sheil, D., Sidiyasa, K., Suzuki, E., van Valkenburg, J.L.C.H., Webb, C.O., Wich, S., Yoneda, T., Zakaria, R., Nicole Zweifel N. (2011). Soils on exposed Sunda Shelf shaped biogeographic patterns in the equatorial forests of Southeast Asia. *Proceedings of the National Academy of Sciences USA*, 108(30), 12343–12347.

Thomas, E. (2008). Descent into the Icehouse. *Geology*, 36(2), 191–192.

Toumoulin, A., Donnadieu, Y., Ladant, J.-B., Batenburg, S.J., Poblete, F., Dupont-Nivet, G. (2020). Quantifying the effect of the Drake passage opening on the Eocene ocean. *Paleoceanography and Paleoclimatology*, 35(8), e2020PA003889.

Uno, K.T., Cerling, T.E., Harris, J.M., Kunimatsu, Y., Leakey, M.G., Nakatsukasa, M., Nakaya, H. (2011). Late Miocene to Pliocene carbon isotope record of differential diet change among East African herbivores. *Proceedings of the National Academy of Sciences USA*, 108(16), 6509–6514.

Vine, F.J. and Matthews, D.H. (1963). Magnetic anomalies over oceanic ridges. *Nature*, 199(4897), 947–949.

Wallace, A.R. (1863). On the physical geography of the Malay Archipelago. *Journal of the Royal Geographical Society*, 33, 1863, 217–234.

Wallis, G., Waters, J., Upton, P., Craw, D. (2016). Transverse alpine speciation driven by glaciation. *Trends in Ecology and Evolution*, 31(12), 916–926 [Online]. Available at: https://doi.org/10.1016/j.tree.2016.08.009.

Wang P., Yao H., Gilbert K.J., Lu Q., Hao Y., Zhang Z., Wang N. (2018). Glaciation-based isolation contributed to speciation in a Palearctic alpine biodiversity hotspot: Evidence from endemic species. *Molecular Phylogenetics and Evolution*, 129, 315–324 [Online]. Available at: doi: 10.1016/j.ympev.2018.09.006. PMID: 30218774.

Webb, A.A.G., Guo, H., Clift, P.D., Husson, L., Müller, T., Costantino, D., Yin, A., Xu, Z., Cao, H., Qin Wang, Q. (2017). The Himalaya in 3D: Slab dynamics controlled mountain building and monsoon intensification. *Lithosphere*, 9(4), 637–651 [Online]. Available at: doi: https://doi.org/10.1130/L636.1.

Wegener, A.L. (1922). *Die entstehung der kontinente und ozeane*. Friedrich Vieweg and Sohn, Braunschweig.

Wen, J., Nie, Z.-L., Ickert-Bond, S.-M. (2016). Intercontinental disjunctions between eastern Asia and western North America in vascular plants highlight the biogeographic importance of the Bering land bridge from Late Cretaceous to Neogene. *Journal of Systematics and Evolution*, 54(5), 469–90.

Westerhold, T., Marwan, N., Drury, A.J., Liebrand, D., Agnini, C., Anagnostou, E., Barnet, J.S.K., Bohaty, S.M., Vleeschouwer, D.D., Florindo, F., Frederichs, T., Hodell, D.A., Holbourn, A.E., Kroon, D., Lauretano, V., Littler, K., Lourens, L.J., Lyle, M., Pälike, H., Röhl, U., Tian, J., Wilkens, R.H., Wilson, P.A., Zachos, J.C. (2020). An astronomically dated record of Earth's climate and its predictability over the last 66 million years. *Science*, 369(6509), 1383–1387.

Woodburne, M.O. (2010). The Great American Biotic Interchange: Dispersals, tectonics, climate, sea level and holding pens. *Journal of Mammalian Evolution*, 17(4), 245–264.

Zachos, J.C., Pagani, M.O., Sloan, L.C., Thomas, E., Billups, K. (2001). Trends, rhythms, and aberrations in global climate 65 Ma to Present. *Science*, 292, 686–693.

5

Island Biogeography

Julia SCHMACK[1] and Matthew BIDDICK[2]

[1] *Centre for Biodiversity and Biosecurity, University of Auckland, New Zealand*

[2] *School of Life Sciences, Technical University of Munich, Germany*

Traditionally, the study of island biogeography seeks to understand the factors affecting species richness, diversification and distribution on islands. However, since its inception in the 1960s, the discipline has expanded to include many related subjects, including natural history, trait evolution, filtering processes, functional ecology and conservation biology. Other types of isolated communities are also now considered, such as those inhabiting mountain summits, sea mounts or forest fragments. This expansion was in many ways inevitable, not only because our understanding of insular communities has improved considerably in the last 60 years, but also because biogeographers quickly realized that understanding the geological and natural history of species is key to understanding their present-day dynamics.

Island biogeography's roots are firmly embedded in population and community ecology. Long before the equilibrium theory of island biogeography (henceforth "ToIB") was derived, botanists and zoologists alike had documented the apparently unwavering tendency for species richness to increase log-linearly with sample area (Arrhenius 1921; Gleason 1922; Schoener 1976; Tjørve *et al.* 2018). The "species–area relationship", as it is now known, has been documented in countless ecological communities globally (Preston 1960; Rosenzweig 1995). While the exact shape and drivers of this relationship are still debated (Lomolino 2000b, 2001; Williamson *et al.* 2001, 2002; Burns *et al.* 2009), it is generally accepted that larger areas house more species because they offer a greater pool of resources for

Biogeography,
coordinated by Eric GUILBERT.

occupants. Richness patterns on isolated islands, however, are disjunct from those on the mainland. This is because species are constantly immigrating and going extinct, processes that inherently vary with island area and isolation.

5.1. The equilibrium theory of island biogeography

MacArthur and Wilson's (1967) ToIB (Figure 5.1) predicts that the number of species inhabiting an island results primarily from 1) the rate at which species colonize it (*immigration*); 2) the rate at which established species become extinct (*extinction*); and 3) the rate at which established species diversify (*speciation*). Immigration rates are expected to decline with increasing species richness, due to the reduction in available niche space. Additionally, the pool of potential colonizers from the mainland reduces as they establish on the island. Extinction rates, on the other hand, decrease with increasing species richness. When the island is near empty, extinction rates are low as few species are available to become extinct. As the island fills, populations become smaller and more prone to extinction via stochastic events or resource limitation.

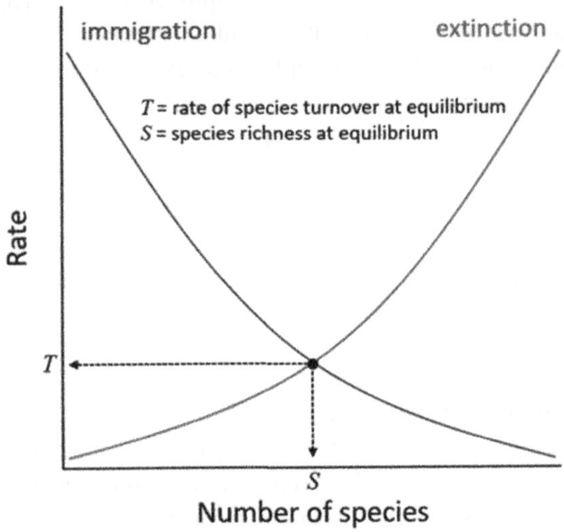

Figure 5.1. *The theory of island biogeography predicts that species richness on islands results from a balance between immigration and extinction. It illustrates how both processes vary as a function of the number of occupant species. Dynamic equilibrium is found at the intersection of these two curves, at which point species richness is approximately maintained but species composition changes through time. For a color version of this figure, see www.iste.co.uk/guilbert/biogeography.zip*

MacArthur and Wilson's theory also incorporates how these two processes vary as a function of island area and isolation (Figure 5.2). First, the *island distance effect* describes how immigration rates should decrease with increasing distance to the mainland. Distant islands are more difficult to colonize as they require greater powers of dispersal to reach, which only a subset of the mainland species pool possess. Second, the *island area effect* describes how rates of extinction should increase with decreasing island area. Small islands are more vulnerable and exposed to stochastic events, such as extreme weather that can wipe out small populations of island inhabitants (Gonzalez and Chaneton 2002). These effects manifest themselves by modulating the immigration and extinction curves for each island, depending on their isolation and area.

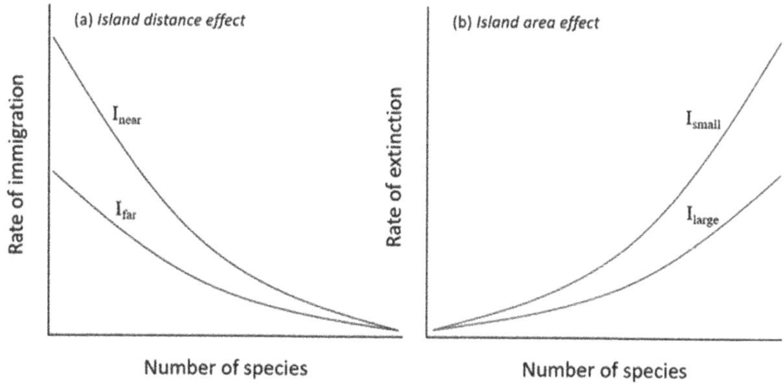

Figure 5.2. *Panel (a) illustrates how island isolation (i.e. distance to the mainland source pool) affects rates of immigration. Panel (b) illustrates how island area affects rates of extinction. For a color version of this figure, see www.iste.co.uk/guilbert/biogeography.zip*

Finally, the ToIB predicts that a dynamic equilibrium exists (i.e. where the immigration and extinction curves intersect), wherein species diversity is approximately maintained but the species comprising this diversity turn over with time (Figure 5.3). Equilibrium is determined not only by the interplay of immigration and extinction, but also by the geographic attributes of the island. Accordingly, small, remote islands are predicted to house fewer species than are large, nearby islands. Between these two extremes, large, remote islands are predicted to house greater species richness than small, nearby islands.

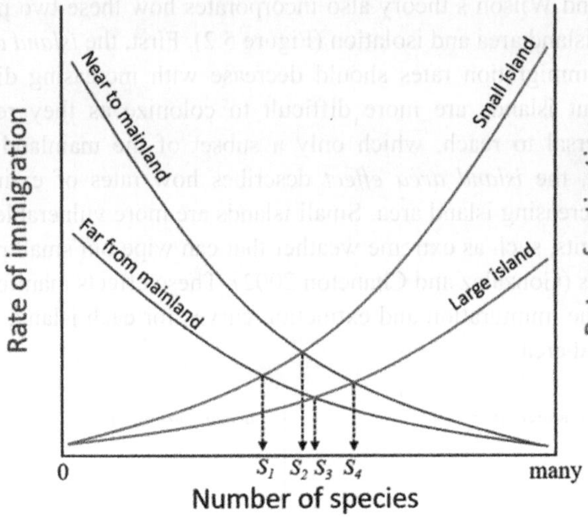

Figure 5.3. *The theory of island biogeography as first proposed by MacArthur and Wilson (1967). For a color version of this figure, see www.iste.co.uk/guilbert/biogeography.zip*

MacArthur and Wilson (1967) cite data from the Krakatau islands in support of their theories' predictions. They highlight how, following an eruption in 1883, the recolonization of birds steadily increased over the course of 25–36 years before reaching equilibrium. Several years later, Edward Wilson paired up with his then PhD student, Daniel Simberloff, to establish an experiment on the mangrove islands of the Florida Keys. With the aid of methyl bromide – a gas that, when dosed correctly, is fatal to arthropods yet harmless to plants – they defaunated the islands of their arthropod fauna, mimicking what they had observed on Krakatau (Simberloff and Wilson 1969, 1970). As predicted, the islands filled with new arthropod species over time and reached a faunistic equilibrium.

More than half a century after its inception, the ToIB continues to find widespread support (Rosenzweig 1995; Whittaker and Fernández-Palacios 2007; see Warren *et al.* 2015 for discussion). Most recently, Valente *et al.* (2020) have provided an exceptional test of the idea of faunistic equilibrium in island birds. By pairing genetic and distributional data, Valente *et al.* (2020) were able to estimate rates of colonization, extinction and speciation. Their results confirm several key predictions of the ToIB, including that 1) colonization rates decrease with increasing insularity; 2) extinction rates decrease with increasing island area; and 3) speciation rates increase with increasing insularity and island area.

However, the ToIB is not without criticism. Its primary assumptions and utility have been questioned (Sauer 1969; Lack 1970), and several authors have suggested that a replacement theory is long overdue (Brown and Lomolino 2000; Lomolino 2000a; Heaney 2007). The assumption of functional equivalence is arguably its most tenuous assumption. It assumes that all species possess equivalent powers of dispersal and are equally likely to speciate and become extinct. While almost certainly violated, this assumption fulfils a useful purpose. Assumptions of equivalence, much like null models, are useful tools for exploring what we should expect were the additional complexity of functional differences not at play (Gotelli and Graves 1996). Thus, while the assumptions of the ToIB are sometimes tenuous, their resulting predictions are invaluable and have played a seminal role in the emergence of novel research fronts, such as metacommunity dynamics (Holyoak *et al.* 2005) and macroecology (Hubbell 2001).

One of the reasons for the ToIB's success is its simplicity. A side effect of this simplicity though is that it does not delineate between natives and exotics, limiting its utility as a conservation tool (but see Chapter 12). This is particularly unfortunate given the disproportionate susceptibility of islands to invasion relative to mainlands (see section 5.5, Loope and Mueller-Dombois 1989; MacDonald and Cooper 1995; Hulme 2004; Castro *et al.* 2010; Kueffer *et al.* 2010; Shaw *et al.* 2010). Burns (2015) provides a modified ToIB that distinguishes between the extinction dynamics of natives and exotics. Exotics are expected to suffer higher extinction rates relative to natives (represented by steeper extinction curves with higher intercepts), as they have not yet had the time to naturalize and establish equilibrium population sizes that vary with island area. With time, however, exotics become naturalized and converge on the same biogeographic dynamics as natives. A large dataset of plants on islands off the coast of New Zealand supports Burns's (2020) ToIB for exotics well. Future work will confirm the utility of its predictions for other taxa and at larger geographic scales.

The ToIBs also do not incorporate the geological ontogeny of oceanic islands. While MacArthur and Wilson revolutionized how we think about ecological communities by portraying them as dynamic systems, their theory conceptualizes islands as static units. In reality, islands are dynamic and ever-changing habitats. For example, oceanic islands emerge from the ocean, build up in altitude and area over time and then erode back into the sea. Accordingly, the core processes underpinning the ToIB (immigration, extinction and speciation) are in tandem with the island ontogeny. Whittaker *et al.* (2008) address this shortfall with their general dynamic model (GDM) of island biogeography. The GDM is founded on the same principal elements as the ToIB; however, it benefits from incorporating how these elements covary with those of the island.

Immediately following emergence from the sea, the island increases in altitude and area. Once the initial burst of volcanic activity is over, island height and area stagnate before slowly decreasing as erosive processes diminish the islands' landmass (this period is probably several orders of magnitude longer than that of emergence and building). Topographic complexity peaks shortly after the peak altitudinal range and area as erosion carves out the terrestrial surface of the island. The authors point out that this is a simplified model of island ontogeny and many oceanic islands will have more complex life histories. Nevertheless, by taking island ontogeny into account in this way, Whittaker *et al.* (2008) derive a more holistic theory of island biogeography (Figure 5.4).

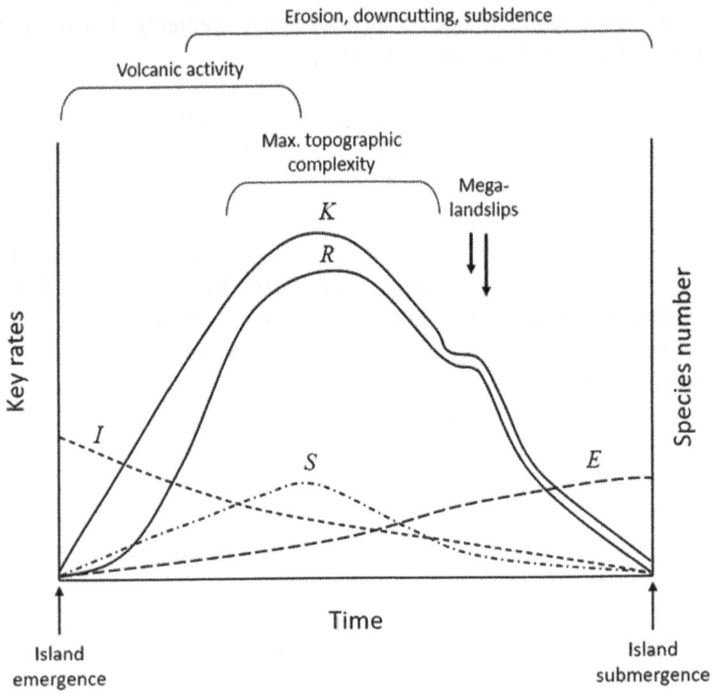

Figure 5.4. *The general dynamic model of island biogeography as derived by Whittaker* et al. *(2008)*

5.2. Insularity and the evolution of emblematic biotas

On his voyage across the world on the HMS *Beagle*, Darwin was struck by the peculiar animals inhabiting the islands he visited. Many of them resembled the continental animals with which he was more accustomed. Yet at the same time, they were somehow different. Some animals like tortoises and iguanas were unusually large. Other animals like elephants and penguins were remarkably small. While exploring the Galapagos, a colleague informed Darwin that he could tell which island a tortoise was from by the shape of its shell. Further still, there appeared to be a type of finch for every island in the archipelago. These observations challenged Darwin's understanding of life on Earth. He had been taught that all species were divinely created – fixed and unchanging with time. Yet, the island animals he observed were clearly modified versions of those from the mainland. This was directly at odds with a static conception of life on Earth. Only decades later would Darwin's observations prove central to developing his theory of evolution via natural selection.

We now understand that the unique conditions of isolated islands cause species to evolve in predictable ways (Whittaker and Fernández-Palacios 2007, Losos and Ricklefs 2009). These evolutionary changes are collectively known as the "island syndrome". The *island syndrome* is a term used to describe the morphological, ecological and behavioral traits typically exhibited by animals endemic to isolated islands (Baeckens and Van Damme 2020). Birds illustrate several aspects of the island syndrome well, often becoming large, losing the ability to fly and showing no obvious fear of humans or other mainland predators. In fact, nearly every isolated archipelago on Earth housed an assemblage of endemic birds with exactly these traits, including the elephant birds of Madagascar (Aepyornithidae), the moa of New Zealand (Dinornithiformes) and the pigeons of the Mascarene islands (Columbidae). Unfortunately, these attributes were not lost on European sailors. The exact mechanisms causing birds to independently converge on these attributes on islands are debated. One possibility is that larger size is favored on islands, and the ability to fly is lost consequently.

Foster (1963, 1964) noticed that many rodents and marsupials increase in size on islands, while carnivores, lagomorphs and artiodactyls tend to decrease in size. Insectivores, on the other hand, show no consistent trend in body size evolution on islands. A decade later, Van Valen (1973) named the phenomenon the *island rule*, proposing it to be accepted as a new evolutionary law, one with fewer exceptions than any other ecotypic rule in nature. The island rule describes the repeated evolution of small animal giants and large animal dwarves, and has been demonstrated in various animal groups, including rodents (Lomolino 1985, 2005; Lomolino *et al.* 2013; Nolfo-Clements *et al.* 2017), marsupials (Lomolino 1985, 2005; Lomolino *et al.* 2013), primates (Brown *et al.* 2004; Welch 2009), carnivores (Lomolino 1985, 2005; Rick *et al.* 2009; Lomolino *et al.* 2013), artiodactyls (Western 1979; Long *et al.* 2019), snakes (Vanek and Burke 2020), birds (Clegg and Owens 2002), dinosaurs (Benton *et al.* 2010) and even deep sea gastropods

(McClain *et al.* 2006). However, many authors have also not found support for its predictions (Meiri 2007; Meiri *et al.* 2006, 2008, 2009; Itescu *et al.* 2014; Itescu *et al.* 2018; Rebouças *et al.* 2018), generating scepticism as to its validity as an evolutionary rule (see Lokatis and Jeschke 2018). Despite more than 50 years of island rule research, a mechanistic explanation for why species converge on intermediate body sizes on islands has yet to be found.

Five primary hypotheses have been put forth to explain the island rule in animals. The first is immigrant selection (Lomolino 1984, 1985). For organisms whose dispersal abilities increase with body size, it is suggested that the filtering effect of colonizing isolated islands naturally selects for the largest individuals, and that the signal of this process should remain, at least in the early stages of colonization. However, this only explains why islands would harbor gigantic forms of mainland organisms. Further, it is not applicable to organisms whose dispersal abilities are unrelated to body size, such as plants, which also exhibit size changes on islands (Carlquist 1974; Carlquist *et al.* 2003; Burns *et al.* 2012; Burns 2016; Biddick *et al.* 2019a) and have recently been shown to obey the island rule (Biddick and Burns 2019; Biddick *et al.* 2019b).

After colonization, a suite of selective pressures are thought to jointly drive convergence on medium body sizes (Sondaar 1977; Heaney 1978; Lomolino 1985). Because islands are more species depauperate than mainland areas of equivalent size (MacArthur and Wilson 1967), interspecific competition for niche space should be reduced. Consequently, larger size in small-bodied organisms may confer a fitness advantage by enabling them to exploit more resources, particularly if larger size enables them to handle both small and large food items. On the other hand, because resources are more limited on islands than on mainlands, smaller size in large-bodied organisms may evolve to reduce their net energetic requirement (McClain *et al.* 2006). Islands also lack many mammalian predators, as they do not possess the dispersal powers required to reach them. This reduction in predation pressure is thought to enable small-bodied prey to evolve larger size, while larger bodied organisms like ungulates may dwarf as they no longer require the physical advantages of large size to fend off predators (Sondaar 1977; see also the discussion in Lomolin 1985). Finally, the combined effects of reduced resource availability and predation pressure are thought to modify life history traits, such as age at maturity (Palkovacs 2003).

While the above hypotheses are ecologically interesting, none have been confirmed empirically, nor can they account for convergent size changes in all types of organisms. Further, the demonstration of the island rule in plants suggests that the most parsimonious explanation for it would be one that is not restricted to animals. Such an explanation may be found in neutral processes that affect all kinds of island colonists, such as evolutionary drift, which has been overlooked in island rule research.

5.3. Island biogeography in the Anthropocene

Human impact is shaping ecosystems on all spatial scales, and most of Earth's natural systems carry our footprint. Naturally, the diversity and distribution of species across the globe are modulated by natural barriers. However, human-mediated dispersal of species has bypassed these natural barriers and facilitated the dispersal of exotics to even the most remote island ecosystems. Extensive transport networks like shipping and airline traffic accelerate the homogenization of Earth's biota (e.g. Rahel 2002; Holway 2006; Hulme 2009). For aquatic ecosystems, both marine and estuarine, the primary invasion pathway is the ballast water of commercial ships, which have been known to carry some 7,000–10,000 species at any one time (Drake and Lodge 2004; Wonham *et al.* 2005). Meanwhile, for terrestrial ecosystems, airline baggage and cargo can harbor exotic plants and animals, as well as infectious diseases and pathogens (Hulme 2009; Tatem 2009). Globalization not only bypasses natural barriers, but it is also the leading cause of habitat disturbance.

The continuous growth of the human population has led to a dramatic reduction in native vegetation cover on continents and islands, and large proportions of the Earth's primary forests have been replaced by human infrastructure and agriculture (Laurance 1999; Tilman *et al.* 2002; Defries *et al.* 2004; Song *et al.* 2018). In addition to the effect of globalization, habitat alteration can significantly change biotic conditions and make ecosystems less favorable for native species than for exotic newcomers (Stachowicz *et al.* 2002; Ward and Masters 2007). Furthermore, countless deliberate introductions of exotic species have taken place throughout history (Mack and Erneberg 2002; Long 2003; Blackburn *et al.* 2007). Human impact is therefore responsible for most recent changes to species distributions, community assemblages and extinctions (Pimm *et al.* 1995; Sala *et al.* 2000; Wilson 2002; Di Marco *et al.* 2018; Andermann *et al.* 2020).

Islands worldwide have been subject to significant land use change, the biotic and abiotic consequences of which have been documented on numerous islands, including many of the world's most important biodiversity hotspots (Rabor 1959; Brooks *et al.* 1997, 2003; Sodhi *et al.* 2004, 2006; Takahashi *et al.* 2017; Gerzabek *et al.* 2019; Supriatna *et al.* 2020). The Azorean archipelago, for instance, is situated 1,400 km from the nearest mainland and holds a considerable number of terrestrial and marine taxa, including many endemics (Borges 2016). Since the arrival of humans in 1432, the islands have lost more than 90% of their primary forest to expanding human activities, including an intensive milk production scheme that comprises a third of Portugal's overall dairy production (De Almeide *et al.* 2020). Once almost entirely covered by evergreen broadleaf *Laurisilva* forest, today only 2.5% of the total area of the Azores is covered by native forest (Triantis *et al.* 2010). Consequently, the biodiversity of the archipelago has been significantly reduced (Borges *et al.* 2019) and most of the endemic forest–dependent taxa belonging to the orders Coleoptera, Hemiptera and Aranea are expected to go extinct in time (Triantis *et al.* 2010).

Human land use can affect island endemics in a more profound way than continental species (Ricketts *et al.* 2005; Kier *et al.* 2009), potentially because endemics often lack traits to cope with rapid environmental changes (Shea and Chesson 2002). In fact, island ecosystems have hosted 61% of known species extinctions and are home to 37% of all critically endangered species (Tershy *et al.* 2015). This phenomenon is also evident when extinction rates are considered. For example, extinction rates (on a per unit area basis) are 187 times higher for birds and 177 times higher for mammals on islands than on continents (Loehle and Eschenbach 2011). Although fossil records provide evidence for various natural extinction events on islands (e.g. Louys *et al.* 2007), there is consensus among the scientific community that human impact is the main driver of the recent extinctions of many island endemics (Diamond 1969; Chown *et al.* 1998; Blackburn *et al.* 2004; Steadman *et al.* 2005; Wood *et al.* 2017).

5.3.1. *Biological invasions*

Range expansions are an important component of Earth's history and the formation of biodiversity patterns. Paleoecological records of natural invasions beg the question: to what extent do prehistoric invasions differ from those that we record today (Vermeij 2005)? Human-mediated dispersal events exceed natural dispersal events in both number and frequency and include many species that would otherwise be unlikely or unable to disperse long distances (Ricciardi 2007). Only occasionally can the number of naturally dispersed individuals be high, for instance, when geographical or physical barriers are removed, permitting a flux of dispersers between formerly isolated areas (Vermeij 2005; Lam *et al.* 2018; Stigall 2019).

Exotic species play an important role in our everyday life. For instance, many of our most important food sources are derived from introduced plants and animals and form the backbone of today's economy. However, some exotics expand rapidly and cause ecological and economic damage. They are subsequently considered *invasive*, regardless of whether they were introduced deliberately or accidentally (Mooney and Cleland 2001; Davis 2009). While invasion success is determined by multiple factors, the *invasiveness* of the species and *invasibility* of the ecosystem have emerged as two of the most important (Richardson and Pyšek 2006; Hui *et al.* 2016; Feng *et al.* 2019). Biogeographic factors can affect both invasiveness and invasibility, and islands have therefore historically served as model ecosystems for studying invasion dynamics (Lonsdale 1999; Wardle 2002; Kueffer *et al.* 2010).

The *invasiveness* of a species can be considered as its propensity to establish and spread in novel environments. Introduction history, species traits and ecological and evolutionary processes are all important determinants of invasiveness (van Kleunen *et al.* 2010). Regarding introduction history, the probability of establishment is predicted to rise with increasing propagule pressure (Lockwood *et al.* 2005; Simberloff 2009; Blackburn *et al.* 2013; but see also Nuñez *et al.* 2011). Traits that enable a species to cope with new environments, such as the social behavior of many Hymenoptera, can also enhance their invasiveness (Moller 1996; Blackburn *et al.* 2009), even when propagule pressure is low. For instance, a single mated *Vespula* wasp queen can establish successful populations in new environments. Both *Vespula germanica* and *Vespula vulgaris* – two highly invasive wasps in New Zealand – likely arrived as hibernating queens on merchandise vessels from Europe (e.g. in timber, Beggs *et al.* 2011), and have established on many of New Zealand's offshore islands despite considerable genetic bottlenecks (Beggs *et al.* 2011; Schmack *et al.* 2019; Schmack *et al.* 2020). Invaders may also experience a release from their natural enemies (predators, competitors and pathogens) on islands, further bolstering their chances of success (Crawley 1987; Keane and Crawley 2002, but see also Colautti *et al.* 2014). Finally, the success of an invasive species may depend on their potential to adapt to novel conditions (Hufbauer *et al.* 2012; Turner *et al.* 2014).

Invasive species may also access prey resources more efficiently and encounter less pressure from resident predators, herbivores and pathogens themselves ("enemy release" hypothesis, Williamson and Griffiths 1996; Keane and Crawley 2002; Sih *et al.* 2010) and use phytochemicals that are novel to the island biota to inhibit and outcompete their native competitors ("novel weapons" hypothesis, Callaway *et al.* 2008; Kim and Lee 2011; Pinzone *et al.* 2018). In addition, species from regions with more phylogenetic diversity have experienced more competition and may therefore be superior to those species found in regions of low phylogenetic diversity, such as islands ("evolutionary imbalance" hypothesis, Mack 2003; Fridley and Sax 2014).

Invasibility is driven by the characteristics of the recipient ecosystem (Hui *et al.* 2016). One common predictor of invasibility is species richness: species rich communities have fewer unoccupied niches available for invaders to exploit ("biotic resistance hypothesis", Elton 1958). Islands, however, are species depauperate relative to continental habitats of equivalent size (D'Antonio and Dudley 1995), leaving more ecological niches vacant, rendering them less resistant to invasion (Elton 1958). In theory, large islands should be more resistant to invasion than small islands as they support more species ("species–area relationship", Connor and McCoy 1979). However, the disproportionate prevalence of human settlements on large islands (e.g. Spatz *et al.* 2017) can obscure the effects of island size and isolation on invasion dynamics (Hulme 2009; Rizali *et al.* 2010; Schmack *et al.* 2020).

Islands are hotspots of biodiversity and can represent the only refuge for many endemic flora and fauna (e.g. Kier *et al.* 2009). For instance, island endemics comprise up to one-quarter of Earth's plant diversity (Kreft *et al.* 2008). On the other hand, islands are also particularly vulnerable to biological invasions and their deleterious consequences (Sax and Brown 2000; Donlan and Wilcox 2008; Pyšek *et al.* 2012). Island invaders affect native species through competition and direct predation, transmission of diseases and pathogens, habitat degradation, disruption of ecosystem functions and restructuring of native food webs (Simberloff 2009, 2013; David *et al.* 2017). Mammals rank among the most damaging and widespread invaders worldwide, with rats and house mice established on more than 80% of the world's oceanic islands (Atkinson 1985; Towns *et al.* 2006; McCreless *et al.* 2016). Among mammals, invasive rats, cats and ungulates have driven countless endemic birds, mammals, amphibians and reptiles to extinction (Courchamp *et al.* 2003; Towns *et al.* 2006; Jones *et al.* 2008; Medina *et al.* 2011; Bellard *et al.* 2016; Doherty *et al.* 2016; Spatz *et al.* 2017). Thus, islands are sadly hotspots of both biodiversity and extinction, and in need of urgent conservation action.

Many island endemics are evolutionarily ill-equipped against introduced competitors and predators (Milberg and Tyrberg 1993; Cox and Lima 2006; Rödl *et al.* 2007; Wood *et al.* 2017). Islands often lack certain functional groups like mammalian predators and herbivores. Consequently, island endemics often lack adaptations required to fend off such groups should they invade (Diamond and Case 1986; Cox and Lima 2006). The most poignant example may be the evolution of flightlessness or ground-nesting in island birds, traits developed in the absence of ground dwelling predators (Clout and Craig 1995). The kakapō (*Strigops habroptilus*) is a flightless, nocturnal, ground-nesting parrot that breeds every two to five years, and a relic of New Zealand's ancient avifauna, half of which went extinct following the arrival of humans in the 13th century (Holdaway *et al.* 2001). Covered in cryptically colored feathers, kakapō are brilliantly defended against visual predators (i.e. raptors). However, they are defenseless against introduced mammalian predators that hunt via smell, and their eggs and nestlings are an easy target for ground dwelling predators (Elliott *et al.* 2001).

In a recent study by Moser *et al.* (2018), data from 257 (sub)tropic islands were analyzed to assess differences in the species–isolation relationship of native and exotic species of vascular plants, reptiles, ants, mammals and birds. The study found that consistent with the ToIB (MacArthur and Wilson 1967; Whittaker *et al.* 2008), species richness of natives decreased with island isolation, but that species richness of all exotic taxa, except for birds, increased with island isolation, reiterating that geographic isolation does not prevent the invasion of island ecosystems. Instead, the authors highlight that remoteness is associated with naivety and reduced diversity on islands, increasing their invasibility (Elton 1958; Simberloff 1995; Denslow 2003). They conclude that the ongoing human-mediated introduction of exotics will likely increase overall species richness on islands, but to the expense of island endemics.

The loss of biodiversity is nowhere more evident than on remote islands. However, natural mainland ecosystems are becoming increasingly fragmented, leading to decreases in species diversity and the disappearance of certain functional groups (Saunders *et al.* 1991; Fischer and Lindenmayer 2007; Leal *et al.* 2012; Haddad *et al.* 2015). Consequently, these mainland areas share many ecological characteristics of island ecosystems. Increasing our knowledge of island invasions and their effects on endemic species may greatly benefit biodiversity conservation in both mainland and island ecosystems (Manne *et al.* 1999; Denslow 2003; Whittaker and Fernández-Palacios 2007; Kier *et al.* 2009).

5.3.2. *Anthropogenic climate change*

The impacts of human-induced climate change are substantial for island communities (Glasspool and Sterrer 2009; Triantis and Mylonas 2009; Foufopoulos *et al.* 2011; Fortini *et al.* 2013; Patiño *et al.* 2013; Vorsino *et al.* 2014; Connell 2015; Ferreira *et al.* 2016; Leclerc *et al.* 2020). Rising sea levels are already forcing the human inhabitants of small oceanic islands like Kiribati to relocate to larger islands (Risse 2009; McIver *et al.* 2014). A rise in mean annual temperature, increases in precipitation, more frequent extreme weather events, such as storms, hurricanes and droughts, and altered cyclical weather systems (e.g. monsoon wind, El Niño) all pose a threat (Seneviratne *et al.* 2012; Harter *et al.* 2015). How exactly island communities will respond to these changes is not yet known.

Current models predict a sea level rise of 0.6–2.4 m by 2100 (Nicholls and Cazevane 2010; Bellard *et al.* 2014). This rise will disproportionally affect island ecosystems, particularly those that are small and low-lying (Harter *et al.* 2015). Bellard *et al.* (2014) investigated the effects of this sea level rise on 10 insular biodiversity hotspots and predicted the most significant habitat loss to occur in the Caribbean Islands, Sundaland and the Philippines, potentially threatening some 300 endemics. Further, many island endemics possess poor dispersal abilities and are therefore not well equipped to respond to climate-induced range shifts (Gillespie *et al.* 2008). Worse still, species that do possess sufficient dispersal abilities may be unable to utilize escape routes (e.g. via island hopping) if they are eliminated by rising sea levels (Fordham and Brook 2010). Finally, low genetic diversity renders island endemics less evolutionarily malleable than their genetically diverse mainland counterparts (Diamond and Case 1986; Frankham 1997; Hermant *et al.* 2013).

Climate change-induced range shifts will be particularly challenging for species with specialized habitat and dietary requirements, small distributions and slow dispersal rates (Taylor and Kumar 2016; Leclerc *et al.* 2020; Radley *et al.* 2020). Additionally, climate change is expected to intensify the pressure of invasive species (Thuiller 2007; Vilà *et al.* 2007; Walther *et al.* 2009; Willis *et al.* 2010; Wan *et al.* 2017), their impacts on island biota (Benning *et al.* 2002; Hellmann *et al.* 2008; Caujapé-Castells *et al.* 2010; Trueman and d'Ozouville 2010) and the distribution of diseases associated with invasive hosts (e.g. avian malaria, Crowl *et al.* 2008; LaPointe *et al.* 2012; Hulme 2014; de Wit *et al.* 2017). Thus, it appears that the synergistic effects of climate change, habitat fragmentation, biological invasions and disease are likely to have greater impacts on island communities than the additive effects of the stressors alone (Brook *et al.* 2008). Such non-additive effects need to be considered in island biodiversity conservation (see Courchamp *et al.* 2014).

5.4. References

Andermann, T., Faurby, S., Turvey, S.T., Antonelli, A., Silvestro, D. (2020). The past and future human impact on mammalian diversity. *Science Advances*, 6(36), eabb2313.

Arrhenius, O. (1921). Species and area. *Journal of Ecology*, 9, 95–99.

Atkinson, I.A.E. (1985). The spread of commensal species of *Rattus* to oceanic islands and their effects on island avifaunas. In *Conservation of Island Birds: Case Studies for the Management of Threatened Island Species*, Moors, P.J. (ed.). International Council for Bird Preservation, Cambridge.

Baeckens, S. and van Damme, R. (2020). The island syndrome. *Current Biology*, 30, R338.

Beggs, J.R., Brockerhoff, E.G., Corley, J.C., Kenis, M., Masciocchi, M., Muller, F., Villemant, C. (2011). Ecological effects and management of invasive alien Vespidae. *BioControl*, 56(4), 505–526.

Bellard, C., Leclerc, C., Courchamp, F. (2014). Impact of sea level rise on the 10 insular biodiversity hotspots. *Global Ecology and Biogeography*, 23(2), 203–212.

Bellard, C., Cassey, P., Blackburn, T.M. (2016). Alien species as a driver of recent extinctions. *Biology Letters*, 12(2), 20150623.

Benning, T.L., LaPointe, D., Atkinson, C.T., Vitousek, P.M. (2002). Interactions of climate change with biological invasions and land use in the Hawaiian Islands: Modeling the fate of endemic birds using a geographic information system. *Proceedings of the National Academy of Sciences USA*, 99(22), 14246–14249.

Benton, M.J., Csiki, Z., Grigorescu, D., Redelstorff, R., Sander, P.M., Stein, K., Weishampel, D.B. (2010). Dinosaurs and the island rule: The dwarfed dinosaurs from Haţeg Island. *Palaeogeography, Palaeoclimatology, Palaeoecology*, 293, 438–454.

Biddick, M. and Burns, K.C. (2019). Support for the island rule does not hide morphological disparity in insular plants. *Proceedings of the National Academy of Sciences*, 412–424.

Biddick, M., Hutton, I., Burns, K.C. (2019a). Independent evolution of allometric traits: A test of the allometric constraint hypothesis in island vines. *Biological Journal of the Linnean Society*, 126, 203–211.

Biddick, M., Hendricks, A., Burns, K.C. (2019b). Plants obey (and disobey) the island rule. *Proceedings of the National Academy of Sciences USA*, 116, 17632–17634.

Blackburn, T.M. and Cassey, P. (2007). Patterns of non-randomness in the exotic avifauna of Florida. *Diversity and Distributions*, 13(5), 519–526.

Blackburn, T.M., Cassey, P., Duncan, R.P., Evans, K.L., Gaston, K.J. (2004). Avian extinction and mammalian introductions on oceanic islands. *Science*, 305(5692), 1955–1958.

Blackburn, T.M., Lockwood, J.L., Cassey, P. (2009). *Avian Invasions: The Ecology and Evolution of Exotic Birds*, Volume 1. Oxford University Press, Oxford.

Blackburn, T.M., Prowse, T.A., Lockwood, J.L., Cassey, P. (2013). Propagule pressure as a driver of establishment success in deliberately introduced exotic species: Fact or artefact? *Biological Invasions*, 15(7), 1459–1469.

Borges, P.A. (2016). A list of the terrestrial and marine biota from the Azores. Version 1.3. Universidade dos Açores. Checklist dataset [Online]. Available at: https://doi.org/10.15468/hyvwxi [Accessed via GBIF.org on January 22, 2021].

Borges, P.A., Santos, A.M., Elias, R.B., Gabriel, R. (2019). The Azores archipelago: Biodiversity erosion and conservation biogeography. In *Encyclopedia of the World's Biomes*, Goldstein, M. and DellaSala, D. (eds). Elsevier, Amsterdam.

Brook, B.W., Sodhi, N.S., Ng, P.K.L. (2003). Catastrophic extinctions follow deforestation in Singapore. *Nature*, 424, 420–423.

Brook, B.W., Sodhi, N.S., Bradshaw, C.J. (2008). Synergies among extinction drivers under global change. *Trends in Ecology and Evolution*, 23(8), 453–460.

Brooks, T.M., Pimm, S.L., Collar, N.J. (1997). Deforestation predicts the number of threatened birds in insular Southeast Asia. *Conservation Biology*, 11, 382–394.

Brown, J.H. and Kodric-Brown, A. (1977). Turnover rates in insular biogeography: Effect of immigration on extinction. *Ecology*, 58, 445–449.

Brown, J.H. and Lomolino, M.V. (2000). Concluding remarks: Historical perspective and the future of island biogeography theory. *Global Ecology and Biogeography*, 9, 87–92.

Brown, P., Sutikna, T., Morwood, M.J., Soejono, R.P., Saptomo, E.W., Due, R.A. (2004). A new small-bodied hominin from the Late Pleistocene of Flores, Indonesia. *Nature*, 431, 1055–1061.

Burns, K.C. (2015). A theory of island biogeography for exotic species. *The American Naturalist*, 186, 441–451.

Burns, K.C. (2016). Size changes in island plants: Independent trait evolution in *Alyxia ruscifolia* (Apocynaceae) on Lord Howe Island. *Biological Journal of the Linnean Society*, 119, 847–855.

Burns, K.C., MCHARDY, P.R., Pledger, S. (2009). The small-island effect: Fact or artefact? *Ecography*, 32, 269–276.

Burns, K.C., Herold, N., Wallace, B. (2012). Evolutionary size changes in plants of the south-west Pacific. *Global Ecology and Biogeography*, 21, 819–828.

Callaway, R.M., Cipollini, D., Barto, K., Thelen, G.C., Hallett, S.G., Prati, D., Klironomos, J. (2008). Novel weapons: Invasive plant suppresses fungal mutualists in America but not in its native Europe. *Ecology*, 89(4), 1043–1055.

Carlquist, S. (1974). *Island Biology*. Columbia University Press, New York.

Carlquist, S.J., Baldwin, B.G., Carr, G.D. (2003). *Tarweeds and Silverswords: Evolution of the Madiinae (Asteraceae)*. Missouri Botanical Garden Press, St. Louis, MO.

Castro, S.A., Daehler, C.C., Silva, L., Torres-Santana, C.W., Reyes-Betancort, J.A., Atikinson, R., Jaramillo, P., Guezou, A., Jaksic, F.M. (2010). Floristic homogenization as a teleconnected trend in oceanic islands. *Diversity and Distributions*, 16, 902–910.

Caujape-Castells, J., Tye, A., Crawford, D.J., Santos-Guerra, A., Sakai, A., Beaver, K., Kueffer, C. (2010). Conservation of oceanic island floras: Present and future global challenges. *Perspectives in Plant Ecology, Evolution and Systematics*, 12(2), 107–129.

Chown, S.L., Gremmen, N.J.M., Gaston, K.J. (1998). Ecological biogeography of southern ocean islands: Species-area relationships, human impacts, and conservation. *The American Naturalist*, 152(4), 562–575.

Clegg, S.M. and OWENS, P. (2002). The "island rule" in birds: Medium body size and its ecological explanation. *Proceedings of the Royal Society of London. Series B: Biological Sciences*, 269, 1359–1365.

Clout, M.N. and Craig, J.L. (1995). The conservation of critically endangered flightless birds in New Zealand. *Ibis*, 137, 181–190.

Colautti, R.I., Ricciardi, A., Grigorovich, I.A., MacIsaac, H.J. (2004). Is invasion success explained by the enemy release hypothesis? *Ecology Letters*, 7(8), 721–733.

Connell, J. (2015). Vulnerable islands: Climate change, tectonic change, and changing livelihoods in the Western Pacific. *The Contemporary Pacific*, 27(1), 1–36.

Connor, E.F. and McCoy, E.D. (1979). The statistics and biology of the species-area relationship. *The American Naturalist*, 113(6), 791–833.

Courchamp, F., Chapuis, J.L., Pascal, M. (2003). Mammal invaders on islands: Impact, control and control impact. *Biological Reviews*, 78(3), 347–383.

Courchamp, F., Hoffmann, B.D., Russell, J.C., Leclerc, C., Bellard, C. (2014). Climate change, sea-level rise, and conservation: Keeping island biodiversity afloat. *Trends in Ecology and Evolution*, 29(3), 127–130.

Cox, J.G. and Lima, S.L. (2006). Naiveté and an aquatic–terrestrial dichotomy in the effects of introduced predators. *Trends in Ecology and Evolution*, 21(12), 674–680.

Crawley, M.J. (1987). What makes a community invasible? In *Colonization, Succession and Stability*, Gray, A.J., Crawley, M.J., Edwards, P.J. (eds). Blackwell Scientific Publications, Oxford.

Crowl, T.A., Crist, T.O., Parmenter, R.R., Belovsky, G., Lugo, A.E. (2008). The spread of invasive species and infectious disease as drivers of ecosystem change. *Frontiers in Ecology and the Environment*, 6(5), 238–246.

D'Antonio, C.M. and Dudley, T.L. (1995). *Biological Invasions as Agents of Change on Islands Versus Mainlands*. Springer, Berlin.

David, P., Thebault, E., Anneville, O., Duyck, P.F., Chapuis, E., Loeuille, N. (2017). Impacts of invasive species on food webs: A review of empirical data. *Advances in Ecological Research*, 56, 1–60.

Davis, M.A. (2009). *Invasion Biology*. Oxford University Press on Demand, Oxford.

De Almeida, A.M., Alvarenga, P., Fangueiro, D. (2012). The dairy sector in the Azores Islands: Possibilities and main constraints towards increased added value. *Tropical Animal Health and Production*, 53, 40.

De Wit, L.A., Croll, D.A., Tershy, B., Newton, K.M., Spatz, D.R., Holmes, N.D., Kilpatrick, A.M. (2017). Estimating burdens of neglected tropical zoonotic diseases on islands with introduced mammals. *The American Journal of Tropical Medicine and Hygiene*, 96(3), 749–757.

Defries, R.S., Foley, J.A., Asner, G.P. (2004). Land-use choices: Balancing human needs and ecosystem function. *Frontiers in Ecology and the Environment*, 2, 249–257.

Denslow, J.S. (2003). Weeds in paradise: Thoughts on the invasibility of tropical islands. *Annals of the Missouri Botanical Garden*, 90, 119–127.

Di Marco, M., Venter, O., Possingham, H.P., Watson, J.E. (2018). Changes in human footprint drive changes in species extinction risk. *Nature Communications*, 9(1), 1–9.

Diamond, J.M. (1969). Avifaunal equilibria and species turnover rates on the Channel Islands of California. *Proceedings of the National Academy of Sciences USA*, 64(1), 57–63.

Diamond J. and Case T.J. (1986). Overview: Introductions, extinctions, exterminations, and invasions. In *Community Ecology*, Diamond, J. and Case T.J. (eds). Harper and Row, Manhattan, New York.

Doherty, T.S., Glen, A.S., Nimmo, D.G., Ritchie, E.G., Dickman, C.R. (2016). Invasive predators and global biodiversity loss. *Proceedings of the National Academy of Sciences USA*, 113(40), 11261–11265.

Donlan, C.J. and Wilcox, C. (2008). Diversity, invasive species and extinctions in insular ecosystems. *Journal of Applied Ecology*, 45, 1114–1123.

Drake, J.M. and Lodge, D.M. (2004). Global hot spots of biological invasions: Evaluating options for ballast–water management. *Proceedings of the Royal Society of London. Series B: Biological Sciences*, 271(1539), 575–580.

Elliott, G.P., Merton, D.V., Jansen, P.W. (2001). Intensive management of a critically endangered species: The kakapo. *Biological Conservation*, 99(1), 121–133.

Elton, C.S. (1958). *The Ecology of Invasions by Animals and Plants*. University of Chicago Press, Chicago.

Feng, Y., Fouqueray, T.D., van Kleunen, M. (2019). Linking Darwin's naturalisation hypothesis and Elton's diversity–invasibility hypothesis in experimental grassland communities. *Journal of Ecology*, 107(2), 794–805.

Ferreira, M.T., Cardoso, P., Borges, P.A., Gabriel, R., de Azevedo, E.B., Reis, F., Elias, R.B. (2016). Effects of climate change on the distribution of indigenous species in oceanic islands (Azores). *Climatic Change*, 138(3), 603–615.

Fischer, J. and Lindenmayer, D.B. (2007). Landscape modification and habitat fragmentation: A synthesis. *Global Ecology and Biogeography*, 16(3), 265–280.

Fordham, D.A. and Brook, B.W. (2010). Why tropical island endemics are acutely susceptible to global change. *Biodiversity and Conservation*, 19(2), 329–342.

Fortini, L.B., Price, J., Jacobi, J., Vorsino, A., Burgett, J., Brinck, K.W., Paxton, E.H. (2013). A landscape-based assessment of climate change vulnerability for all native Hawaiian plants. Technical Report, HCSU-044 [Online]. Available: http://hilo.hawaii.edu/hcsu/documents/TR44_Fortini_plant_vulnerability_assessment.pdf [Accessed January 29, 2021].

Foster, J.B. (1963). *The Evolution of the Native Land Mammals of the Queen Charlotte Islands and the Problem of Insularity*. University of British Columbia, Vancouver.

Foster, J.B. (1964). Evolution of mammals on islands. *Nature*, 202, 234–235.

Foufopoulos, J., Kilpatrick, A.M., Ives, A.R. (2011). Climate change and elevated extinction rates of reptiles from Mediterranean islands. *The American Naturalist*, 177(1), 119–129.

Frankham, R. (1997). Do island populations have less genetic variation than mainland populations? *Heredity*, 78, 311–327.

Fridley, J.D. and Sax, D.F. (2014). The imbalance of nature: Revisiting a Darwinian framework for invasion biology. *Global Ecology and Biogeography*, 23(11), 1157–1166.

Gerzabek, M.H., Bajraktarevic, A., Keiblinger, K., Mentler, A., Rechberger, M., Tintner, J., Wriessnig, K., Gartner, M., Valenzuela, X.S., Troya, A., Couenberg, P.M., Jäger, H., Carrión, J.E., Zehetner, F. (2019). Agriculture changes soil properties on the Galápagos Islands – Two case studies. *Soil Research*, 57, 201–214.

Gillespie, R.G., Claridge, E.M., Roderick, G.K. (2008). Biodiversity dynamics in isolated island communities: Interaction between natural and human-mediated processes. *Molecular Ecology*, 17(1), 45–57.

Glasspool, A.F. and Sterrer, W. (2009). Bermuda. In *Encyclopedia of Islands*, Gillespie, R. and Clague, D. (eds). University of California Press, Berkeley, CA.

Gleason, H.A. (1922). On the relation between species and area. *Ecology*, 3, 158–162.

Gonzalez, A. and Chaneton, E.J. (2002). Heterotroph species extinction, abundance and biomass dynamics in an experimentally fragmented microecosystem. *Journal of Animal Ecology*, 71, 594–602.

Gotelli, N.J. and Graves, G.R. (1996). *Null Models in Ecology*. Smithsonian Institution Press, Washington, DC.

Haddad, N.M., Brudvig, L.A., Clobert, J., Davies, K.F., Gonzalez, A., Holt, R.D., Townshend, J.R. (2015). Habitat fragmentation and its lasting impact on Earth's ecosystems. *Science Advances*, 1(2), e1500052.

Harter, D.E., Irl, S.D., Seo, B., Steinbauer, M.J., Gillespie, R., Triantis, K.A., Beierkühnlein, C. (2015). Impacts of global climate change on the floras of oceanic islands – Projections, implications and current knowledge. *Perspectives in Plant Ecology, Evolution and Systematics*, 17(2), 160–183.

Heanley, L.R. (1978). Island area and body size of insular mammals: Evidence from the tri-colored squirrel (*Callosciurus prevosti*) of Southeast Asia. *Evolution*, 32, 29–44.

Heanley, L.R. (2007). Is a new paradigm emerging for oceanic island biogeography? *Journal of Biogeography*, 34(5), 753–757.

Hellmann, J.J., Byers, J.E., Bierwagen, B.G., Dukes, J.S. (2008). Five potential consequences of climate change for invasive species. *Conservation Biology*, 22(3), 534–543.

Hermant, M., Prinzing, A., Vernon, P., Convey, P., Hennion, F. (2013). Endemic species have highly integrated phenotypes, environmental distributions and phenotype–environment relationships. *Journal of Biogeography*, 40(8), 1583–1594.

Holdaway, R.N., Worthy, T.H., Tennyson, A.J. (2001). A working list of breeding bird species of the New Zealand region at first human contact. *New Zealand Journal of Zoology*, 28(2), 119–187.

Holway, D.A. and Suarez, A.V. (2006). Homogenization of ant communities in mediterranean California: The effects of urbanization and invasion. *Biological Conservation*, 127(3), 319–326.

Holyoak, M., Leibold, M.A., Holt, R.D. (2005). *Metacommunities: Spatial Dynamics and Ecological Communities*. University of Chicago Press, Chicago, IL.

Hubbell, S.P. (2001). *The Unified Neutral Theory of Biodiversity and Biogeography (MPB-32)*. Princeton University Press, Princeton, NJ.

Hufbauer, R.A., Facon, B., Ravigne, V., Turgeon, J., Foucaud, J., Lee, C.E., Estoup, A. (2012). Anthropogenically induced adaptation to invade (AIAI): Contemporary adaptation to human-altered habitats within the native range can promote invasions. *Evolutionary Applications*, 5(1), 89–101.

Hui, C., Richardson, D.M., Landi, P., Minoarivelo, H.O., Garnas, J., Roy, H.E. (2016). Defining invasiveness and invasibility in ecological networks. *Biological Invasions*, 18(4), 971–983.

Hulme, P. (2004). Islands, invasions and impacts: A Mediterranean perspective. *Island Ecology: Recopilación de las ponencias presentadas en el Symposium de Ecología Insular organizado por la Asociación Española de Ecología Terrestre (AEET) celebrado en Santa Cruz de la Palma (Islas Canarias) del 18 al 24 de noviembre, 2002*, Asociación española de ecología terrestre, AEET, 359–383.

Hulme, P.E. (2009). Trade, transport and trouble: Managing invasive species pathways in an era of globalization. *Journal of Applied Ecology*, 46(1), 10–18.

Hulme, P.E. (2014). Invasive species challenge the global response to emerging diseases. *Trends in Parasitology*, 30(6), 267–270.

Itescu, Y., Karraker, N.E., Raia, P., Pritchard, P.C., Meiri, S. (2014). Is the island rule general? Turtles disagree. *Global Ecology and Biogeography*, 23, 689–700.

Itescu, Y., Schwawrz, R., Donihue, C.M., Slavenko, A., Roussos, S.A., Sagonas, K., Valakos, E.D., Foufopoulos, J., Pafilis, P., Meiri, S. (2018). Inconsistent patterns of body size evolution in co-occurring island reptiles. *Global Ecology and Biogeography*, 27, 538–550.

Jones, H.P., Tershy, B.R., Zavaleta, E.S., Croll, D.A., Keitt, B.S., Finkelstein, M.E., Howald, G.R. (2008). Severity of the effects of invasive rats on seabirds: A global review. *Conservation Biology*, 22(1), 16–26.

Keane, R.M. and Crawley, M.J. (2002). Exotic plant invasions and the enemy release hypothesis. *Trends in Ecology and Evolution*, 17, 164–170.

Kier, G., Kreft, H., Lee, T.M., Jetz, W., Ibisch, P.L., Nowicki, C., Barthlott, W. (2009). A global assessment of endemism and species richness across island and mainland regions. *Proceedings of the National Academy of Sciences USA*, 106(23), 9322–9327.

Kim, Y.O. and Lee, E.J. (2011). Comparison of phenolic compounds and the effects of invasive and native species in East Asia: Support for the novel weapons hypothesis. *Ecological Research*, 26(1), 87–94.

Kreft, H., Jetz, W., Mutke, J., Kier, G., Barthlott, W. (2008). Global diversity of island floras from a macroecological perspective. *Ecology Letters*, 11(2), 116–127.

Kueffer, C., Daehler, C.C., Torres-Santana, C.W., Levergne, C., Meyer, J.Y., Otto, R., Silva, L. (2010). A global comparison of plant invasions on oceanic islands. *Perspectives in Plant Ecology, Evolution and Systematics*, 12, 145–161.

La Pointe, D.A., Atkinson, C.T., Samuel, M.D. (2012). Ecology and conservation biology of avian malaria. *Annals of the New York Academy of Sciences*, 1249(1), 211–226.

Lack, D. (1970). Island birds. *Biotropica*, 29–31.

Lam, A.R., Stigall, A.L., Matzke, N.J. (2018). Dispersal in the Ordovician: Speciation patterns and paleobiogeographic analyses of brachiopods and trilobites. *Palaeogeography, Palaeoclimatology, Palaeoecology*, 489, 147–165.

Laurance, W.F. (1999). Reflections on the tropical deforestation crisis. *Biological Conservation*, 91, 109–117.

Leal, I.R., Filgueiras, B.K., Gomes, J.P., Iannuzzi, L., Andersen, A.N. (2012). Effects of habitat fragmentation on ant richness and functional composition in Brazilian Atlantic forest. *Biodiversity and Conservation*, 21(7), 1687–1701.

Leclerc, C., Courchamp, F., Bellard, C. (2020). Future climate change vulnerability of endemic island mammals. *Nature Communications*, 11(1), 1–9.

Lockwood, J.L., Cassey, P., Blackburn, T. (2005). The role of propagule pressure in explaining species invasions. *Trends in Ecology and Evolution*, 20(5), 223–228.

Loehle, C. and Eschenbach, W. (2012). Historical bird and terrestrial mammal extinction rates and causes. *Diversity and Distributions*, 18(1), 84–91.

Lokatis, S. and Jeschke, J.M. (2018). The island rule: An assessment of biases and research trends. *Journal of Biogeography*, 45, 289–303.

Lomolino, M.V. (1985). Body size of mammals on islands: The island rule reexamined. *The American Naturalist*, 125, 310–316.

Lomolino, M.V. (2000a). A call for a new paradigm of island biogeography. *Global Ecology and Biogeography*, 9, 1–6.

Lomolino, M.V. (2000b). Ecology's most general, yet protean pattern: The species-area relationship. *Journal of Biogeography*, 17–26.

Lomolino, M.V. (2001). The species-area relationship: New challenges for an old pattern. *Progress in Physical Geography*, 25, 1–21.

Lomolino, M.V. (2005). Body size evolution in insular vertebrates: Generality of the island rule. *Journal of Biogeography*, 32, 1683–1699.

Lomolino, M.V., Geer, A.A., Lyras, G.A., Palombo, M.R., Sax, D.F., Rozzi, R. (2013). Of mice and mammoths: Generality and antiquity of the island rule. *Journal of Biogeography*, 40, 1427–1439.

Long, J.L. (2003). *Introduced Mammals of the World*. CSIRO Publishers, Collingwood, Australia.

Long, E.S., Courtney, K.L., Lippert, J.C., Wall-Scheffler, C.M. (2019). Reduced body size of insular black-tailed deer is caused by slowed development. *Oecologia*, 189, 675–685.

Lonsdale, W.M. (1999). Global patterns of plant invasions and the concept of invasibility. *Ecology*, 80(5), 1522–1536.

Loope, L.L. and Mueller-Dombois, D. (1989). Characteristics of invaded islands, with special reference to Hawai'i. In *Biological Invasions: A Global Perspective*, Drake, J.A., Mooney, H., di Castri, F., Groves, R., Kruger, F., Rejmanek, M., Williamson, M. (eds). Wiley, New York.

Losos, J.B. and Ricklefs, R.E. (2009). *The Theory of Island Biogeography Revisited*. Princeton University Press, Princeton, NJ.

Louys, J., Curnoe, D., Tong, H. (2007). Characteristics of Pleistocene megafauna extinctions in Southeast Asia. *Palaeogeography, Palaeoclimatology, Palaeoecology*, 243(1–2), 152–173.

MacArthur, R.H. and Wilson, E.O. (1967). *The Theory of Island Biogeography*. Princeton University Press, Princeton, NJ.

MacDonald, I. and Cooper, J. (1995). *Insular Lessons for Global Biodiversity Conservation with Particular Reference to Alien Invasions*. Springer, Berlin, Heidelberg.

Mack, R.N. (2003). Phylogenetic constraint, absent life forms, and preadapted alien plants: A prescription for biological invasions. *International Journal of Plant Science*, 164, 185–196.

Mack, R.N. and Erneberg, M. (2002). The United States naturalized flora: Largely the product of deliberate introductions. *Annals of the Missouri Botanical Garden*, 176–189.

Manne, L.L., Brooks, T.M., Pimm, S.L. (1999). Relative risk of extinction of passerine birds on continents and islands. *Nature*, 399(6733), 258–261.

McClain, C.R., Boyer, A.G., Rosenberg, G. (2006). The island rule and the evolution of body size in the deep sea. *Journal of Biogeography*, 33, 1578–1584.

McCreless, E.E., Huff, D.D., Croll, D.A., Tershy, B.R., Spatz, D.R., Holmes, N.D., Wilcox, C. (2016). Past and estimated future impact of invasive alien mammals on insular threatened vertebrate populations. *Nature Communications*, 7(1), 1–11.

McIver, L., Woodward, A., Davies, S., Tibwe, T., Iddings, S. (2014). Assessment of the health impacts of climate change in Kiribati. *International Journal of Environmental Research and Public Health*, 11(5), 5224–5240.

Medina, F.M., Bonnaud, E., Vidal, E., Tershy, B.R., Zavaleta, E.S., Donlan, C.J., Nogales, M. (2011). A global review of the impacts of invasive cats on island endangered vertebrates. *Global Change Biology*, 17(11), 3503–3510.

Meiri, S. (2007). Size evolution in island lizards. *Global Ecology and Biogeography*, 16, 702–708.

Meiri, S., Dayan, T., Simberloff, D. (2006). The generality of the island rule reexamined. *Journal of Biogeography*, 33, 1571–1577.

Meiri, S., Cooper, N., Purvis, A. (2008). The island rule: Made to be broken? *Proceedings of the Royal Society of London B: Biological Sciences*, 275, 141–148.

Meiri, S., Dayan, T., Simberloff, D., Grenyer, R. (2009). Life on the edge: Carnivore body size variation is all over the place. *Proceedings of the Royal Society of London B: Biological Sciences*, 276, 1469–1476.

Milberg, P. and Tyrberg, T. (1993). Naïve birds and noble savages – A review of man-caused prehistoric extinctions of island birds. *Ecography*, 16(3), 229–250.

Moller, H. (1996). Lessons for invasion theory from social insects. *Biological Conservation*, 78(1–2), 125–142.

Mooney, H.A. and Cleland, E.E. (2001). The evolutionary impact of invasive species. *Proceedings of the National Academy of Sciences USA*, 98(10), 5446–5451.

Moser, D., Lenzner, B., Weigelt, P., Dawson, W., Kreft, H., Pergl, J., Essl, F. (2018). Remoteness promotes biological invasions on islands worldwide. *Proceedings of the National Academy of Sciences USA*, 115(37), 9270–9275.

Nicholls, R.J. and Cazenave, A. (2010). Sea-level rise and its impact on coastal zones. *Science*, 328(5985), 1517–1520.

Nolfo-Clements, L., Butcher, R., Leite, M., Clements, M. (2017). Evidence of the island rule and microevolution in white-footed mice (*Peromyscus leucopus*) in an urban harbor archipelago. *Mammal Research*, 62, 423–430.

Nuñez, M.A., Moretti, A., Simberloff, D. (2011). Propagule pressure hypothesis not supported by an 80-year experiment on woody species invasion. *Oikos*, 120(9), 1311–1316.

Palkovacs, E.P. (2003). Explaining adaptive shifts in body size on islands: A life history approach. *Oikos*, 103, 37–44.

Patiño, J., Medina, R., Vanderpoorten, A., González-Mancebo, J.M., Werner, O., Devos, N., Ros, R.M. (2013). Origin and fate of the single-island endemic moss *Orthotrichum handiense*. *Journal of Biogeography*, 40(5), 857–868.

Pimm, S.L., Russell, G.J., Gittleman, J.L., Brooks, T.M. (1995). The future of biodiversity. *Science*, 269, 347–350.

Pinzone, P., Potts, D., Pettibone, G., Warren, R. (2018). Do novel weapons that degrade mycorrhizal mutualisms promote species invasion? *Plant Ecology*, 219(5), 539–548.

Preston, F. (1960). Time and space and the variation of species. *Ecology*, 41, 611–627.

Pyšek, P., Jarošík, V., Hulme, P.E., Pergl, J., Hejda, M., Schaffner, U., Vilà. M. (2012). A global assessment of invasive plant impacts on resident species, communities and ecosystems: The interaction of impact measures, invading species' traits and environment. *Global Change Biology*, 18, 1725–1737.

Rabor, D. (1959). The impact of deforestation on birds of Cebu, Philippines, with New Records for That Island. *The Auk*, 76(1), 37–43.

Radley, P.M., van Etten, E.J., Blake, D., Davis, R.A. (2020). Breeding and feeding habitat selection by an island endemic bird may increase its vulnerability to climate change. *Biotropica*, 53(2), 422–432.

Rahel, F.J. (2002). Homogenization of freshwater faunas. *Annual Review of Ecology and Systematics*, 33(1), 291–315.

Rebouças, R., Da Silva, H.R., Solé, M. (2018). Frog size on continental islands of the coast of Rio de Janeiro and the generality of the Island Rule. *PloS One*, 13, e0190153.

Richardson, D.M. and Pyšek, P. (2006). Plant invasions: Merging the concepts of species invasiveness and community invasibility. *Progress in Physical Geography*, 30(3), 409–431.

Rick, T.C., Erlandson, J.M., Vellanoweth, R.L., Braje, T.J., Collins, P.W., Guthrie, D.A., Stafford Jr. T.W. (2009). Origins and antiquity of the island fox (*Urocyon littoralis*) on California's Channel Islands. *Quaternary Research*, 71, 93–98.

Ricketts, T.H., Dinerstein, E., Boucher, T., Brooks, T.M., Butchart, S.H., Hoffmann, M., Wikramanayake, E. (2005). Pinpointing and preventing imminent extinctions. *Proceedings of the National Academy of Sciences USA*, 102(51), 18497–18501.

Risse, M. (2009). The right to relocation: Disappearing island nations and common ownership of the earth. *Ethics and International Affairs*, 23(3), 281–300.

Rizali, A., Lohman, D.J., Buchori, D., Prasetyo, L.B., Triwidodo, H., Bos, M.M., Schulze, C.H. (2010). Ant communities on small tropical islands: Effects of island size and isolation are obscured by habitat disturbance and "tramp" ant species. *Journal of Biogeography*, 37(2), 229–236.

Rödl, T., Berger, S., Michael Romero, L., Wikelski, M. (2007). Tameness and stress physiology in a predator-naive island species confronted with novel predation threat. *Proceedings of the Royal Society of London. Series B: Biological Sciences*, 274(1609), 577–582.

Rosenzweig, M.L. (1995). *Species Diversity in Space and Time*. Cambridge University Press, Cambridge.

Sala, O.E., Chapin, F.S., Armesto, J.J., Berlow, E., Bloomfield, J., Dirzo, R., Wall, D.H. (2000). Global biodiversity scenarios for the year 2100. *Science*, 287(5459), 1770–1774.

Sauer, J.D. (1969). Oceanic islands and biogeographical theory: A review. *Geographical Review*, 59, 582–593.

Saunders, D.A., Hobbs, R.J., Margules, C.R. (1991). Biological consequences of ecosystem fragmentation: A review. *Conservation Biology*, 5, 18–32.

Sax, D.F. and Brown, J.H. (2000). The paradox of invasion. *Global Ecology and Biogeography*, 9, 363–372.

Schmack, J.M., Brenton-Rule, E.C., Veldtman, R., Wenseleers, T., Beggs, J.R., Lester, P.J., Bulgarella, M. (2019). Lack of genetic structuring, low effective population sizes and major bottlenecks characterise common and German wasps in New Zealand. *Biological Invasions*, 21(10), 3185–3201.

Schmack, J.M., Schleuning, M., Ward, D.F., Beggs, J.R. (2020). Biogeography and anthropogenic impact shape the success of invasive wasps on New Zealand's offshore islands. *Diversity and Distributions*, 26(4), 441–452.

Schoener, T. (1976). The species-area relation within archipelagos: Models and evidence from island land birds. *16th International Ornithological Congress*, Canberra, Australia, August 12–17, 1974, 1976. Australian Academy of Sciences, 629–642.

Seneviratne, S.I., Nicholls, N., Easterling, D., Goodess, C.M., Kanae, S., Kossin, J., Luo, Y., Marengo, J., McInnes, K., Rahimi, M. (2012). Changes in climate extremes and their impacts on the natural physical environment. In *Managing the Risks of Extreme Events and Disasters to Advance Climate Change Adaptation: A Special Report of Working Groups I and II of the Intergovernmental Panel on Climate Change*, Field, C.B., Barros, B., Stocker, T.F., Qin, D., Dokken, D.J., Ebi, K.L., Mastrandrea, M.D., Mach, K.J., Plattner, G.-K., Allen, S.K., Tignor, M., Midgley, P.M. (eds). Cambridge University Press, Cambridge.

Shaw, J.D., Spear, D., Greve, M., Chown, S.L. (2010). Taxonomic homogenization and differentiation across Southern Ocean Islands differ among insects and vascular plants. *Journal of Biogeography*, 37, 217–228.

Shea, K. and Chesson, P. (2002). Community ecology theory as a framework for biological invasions. *Trends in Ecology and Evolution*, 17(4), 170–176.

Sih, A., Bolnick, D.I., Luttbeg, B., Orrock, J.L., Peacor, S.D., Pintor, L.M., Vonesh, J.R. (2010). Predator–prey naïveté, antipredator behavior, and the ecology of predator invasions. *Oikos*, 119(4), 610–621.

Simberloff, D. (1995). Why do introduced species appear to devastate islands more than mainland areas? *Pacific Science*, 49, 87–97.

Simberloff, D. (2009). The role of propagule pressure in biological invasions. *Annual Review of Ecology, Evolution, and Systematics*, 40, 81–102.

Simberloff, D. (2013). *Invasive Species: What Everyone Needs to Know*. Oxford University Press, Oxford.

Simberloff, D.S. and Wilson, E.O. (1969). Experimental zoogeography of islands: The colonization of empty islands. *Ecology*, 50, 278–296.

Simberloff, D.S. and Wilson, E.O. (1970). Experimental zoogeography of islands: A two-year record of colonization. *Ecology*, 51, 934–937.

Sodhi, N.S., Koh, L.P., Brook, B.W., Ng, P.K.L. (2004). Southeast Asian biodiversity: The impending disaster. *Trends in Ecology and Evolution*, 19, 654–660.

Sodhi, N.S., Lee, T.M., Koh, L.P., Prawiradilaga, D.M. (2006). Long-term avifaunal impoverishment in an isolated tropical woodlot. *Conservation Biology*, 20(3), 772–779.

Sondaar, P.Y. (1977). Insularity and its effect on mammal evolution. In *Major Patterns in Vertebrate Evolution*, Hecht, E. (ed.). Springer, New York.

Song, X.P., Hansen, M.C., Stehman, S.V., Potapov, P.V., Tyukavina, A., Vermote, E.F., Townshend, J.R. (2018). Global land change from 1982 to 2016. *Nature*, 560(7720), 639–643.

Spatz, D.R., Zilliacus, K.M., Holmes, N.D., Butchart, S.H., Genovesi, P., Ceballos, G., Croll, D.A. (2017). Globally threatened vertebrates on islands with invasive species. *Science Advances*, 3(10), e1603080.

Stachowicz, J.J., Fried, H., Osman, R.W., Whitlatch, R.B. (2002). Biodiversity, invasion resistance, and marine ecosystem function: Reconciling pattern and process. *Ecology*, 83(9), 2575–2590.

Steadman, D.W., Martin, P.S., MacPhee, R.D., Jull, A.T., McDonald, H.G., Woods, C.A., Hodgins, G.W. (2005). Asynchronous extinction of late Quaternary sloths on continents and islands. *Proceedings of the National Academy of Sciences USA*, 102(33), 11763–11768.

Stigall, A.L. (2019). The invasion hierarchy: Ecological and evolutionary consequences of invasions in the fossil record. *Annual Review of Ecology, Evolution, and Systematics*, 50, 355–380.

Supriatna, J., Shekelle, M., Fuad, H.A., Winarni, N.L., Dwiyahreni, A.A., Farid, M., Zakaria, Z. (2020). Deforestation on the Indonesian island of Sulawesi and the loss of primate habitat. *Global Ecology and Conservation*, 24, e01205.

Takahashi, A., Kumagai, T., Kanamori, H., Fujinami, H., Hiyama, T., Hara, M. (2017). Impact of tropical deforestation and forest degradation on precipitation over Borneo Island. *Journal of Hydrometeorology*, 18(11), 2907–2922.

Tatem, A.J. (2009). The worldwide airline network and the dispersal of exotic species: 2007–2010. *Ecography*, 32(1), 94–102.

Taylor, S. and Kumar, L. (2016). Global climate change impacts on Pacific islands terrestrial biodiversity: A review. *Tropical Conservation Science*, 9(1), 203–223.

Tershy, B.R., Shen, K.W., Newton, K.M., Holmes, N.D., Croll, D.A. (2015). The importance of islands for the protection of biological and linguistic diversity. *Bioscience*, 65(6), 592–597.

Thuiller, W. (2007). Climate change and the ecologist. *Nature*, 448(7153), 550–552.

Tilman, D., Cassman, K.G., Matson, P.A., Naylor, R., Polasky, S. (2002). Agricultural sustainability and intensive production practices. *Nature*, 418, 671–677.

Tjørve, E., Calf Tjørve, K.M., Šizlingová, E., Šizling, A.L. (2018). Great theories of species diversity in space and why they were forgotten: The beginnings of a spatial ecology and the Nordic early 20th-century botanists. *Journal of Biogeography*, 45, 530–540.

Towns, D.R., Atkinson, I.A., Daugherty, C.H. (2006). Have the harmful effects of introduced rats on islands been exaggerated? *Biological Invasions*, 8(4), 863–891.

Triantis, K.A. and Mylonas, M. (2009). Greek islands, biology. *Encyclopedia of Islands*, 388–392.

Triantis, K.A., Borges, P.A., Ladle, R.J., Hortal, J., Cardoso, P., Gaspar, C., Whittaker, R.J. (2010). Extinction debt on oceanic islands. *Ecography*, 33(2), 285–294.

Trueman, M. and d'Ozouville, N. (2010). Characterizing the Galapagos terrestrial climate in the face of global climate change. *Galapagos Research*, 67, 26–37.

Turner, K.G., Hufbauer, R.A., Rieseberg, L.H. (2014). Rapid evolution of an invasive weed. *New Phytologist*, 202(1), 309–321.

Valente, L., Phillimore, A.B., Melo, M., Warren, B.H., Clegg, S.M., Havenstein, K., Tiedemann, R., Illera, J.C., Thébaud, C., Aschenbach, T. (2020). A simple dynamic model explains the diversity of island birds worldwide. *Nature*, 579, 92–96.

Van Kleunen, M., Dawson, W., Schlaepfer, D., Jeschke, J.M., Fischer, M. (2010). Are invaders different? A conceptual framework of comparative approaches for assessing determinants of invasiveness. *Ecology Letters*, 13(8), 947–958.

Van Valen, L. (1973). A new evolutionary law. *Evolutionary Theory*, 1, 1–30.

Vanek, J.P. and Burke, R.L. (2020). Insular dwarfism in female Eastern hog-nosed snakes (*Heterodon platirhinos*; Dipsadidae) on a barrier island. *Canadian Journal of Zoology*, 98, 157–164.

Vermeij, G. (2005). *Invasion as Expectation: A Historical Fact of Life. Species Invasions: Insights Into Ecology, Evolution and Biogeography*, Sax, D.F., Stachowicz, J.J., Gaines, S.D. (eds). Sinauer Press, Sunderland, MA.

Vilà, M., Corbin, J.D., Dukes, J.S., Pino, J., Smith, S.D. (2007). Linking plant invasions to global environmental change. In *Terrestrial Ecosystems in a Changing World*, Canadell, J.G., Pataki, D.E., Pitelka, L.F. (eds). Springer, Berlin.

Vorsino, A.E., Fortini, L.B., Amidon, F.A., Miller, S.E., Jacobi, J.D., Price, J.P., Koob, G.A. (2014). Modeling Hawaiian ecosystem degradation due to invasive plants under current and future climates. *PloS One*, 9(5), e95427.

Walther, G.R., Roques, A., Hulme, P.E., Sykes, M.T., Pyšek, P., Kühn, I., Settele, J. (2009). Alien species in a warmer world: Risks and opportunities. *Trends in Ecology and Evolution*, 24(12), 686–693.

Wan, J.Z., Wang, C.J., Tan, J.F., Yu, F.H. (2017). Climatic niche divergence and habitat suitability of eight alien invasive weeds in China under climate change. *Ecology and Evolution*, 7(5), 1541–1552.

Ward, N.L. and Masters, G.J. (2007). Linking climate change and species invasion: An illustration using insect herbivores. *Global Change Biology*, 13(8), 1605–1615.

Wardle, D.A. (2002). Islands as model systems for understanding how species affect ecosystem properties. *Journal of Biogeography*, 29(5–6), 583–591.

Warren, B.H., Simberloff, D., Ricklefs, R.E., Aguilée, R., Condamine, F.L., Gravel, D., Morlon, H., Mouquet, N., Rosindell, J., Casquet, J. (2015). Islands as model systems in ecology and evolution: Prospects fifty years after MacArthur-Wilson. *Ecology Letters*, 18, 200–217.

Western, D. (1979). Size, life history and ecology in mammals. *African Journal of Ecology*, 17, 185–204.

Whittaker, R.J. and Fernández-Palacios, J.M. (2007). *Island Biogeography: Ecology, Evolution, and Conservation*. Oxford University Press, Oxford.

Whittaker, R.J., Triantis, K.A., Ladle, R.J. (2008). A general dynamic theory of oceanic island biogeography. *Journal of Biogeography*, 35, 977–994.

Williamson, M. and Griffiths, B. (1996). *Biological Invasions*. Springer Science and Business Media, Berlin.

Williamson, M., Gaston, K.J., Lonsdale, W. (2001). The species–area relationship does not have an asymptote! *Journal of Biogeography*, 28, 827–830.

Williamson, M., Gaston, K.J., Lonsdale, W. (2002). An asymptote is an asymptote and not found in species–area relationships. *Journal of Biogeography*, 29, 1713–1713.

Willis, C.G., Ruhfel, B.R., Primack, R.B., Miller-Rushing, A.J., Losos, J.B., Davis, C.C. (2010). Favorable climate change response explains non-native species' success in Thoreau's woods. *PloS One*, 5(1), e8878.

Wlech, J.J. (2009). Testing the island rule: Primates as a case study. *Proceedings of the Royal Society of London. Series B: Biological Sciences*, 276, 675–682.

Wonham, M.J., Lewis, M.A., MacIsaac, H.J. (2005). Minimizing invasion risk by reducing propagule pressure: A model for ballast-water exchange. *Frontiers in Ecology and the Environment*, 3(9), 473–478.

Wood, J.R., Alcover, J.A., Blackburn, T.M., Bover, P., Duncan, R.P., Hume, J.P., Wilmshurst, J.M. (2017). Island extinctions: Processes, patterns, and potential for ecosystem restoration. *Environmental Conservation*, 44(4), 348–358.

6

Cave Biogeography

Arnaud FAILLE
Stuttgart State Museum of Natural History, Germany

The subterranean environment comprises voids of any size in which life can develop in aphotic, aseasonal and largely oligotrophic conditions. A small proportion of living organisms have been able to evolve and adapt to such conditions. Some of them have become strictly dependent on this harsh environment, at the price of a set of profound biological adaptations. Their physiological, biological and morphological adaptations have converged towards a drastic reduction in dispersal ability and between-population gene flow. Combined with the extreme fragmentation of many subterranean habitats, these traits make the subterranean fauna a model of special interest compared to the surface fauna for exploring the impact of habitat fragmentation on speciation, diversification and evolutionary radiation, in particular for defining and dating barriers to dispersal based on geological and hydrogeological evidence.

6.1. Physical characteristics of subterranean environments

The study of insular habitats has been at the origin of major biogeographic and evolutionary theories (Wallace 1880; MacArthur and Wilson 1967; Whittaker *et al.* 2017). Hypogean environments offer a similar overall insular configuration, with similar limitations to dispersal (Culver 1970; Barr and Holsinger 1985). These fragmented landscapes are the result of a long uninterrupted evolution through geological times, giving a temporal framework to study the processes of adaptation, speciation and radiation because of the low dispersal ability of subterranean

Biogeography,
coordinated by Eric GUILBERT.

invertebrates and their seclusion into stable geological units, much more stable in the long term than surface habitats. Such subterranean environments are diverse around the world, present in all climatic zones, at all altitudes from deep sea to mountain tops and developed in various geological settings, well beyond the classical lava caves and karstic areas. They share major stringent constraints, putting hypogean organisms under very strong unidirectional selection. These environments are simple and homogeneous through space and time, *de facto* reducing the physical factors to take into consideration in evolutionary studies (Jeannel 1926). Several parameters remain almost unchanged through time, including the absence of light, constant temperatures, high humidity and an irregular and often scarce food supply, which has generated a severely truncated biodiversity (Gibert and Deharveng 2002) with low numbers of species and lineages and low population densities. These characteristics make easier the comparison between populations for detecting local adaptations, both in the short (population genetics) and long (study of the convergence and parallelism in various species or lineages) term. Subterranean environments are not limited to caves. They also comprise more shallow habitats, deep soil, the "Milieu souterrain superficiel", and phreatic and hyporheic habitats (Juberthie *et al.* 1980; Giachino and Vailati 2010; Ortuño *et al.* 2013; Pipan and Culver 2019). One of the hypothesized outcomes of physical characteristics of subterranean habitats is the prevalence of radiative diversifications: a reduced number of migrations (and of successful adaptation), events leading to few highly diverse lineages, each with many species. For all the reasons listed above and particularly their fragmentation and durability, subterranean systems, like oceanic islands, provide exceptional opportunities for the study of spatio-temporal patterns that shape biodiversity.

6.2. Diversity and adaptations of the cave fauna

6.2.1. *Underground evolution*

The isolation and often the fragmentation of the geological cover in which they live generate a high degree of endemism among obligate subterranean organisms. This allows us to reconstruct the evolution of lineage dispersion through time, and to identify which temporal events had an impact in this dispersal. Contrary to epigean habitats like forest or aquatic environments, hypogean habitats are slowly changing at the geological scale, in their geographical pattern, interconnections and structures. Within these subterranean habitats, organisms with various biologies and dispersal abilities are interacting, allowing the same hypotheses to be tested on functionally different biological models, thus refining and strengthening the statistical value of conclusions.

6.2.1.1. *Features and biogeographical interest of the subterranean fauna*

The evolution and origin of cave life have fascinated evolutionists and biologists for a long time (Lamarck 1809; Darwin 1859; Racovitza 1907). The organisms living underground share a highly modified morphology and biology, combining adaptations like loss of some traits (eye degeneration, depigmentation), with enhancement of others (mechanical and chemical sensory organs, body shape modifications), the so-called "troglomorphies" (Christiansen 1962). The most troglomorphic species are often considered to have colonized the underground environment the earliest, although the subterranean colonization and subsequent appearance of troglomorphic features can be fast (Howarth 1983; Wessel *et al.* 2013). Recent studies on the cave fish *Astyanax mexicanus* (De Filippi, 1853) indicate that cave-related traits can appear even within a single generation through phenotypic plasticity (Bilandžija *et al.* 2020). Morphological adaptations are generally associated with changes in life cycle and metabolism: subterranean species tend to develop a K-strategy, that is, fewer but much larger eggs, longer life span and in some extreme cases reduction of the number and duration of larval instars (Deleurance 1958). Stenothermy, stenhygroby, low dispersal ability and flightlessness are known to enforce isolation of populations and to promote speciation (see, for example, Ikeda *et al.* 2012; Polato *et al.* 2018). The long-term isolation of subterranean populations in well-delimited geological units with few or no gene flow has led to a high level of endemism, as well as to strong genetic structuring at the population scale, mostly shaped by fragmentation of the geological cover (Barr 1968; Caccone 1985). Depending on the evolutionary level of interest, cave animals thus provide excellent models to study recent pleistocene splits as well as ancient paleogeological events. In this respect, cave faunas keep a signal of ancient distributions and tectonics, whereas most other surface groups have lost this signal due to subsequent dispersal events (see, for example, Faille *et al.* 2014). For the same reasons, subterranean groups are also particularly promising when studying the impact of past and predicted climatic changes on biodiversity (Sánchez-Fernández *et al.* 2016; Mammola 2019).

6.2.2. *Diversity*

6.2.2.1. *Taxonomic richness*

A limited number of taxonomic groups have successfully colonized subterranean habitats despite their harsh environmental conditions. A few are vertebrates, essentially aquatic ones like the emblematic olm *Proteus anguineus* Laurenti, 1768, from caves of the Dinarides. The majority of the vertebrate species found underground belong to fishes, with about 200 obligate species mostly belonging to the two orders Cypriniformes and Siluriformes (Niemiller *et al.* 2019). These species are mainly found in tropical areas, the bulk of the diversity being in China, south of the Yangtze River (Ma *et al.* 2019). Nevertheless, the large majority of subterranean biodiversity is dominated by aquatic and terrestrial crustaceans, arachnids, myriapods, springtails and insects. Apart from Isopoda, the most diverse group of Crustacea in terrestrial subterranean ecosystems with more than 300 species worldwide (Taiti 2004), most of the subterranean Crustacea are aquatic and belong to the superorder Peracarida (more than 1,800 species, half of which Amphipoda) and the subclass Copepoda (ca 1,000 species) (Stoch and Galassi 2010; Deharveng and Bedos 2018; Mejía-Ortíz 2019). Among Hexapoda, the group with the highest diversity in subterranean environments is the order Coleoptera with ca 2,500 described species, followed by Collembola (ca 500 troglobitic (= cavernicolous) species described so far). Other groups are Diplopoda, spiders, Orthoptera, Diplura, Hemiptera, Zygentoma, Blattodea and Psocoptera (Deharveng and Bedos 2018). Diplura is the group with the highest ratio of troglobitic species to total species (Sendra *et al.* 2020). The quasi-absence underground of several of the richest groups of living beings (plants, Diptera, Hymenoptera, Lepidoptera) is noticeable, and raises major eco-evolutionary questions.

Most of the animals that successfully colonized the terrestrial subterranean environments belong to groups living in the soil and leaf litter, where their epigean relatives are often highly diversified; many are also narrow endemics. These edaphic species share characters of the cave fauna, such as eyelessness, depigmentation and apterism in insects, and stenhygroby. Cave animals appear therefore to belong to groups preadapted to an underground life (beetles, Collembola, Myriapoda, Isopoda, Arachnida), as observed in fishes, in which permanent darkness imposes stronger negative selection on groups of fishes adapted to surface habitats (Vandel 1964; Barr 1968; Torres-Dowdall *et al.* 2018).

Among insects, Coleoptera have been the most successful arthropod group, in terms of species richness, to colonize subterranean habitats, but only two of their numerous families are highly diverse in this environment: the Leiodidae Leptodirini, a Western Palearctic tribe of saprophagous beetles, and the Carabidae of the tribe Trechini, which underwent subterranean radiations in several parts of the world (Jeannel 1911, 1928; Casale *et al.* 1998). Subterranean species of both groups share morphological modifications considered as adaptations to a subterranean lifestyle: loss of metathoracic wings, eyes and body pigments, and parallel changes in body shape and size. The third most diverse group of subterranean Coleoptera is aquatic: the Dytiscidae of the subfamily Hydroporinae, with more than 100 species in the calcrete aquifers of inland arid zones of Australia (Leys *et al.* 2003; Balke *et al.*

2004; Watts and Humphreys, 2009). These species have developed troglomorphic traits similar to those found in the terrestrial groups.

6.2.2.2. *World patterns of distribution of the cave fauna*

Although this global pattern is changing fast as the result of new discoveries, most of the described subterranean diversity is still found in a belt located between 42° and 46° in Europe and 34° in North America (Culver *et al.* 2006). The richest areas of subterranean biodiversity are located in the Western Palearctic area, between the Pyrenees (France, Spain) and the Black Sea (Figure 6.1) (Culver and Pipan 2019; Iannella *et al.* 2020). Interestingly, most of the relict terrestrial species, that is, species without a close epigean relative in their distribution area, are localized in the Western Mediterranean region, particularly the Iberian Peninsula and the Dinaric Alps (Deharveng *et al.* 2019).

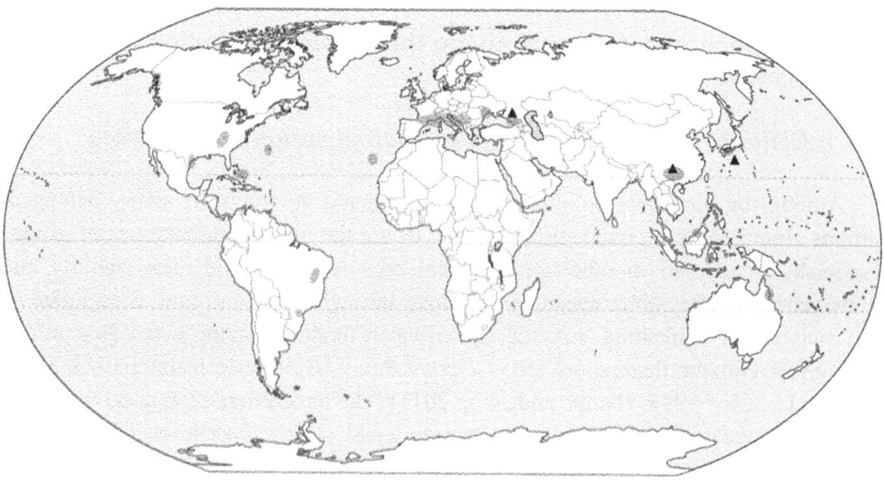

Figure 6.1. *Map of world hotspots of subterranean biodiversity. This includes the sites with more than 20 troglobitic or stygobitic species recorded in Culver and Pipan (2019) and Deharveng and Bedos (2019); the triangles indicate the emerging hotspots of Shikoku (Japan), the Western Caucasus and Southern China. For a color version of this figure, see www.iste.co.uk/guilbert/biogeography.zip*

The Dinarides host four of the six richest cave systems in the world (>40 subterranean species). In North America, the hotspots of subterranean diversity are located in Texas, Alabama and Kentucky (Christman *et al.* 2005; Culver and Pipan 2009). The other identified hotspots, hosting more than 20 troglobitic species each, are located in Bermuda, the Canary Islands, Northern and Western Australia and Indonesia (Culver and Pipan 2019; Deharveng and Bedos 2019). The Japanese island of Shikoku as well as the Caucasus are expected to become hotspots of subterranean biodiversity, and new hotspots are rapidly emerging in China, Brazil and Southeast Asia (Souza Silva and Ferreira 2016; Tian *et al.* 2016; Culver and Pipan 2019; Deharveng and Bedos 2019) (Figure 6.1). These areas, which were completely unknown for cave biodiversity until recently, host a highly diverse subterranean fauna. A striking example is the remarkable diversification of the ground beetle tribe Trechini in Southern China, mostly in the Guizhou and Guangxi provinces, with about 170 species described in ca 30 years, a number continuously increasing and apparently not close to reaching a plateau (Huang *et al.* 2020). Among the thoroughly studied zoological groups, southern China is also a biodiversity hotspot for Diplopoda (Golovatch 2015). Tropical subterranean ecosystems, although on average much less sampled than temperate ones, host a rich fauna, especially of species linked to guano (Deharveng and Bedos 2019).

6.3. Vicariance and dispersal shape the global distribution patterns of cave animals

6.3.1. *Disjunct distributions and the relictual status of cave biota*

Among the biota that colonized the underground environment, many belong to groups absent from the local surface fauna, or are the only representatives of groups especially diversified in other geographical areas of the world. The stability and homogeneity of the subterranean environment through space and time is regarded as the main factor explaining this, the subterranean medium having acted as a refuge during past climatic fluctuations (the *Climatic Relict Hypothesis*, Jeannel 1943, 1959; Peck and Finston 1993; Hampe and Jump 2011). The underground fauna is therefore a precious source of unique information regarding past surface biocenoses. Underground species are also especially important to understand the evolution of lineages. Many of these species belong to highly derived, relictual genera with one or few species only and a very narrow distribution. They are usually morphologically and genetically very isolated and it is often difficult to trace back their origin (Assmann *et al.* 2010).

6.3.1.1. *Terrestrial groups*

Many examples of relictual distribution patterns can be found in the subterranean terrestrial fauna. Examples can be found in many groups, such as the tertiary relict and only stygobitic bivalve genus *Congeria* from the Dinaric karst (Bilandžija *et al.* 2013) and Diplopoda (e.g. the genera *Marboreuma* in the Pyrenees, *Cantabrodesmus* in Cantabria or *Caucasodesmus* in the northern Caucasus (Golovatch 2009)), among many others. Many similar cases are found in Coleoptera, a well-known example being the monotypic genus *Dalyat* from Southeastern Spain, the only European representative of the subfamily Promecognathinae, a group with disjunct distribution with representatives nowadays restricted to South Africa and Western North America. Gondwanan vicariance is regarded as the explanation of this striking pattern, the current disjunct distribution resulting from the isolation of the Iberian plate from Pangaea in the late Jurassic to early Cretaceous (ca 145 Mya) (Ribera *et al.* 2005) (Figure 6.2(a)).

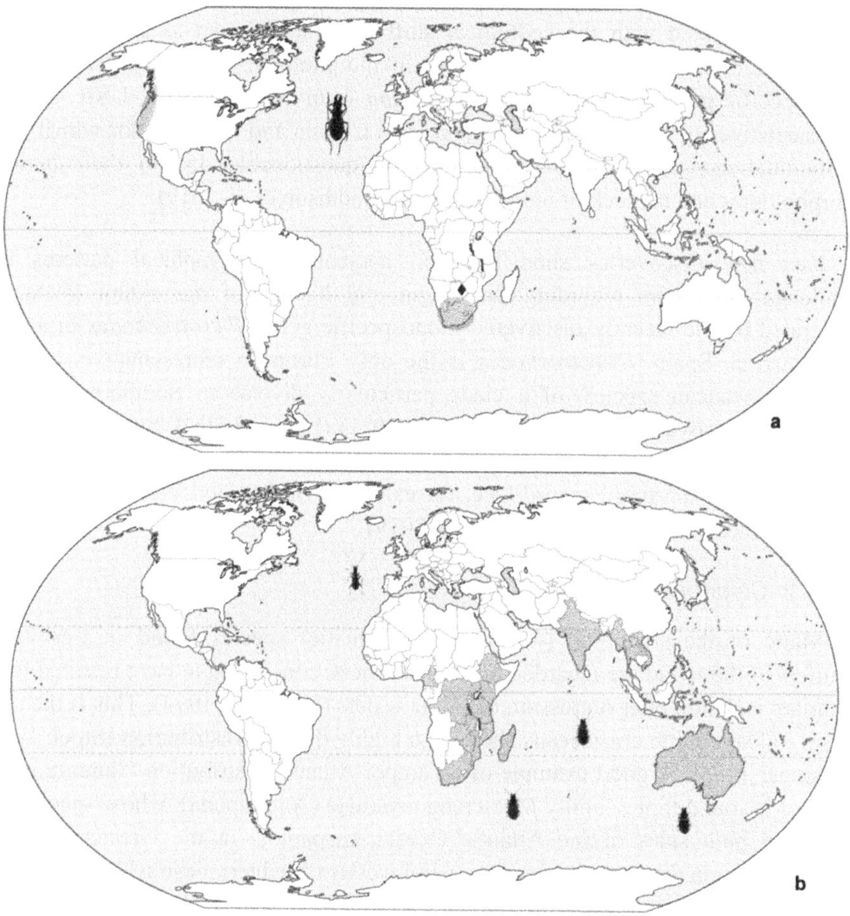

Figure 6.2. *Distribution patterns of some beetle groups with cave species: (a) Promecognathinae; asterisk: location of the monotypic subterranean genus* Dalyat*; rhombus: Cretaceous fossil genus* Palaeoaxinidium. *(b) Trechodina; known distribution of epigean species of the clade Trechodes; asterisk: monotypic genus* Iberotrechodes, *the only subterranean species of the clade. For a color version of this figure, see www.iste.co.uk/guilbert/biogeography.zip*

Other examples of such highly disjunct distributions are naturally found in the epigean fauna, especially in high mountain groups. This is the case of the sister relationship – evidenced by use of molecular markers – between the carabid genera *Himalotrechodes* from the Himalayas and *Pachytrechodes* from the Udzungwa and Uluguru Mountains in Tanzania, a disjunction of more than 6,500 km in a straight line (Faille *et al.* 2021). Such examples are especially common in the cave fauna, but are associated with the ecological shift from an epigean to an underground lifestyle and striking adaptative changes in morphology. This is the case of the monospecific ground beetle genus *Horologion* from West Virginia, USA, and the representatives of the subtribe Lovriciina from Croatia and Bulgaria, for which even subfamilial assignment among Trechinae is questionable due to their derived morphologies and the lack of molecular data (Maddison *et al.* 2019).

Key new discoveries shed light on ancient biogeographical patterns but challenge our views regarding the origin and history of the extant fauna, as illustrated by the recently discovered monospecific genus *Iberotrechodes* in a cave in Cantabria, Spain. *Iberotrechodes* is the only European representative, and the only subterranean species, of a clade particularly diverse in Southern and East Africa, Madagascar and India (Faille *et al.* 2021) (Figure 6.2(b)). Such unexpected biogeographical patterns are rare in the extant beetle fauna, whereas they are well documented in the fossil record (see, for example, Barden and Ware 2017; Poinar 2018; Brunke *et al.* 2019; Mashimo *et al.* 2019).

6.3.1.2. *Stygobitic groups*

Many of the stygobitic (=subterranean aquatic) species found in freshwater aquifers of the world are regarded as marine relicts, considered to have resulted from vicariant isolation after regression of saline waters (Holsinger 1994). This is the case of several stygobitic crustaceans showing a highly disjunct distribution (Stoch 1993; Holsinger 1994). A good example of an amphi-Atlantic distribution – among many others – is found in the family Metacrangonyctidae (Amphipoda), whose species are found on both sides of the Atlantic Ocean: Hispaniola in the Greater Antilles, Fuerteventura in the Canary Islands, several western Mediterranean islands (Balearic Islands, Elba), North Africa, the Middle East and Southwestern Oman (Bauzà-Ribot *et al.* 2012; Jaume and Vonk 2012; Pons *et al.* 2019) (Figure 6.3). The evolutionary history of the group extends back to the early Cretaceous. Vicariance by plate tectonics remains the main explanatory factor for the amphi-Atlantic distribution displayed by many groups of subterranean Crustacea (see, for example, Jaume 2008). A shared biogeographical pattern, of similar age and resulting from vicariance by continental drift, is found in amphipods of the family Crangonyctidae (Copilaş-Ciocianu *et al.* 2019). Another well-known example is the order Thermosbaenacea, whose current distribution pattern was also found to be caused by the continental drift. The current distribution of this remarkable group of crustaceans reflects the area covered by the Tethys Sea or its coastlines. It is therefore regarded as a relict of a widespread shallow-water marine Tethyan fauna having sheltered in interstitial groundwater following marine regressions, as observed in other groups like the enigmatic crustacean of the class Remipedia, the potential sister group of Hexapoda (Yager 1981; Jaume 2008; Lozano-Fernandez *et al.* 2019).

Figure 6.3. *An example of amphi-Atlantic distribution, the family Metacrangonyctidae (Amphipoda). For a color version of this figure, see www.iste.co.uk/guilbert/biogeography.zip*

The remarkable and isolated aquatic isopod *Cantabroniscus primitivus* Vandel, 1965, endemic to a narrow area in the Cantabrian Mountains, Spain, is considered closely related to the Central American genus *Typhlotricholigioides* Rioja, 1953, from Mexico (Vandel 1965; Bellés 1987). But this hypothesis remains to be tested in a phylogenetic framework in order to verify the degree of affinity between the two genera and confirm that their present-day distribution resulted from an early split of Laurasia; such dating would have an impact on the entire tree of life of Oniscidean isopods, and is considered dubious although no alternative hypothesis explaining this striking pattern has yet been proposed (Broly *et al.* 2013). Another intriguing discovery of biogeographical interest is the recent finding of a stygobitic species of *Exocelina*, a genus of diving beetle, especially diverse in New Guinea and New Caledonia (145 species), in the Malay Peninsula, filling a 4,000 km distribution gap between Melanesia and southern China (Balke and Ribera 2020).

6.3.2. Colonization of the subterranean environment: reassessing biogeographic hypotheses

The age of underground colonization by different groups has long been based on biogeographic and morphological observations only, but the more recent contribution of molecular data provides a phylogenetic framework and allows dating splits. One of the main problems is the lack of reliable fossils to calibrate phylogenetic reconstructions, but paleogeological events can sometimes be used to date divergences. The use of molecular approaches has increased in the last decades, as these methods appear very useful to understand the histories of these groups, alternatively considered as living fossils (Jeannel 1943, Stepien *et al.* 2001) or recent invaders of the subterranean environment (e.g. Howarth 1980). Early contributions to the study of the molecular evolution of subterranean organisms were summarized in Juan *et al.* (2010).

The age of subterranean colonization largely depends on the group in question. Some are assumed to be ancient, as is the case for the Pyrenean beetle lineages Leptodirini (late Oligocene, Cieslak *et al.* 2014) and Trechini (Oligocene–Miocene, Faille *et al.* 2010, 2013), whereas some are regarded as recent invaders of the subterranean environment, as evidenced for American fishes of the genera *Amblyopsis* and *Astyanax* or *Nesticella* spiders (Niemiller *et al.* 2013; Zhang and Li 2013; Fumey *et al.* 2018;). In some cases, the relictual status of the species could have largely preceded the shift to a subterranean lifestyle and should therefore be treated separately from their invasion of subterranean ecosystems, which may have occurred only recently. Moreover, for most cave relicts, a phylogenetic calibrated framework is still lacking.

6.3.2.1. *Impact of glaciations on present-day underground biodiversity*

Pleistocene glaciations were often considered to have had a key impact in shaping the distribution and diversity of the hypogean fauna (Stoch and Galassi 2010; Niemiller *et al.* 2013 and references herein). The impact of the glaciations is sometimes regarded as positive, as they would have allowed cryophilic species to expand their range (Barr 1968).

Most of the recent calibrated phylogenies, however, underline the ancientness of the splits, suggesting that Quaternary glaciations had a reduced impact on cave biota at the species level. A study on the amphipod *Niphargus rhenorhodanensis* Schellenberg, 1937, suggests that Quaternary glaciers may not have influenced the distribution of the species (Lefébure *et al.* 2007). The phylogeographic structure of the cave beetle species *Aphaenops cerberus* (Dieck, 1869) in the French Pyrenees suggests a lack of gene flow among populations, and dating of the splits between populations, failed to detect any impact of the recent (Würm) glaciations (Faille *et al.* 2015). However, a few montane species living as epigean at a high altitude are known to live in caves at a low altitude, which suggests that glaciations had some impact at least on species distribution and cave colonization.

6.3.2.2. *Ongoing processes of underground colonization*

A few species are of particular interest for the study of ongoing subterranean colonization processes. These are the species simultaneously having isolated populations outside of caves and cave populations, with local adaptations to these different environmental conditions. A well-known example is the fish genus *Astyanax* (Bilandžija *et al.* 2020). Another good model for studying ongoing subterranean colonization processes is the ground beetle species *Trechus fulvus* Dejean, 1831. The species is widespread in Western Europe, with various ecologies (lapidicolous in the north of its range, epigean to hypogean in the south of its range), and the existence of multiple isolated populations showing morphological variability, with some populations having apterous individuals with reduced eyes and other populations also including macropterous specimens (Jeannel 1920; Faille *et al.* 2014; Ortuño *et al.* 2017). A few subterranean lineages are highly diverse underground, with many similar species and a usually fragmented distribution, but genetic divergence more recent than that of the geological splits. Such a pattern suggests that dispersion events occur, but rarely: too rarely to maintain gene flow, but often enough to allow range expansion followed by allopatric speciation (Zakšek *et al.* 2019). Stepping stone colonization can occur during narrow temporal windows, as exemplified by the Pyrenean Leptodirini genus *Troglocharinus* (Rizzo *et al.* 2013), as well as occasional dispersion through water, as quoted for Atyidae shrimps (Holsinger 2012). Fragmentation of the distribution and low dispersal abilities lead to strong population structure in cave biota, as evidenced in many groups such as beetles (Faille *et al.* 2015; Boyd *et al.* 2020; Balogh *et al.* 2020) or *Niphargus* amphipods (Lefébure *et al.* 2007). This suggests that both vicariance and dispersal have a key role in shaping the current biogeographic patterns observed in the subterranean fauna (Christman *et al.* 2005; Jurado-Rivera *et al.* 2017).

6.3.2.3. *Forever trapped? Secondary diversification underground and "flourishing" lineages*

In striking contrast with the highly divergent and species-poor relicts quoted above, a few groups of beetles gave rise to flourishing hypogean lineages, contradicting the suggestion that cave evolution is a dead end (Ribera *et al.* 2018). This is the case of the remarkable radiations of diving beetles found in some subterranean aquifers of Australia (Cooper *et al.* 2002, Leijs *et al.* 2012, Langille *et al.* 2020), the ground beetles of the tribe Trechini (see, for example, genera

Aphaenops in Pyrenees, *Dongodytes* in Guangxi, among others) and the Leiodidae of the tribe Leptodirini (e.g. *Anthroherpon* in Dinarid Alps, *Speonomus* in Pyrenees) (Faille *et al.* 2010; Ribera *et al.* 2010; Njunjić *et al.* 2018). In some of these speciose lineages, remarkable life history strategies have developed. In Leiodidae Leptodirini, one of the two main Palearctic subterranean radiations of beetles, the dramatic contraction of the larval phase (in the most modified species, the female lays a single enormous egg, out of which hatches a larva that does not feed and pupates directly without molting) is correlated with an increase in diversification rates (Cieslak *et al.* 2014). A similar convergent evolution is observed in the second flourishing lineage of cave beetles, the Carabidae Trechini (Faille and Pluot-Sigwalt 2015). Such evolutionary novelty, virtually lacking in surface species, appears to be a key innovation boosting the diversification of subterranean biota. A similar case of shift in the larval development associated with a successful colonization of hypogean water is evidenced in some crustaceans, like the freshwater shrimps Atyidae. In this mainly marine family, some species groups successfully colonized rivers and other aquatic habitats. This colonization succeeded due to the switch from a planktonic larva to a sessile larva. Some of these crustacean species successfully colonized the subterranean environment, suggesting that a sessile larva and a complete freshwater life cycle is – in that particular case – a prerequisite for a successful colonization of subterranean ecosystems (Rintelen *et al.* 2012).

6.4. Perspectives in subterranean biogeography

An accurate knowledge of subterranean diversity at the species level, combined with a comprehensive overview of the geological and paleoclimatic histories of the areas of interest, is a prerequisite to the understanding of biogeographic patterns.

Many questions regarding the biogeography of cave biodiversity remain to be addressed, and six questions related to biogeographic issues – among which global biodiversity patterns of subterranean biodiversity, the relative importance of dispersal and the history of subterranean colonizations – were recently listed among the 50 priority questions on cave biology to be addressed in the future (Mammola *et al.* 2020). What are the origins of isolated and relictual taxa? Which explanations can we provide for the large gaps in the distributions of subterranean biota and their relatives? What is the real impact of extinctions on the observed distributions? What major geological (tectonic, orogeny) and climatic changes can we invoke to understand the current biogeographical patterns observed? We are just in the infancy of understanding the mechanisms that led to the isolation and differentiation of subterranean biodiversity. Ways of colonization of the underground environment –

active versus passive – are the topic of many contributions, but for most zoological groups, no phylogenetic framework is yet available.

Limiting factors in our understanding of global biogeographic patterns of the subterranean fauna

There are still many gray areas to be investigated before having a clear vision of the patterns of diversity of the underground fauna on a global scale (Zagmajster *et al.* 2019). Many regions of the world remain underprospected, and many are inaccessible for geographic or political reasons.

Nonetheless, the growing number of recent discoveries suggests that new major hotspots of underground diversity exist on Earth. The fundamental knowledge of the diversity of underground life forms is still fragmented and the inventory of underground species is far from reaching a plateau, especially in areas currently under exploration, such as the immense karsts of southern China, which undoubtedly host a large number of undescribed species (Tian *et al.* 2016).

Generalization of the use of environmental DNA is promising, especially to accurately document the local richness and fine distribution of stygobitic animals. This method has already been successfully used to detect the presence of aquatic biota like the olm, *Proteus anguinus* or the amphipod genus *Stygobromus*, thus extending the known range of these species (Gorički *et al.* 2017; Niemiller *et al.* 2018).

The main challenge for the years to come, in addition to the basic inventory of underground life forms and the filling of gaps in the geographic and taxonomic coverage of subterranean organisms, lies in the development of phylogenetic studies targeting the major subterranean groups. Only a thorough knowledge of the evolutionary histories of various lineages will make it possible to shed light on the impact of paleoclimatic and paleogeographic events on the origin and history of the present-day subterranean biodiversity, as strong hypotheses of phylogenetic relationships are crucial for interpreting the biogeographic significance of observed geographical patterns. For a long time, these hypotheses were based on morphological data, with the risk of over- or under-interpreting the phylogenetic importance of morphological traits. Molecular approaches, by confirming or invalidating previous biogeographical hypotheses and by suggesting new hypotheses, are leading to new developments in subterranean biology. Several recently discovered relictual cave taxa have been keystones to understand the past distributions of living groups, revealing unsuspected sources of regional biodiversity. In turn, they have led to a re-examination of the morphology and the identification of diagnostic morphological characters supporting the clades.

Future investigations should aim at testing the correlations between the size of "geological islands" and their genetic diversity, focusing on interpopulational genetic structuration to infer paths and sequences of colonization. Thorough group-centered studies are needed to date subterranean colonization events and identify the temporal windows which have allowed the dispersal of cave animals. The role of habitat size in shaping the present-day distribution of cave faunas remains largely to be tested. The consideration of ecological parameters and tolerance thresholds, the environmental envelope of the species considered, is an essential prerequisite for understanding the history of subterranean lineages and their capacity to adapt to global changes. These studies combined will allow testing of the many existing narrative hypotheses on the origin and age of cave lineages.

Subterranean animals are remarkable for their narrow endemicity and deserve a particular attention in heritage terms. Therefore, studies on the cave fauna will also have a strong impact on conservation issues, as most subterranean species have restricted distribution areas and are thus potentially more vulnerable than widespread epigean species, though, as stressed above, their habitat is much less fragile than epigean habitats. Our understanding of cave fauna evolution and biogeography has significantly progressed in the last decades, and the combination of information obtained from different model groups, from Crustacea to insects or fishes, and different methodological approaches is providing answers to old questions regarding the subterranean evolutionary processes and biogeographic patterns of the cave biota of the world.

6.5. Acknowledgments

I am thankful to David C. Culver (AU, Washington), Louis Deharveng (MNHN, Paris) and Daniel Whitmore (SMNS, Stuttgart) for very helpful comments on a preliminary version of the manuscript.

6.6. References

Balke, M. and Ribera, I. (2020). A subterranean species of *Exocelina* diving beetle from the Malay Peninsula filling a 4,000 km distribution gap between Melanesia and southern China. *Subterr. Biol.*, 34, 25–37.

Balke, M., Watts, C.H.S., Cooper, S.J.B., Humphreys, W.F., Vogler, A.P. (2004). A highly modified stygobiont diving beetle of the genus *Copelatus* (Coleoptera, Dytiscidae): Taxonomy and cladistic analysis based on mitochondrial DNA sequences. *Syst. Entomol.*, 29, 59–67.

Balogh, A., Ngo, L., Zigler, K.S., Dixon, G. (2020). Population genomics in two cave-obligate invertebrates confirms extremely limited dispersal between caves. *Sci. Rep.*, 10, 17554.

Barden, P. and Ware, J.L. (2017). Relevant relicts: The impact of fossil distributions on biogeographic reconstruction. *Insect Syst. Divers.*, 1(1), 73–80.

Barr, T.C. (1968). Cave ecology and the evolution of troglobites. In *Evolutionary Biology*, Dobzhansky, T., Hecht, M.K., Steere, W.C. (eds). Plenum Press, New York.

Barr, T.C. and Holsinger, J.R. (1985). Speciation in cave faunas. *Annu. Rev. Ecol. Syst.*, 16, 313–37.

Bauzà-Ribot, M.M., Juan, C., Nardi, F., Oromí, P., Pons, J., Jaume, D. (2012). Mitogenomic phylogenetic analysis supports continental-scale vicariance in subterranean thalassoid crustaceans. *Curr. Biol.*, 22, 1–6.

Bellés, X. (1987). *Fauna cavernícola i intersticial de la Península Ibérica i les Illes Balears*. Editorial Moll-CSIC, Palma de Mallorca.

Bilandžija, H., Morton, B., Podnar, M., Ćetković H. (2013). Evolutionary history of relict *Congeria* (Bivalvia: Dreissenidae): Unearthing the subterranean biodiversity of the Dinaric Karst. *Front. Zool.*, 10, 5.

Bilandžija, H., Hollifield, B., Steck, M., Meng, G., Ng, M., Koch, A.D., Gračan, R., Ćetković, H., Porter, M.L., Renner, K.J., Jeffery, W. (2020). Phenotypic plasticity as a mechanism of cave colonization and adaptation. *Elife*, 9:e51830.

Boyd, O.F., Philips, T.K., Johnson, J.R., Nixon, J.J. (2020). Geographically structured genetic diversity in the cave beetle *Darlingtonea kentuckensis* Valentine, 1952 (Coleoptera, Carabidae, Trechini, Trechina). *Subterr. Biol.*, 34, 1–23.

Broly, P., Deville, P., Maillet, S. (2013). The origin of terrestrial isopods (Crustacea: Isopoda: Oniscidea). *Evol. Ecol.*, 27, 461–476.

Brunke, A.J., Żyła, D., Yamamoto, S., Solodovnikov, A. (2019). Baltic amber Staphylinini (Coleoptera: Staphylinidae: Staphylininae): A rove beetle fauna on the eve of our modern climate. *Zool. J. Linn. Soc.*, 187, 166–19.

Caccone, A. (1985). Gene flow in cave arthropods: A qualitative and quantitative approach. *Evolution*, 39, 1223–1234.

Casale, A., Vigna-Taglianti, A., Juberthie, C. (1998). Coleoptera Carabidae. In *Encyclopedia Biospeologica. Tome II*, Juberthie, C. and Decu, V. (eds). Société de Biospéologie, Moulis.

Christiansen, K. (1962). Proposition pour la classification des animaux cavernicoles. *Spelunca*, 2, 76–78.

Christman, M.C., Culver, D.C., Madden, M., White, D. (2005). Patterns of endemism of the eastern North American cave fauna. *J. Biogeogr.*, 32, 1441–1452.

Cieslak, A., Fresneda, J., Ribera, I. (2014). Life history evolution and diversification in Leptodirini cave beetles. *Proc. R. Soc. Lond. B Biol. Sci.*, 281, 20132978.

Cooper, S.J.B., Hinze, S., Leys, R., Watts, C.H.S., Humphreys, W.F. (2002). Islands under the desert: Molecular systematics and evolutionary origins of stygobitic water beetles (Coleoptera: Dytiscidae) from central Western Australia. *Invertebr. Syst.*, 16, 589–598.

Copilaş-Ciocianu, D., Sidorov, D., Gontcharov, A. (2019). Adrift across tectonic plates: Molecular phylogenetics supports the ancient Laurasian origin of old limnic crangonyctid amphipods. *Org. Divers. Evol.*, 19, 191–207.

Culver, D.C. (1970). Analysis of simple cave communities I: Caves as islands. *Evolution*, 24, 463–474.

Culver, D.C. and Pipan, T. (2019). *The Biology of Caves and Other Subterranean Habitats*, 2nd edition. Oxford University Press, Oxford.

Culver, D.C. and Sket, B. (2000). Hotspots of subterranean biodiversity in caves and wells. *J. Caves Karst. Stud.*, 62, 11–17.

Darwin, C. (1859). *On the Origin of Species by Means of Natural Selection*. Murray, London.

Deharveng, L. and Bedos, A. (2018). Diversity of terrestrial invertebrates in subterranean habitats. In *Cave Ecology*, Moldovan, O.T., Kováč, L., Halse, S. (eds). Springer, Cham.

Deharveng, L. and Bedos, A. (2019). Biodiversity in the tropics. In *Encyclopedia of Caves*, 3rd edition, White, W.B., Culver, D.C., Pipan, T. (eds). Academic Press, Waltham.

Deharveng, L., Brehier, F., Bedos, A., Tian, M., Li, Y., Zhang, F., Qin, W., Tan, X. (2009). Mulun and surrounding karsts (Guangxi) host the richest cave fauna of China. *Subterr. Biol.*, 6, 75–79.

Deharveng, L., Gibert, J., Culver, D.C. (2019). Biodiversity in Europe. In *Encyclopedia of Caves*, 3rd edition, White, W.B., Culver, D.C., Pipan, T. (eds). Academic Press, Waltham.

Deleurance, S. (1958). La contraction du cycle évolutif des Coléoptères Bathysciinae et Trechinae en milieu souterrain. *C. R. Seances Acad. Sci.*, 247, 752–753.

Faille, A., Ribera, I., Deharveng, L., Bourdeau, C., Garnery, L., Queinnec, E., Deuve, T. (2010). A molecular phylogeny shows the single origin of the Pyrenean subterranean Trechini ground beetles (Coleoptera: Carabidae). *Mol. Phylogenet. Evol.*, 54, 97–105.

Faille, A., Casale, A., Balke, M., Ribera, I. (2013). A molecular Phylogeny of alpine subterranean Trechini (Coleoptera: Carabidae). *BMC Evol. Biol.*, 13, 248.

Faille, A., Andújar, C., Fadrique, F., Ribera, I. (2014). Late Miocene origin of an Ibero-Maghrebian clade of ground beetles with multiple colonisations of the subterranean environment. *J. Biogeogr.*, 41, 1979–1990.

Faille, A., Tänzler, R., Toussaint, E.F.A. (2015). On the way to speciation: Shedding light on the karstic phylogeography of the micro-endemic cave beetle *Aphaenops cerberus* in the Pyrenees. *J. Hered.*, 106, 692–699.

Faille, A., Balart-García, P., Fresneda, J., Bourdeau, C., Ribera, I. (2021). A remarkable new genus of Iberian troglobitic Trechodina (Coleoptera: Carabidae: Trechinae: Trechini), with a revisited molecular phylogeny of the subtribe. *Ann. Soc. Entomol. Fr.*, 57(2), 85–106.

Fumey, J., Hinaux, H., Noirot, C., Thermes, C., Rétaux, S., Casane, D. (2018). Evidence for late Pleistocene origin of *Astyanax mexicanus* cavefish. *BMC Evol. Biol.*, 18, 43.

Giachino, P.M. and Vailati, D. (2010). *The Subterranean Environment. Hypogean Life, Concepts and Collecting Techniques.* WBA Handbooks, Vol. 3. World Biodiversity Association Onlus, Verona.

Gibert J. and Deharveng L. (2002). Subterranean ecosystems: A truncated functional biodiversity. *BioScience*, 52(6), 473–481.

Golovatch, S.I. (2015). Cave diplopoda of southern China with reference to millipede diversity in Southeast Asia. *ZooKeys*, 510, 79–94.

Gorički, Š., Stanković, D., Snoj, A., Kuntner, M., Jeffery, W.R., Trontelj, P., Pavićević, M., Grizelj, Z., Năpăruş-Aljančič, M., Aljančič, G. (2017). Environmental DNA in subterranean biology: Range extension and taxonomic implications for *Proteus*. *Sci. Rep.*, 7, 91–93.

Hampe, A. and Jump, A.S. (2011). Climate relicts: Past, present, future. *Annu. Rev. Ecol. Evol. Syst.*, 42, 313–333.

Holsinger, J.R. (1994). Pattern and process in the biogeography of subterranean amphipods. *Hydrobiologia*, 287, 131–145.

Holsinger, J.R. (2012). Vicariance and dispersalist biogeography. In *Encyclopedia of Caves*, 2nd edition, Culver, D.C. and White, W.B. (eds). Elsevier Academic Press, Amsterdam.

Howarth, F.G. (1983). Bioclimatic and geologic factors governing the evolution and distribution of Hawaiian cave insects. *Entomologia Generalis*, 8, 17–26.

Huang, S., Tian, M., Faille, A. (2020). Three new species of the aphaenopsian trechine genus *Pilosaphaenops* Deuve and Tian, 2008 from South China Karst (Coleoptera: Carabidae: Trechinae). *Ann. Soc. Entomol. Fr.*, 56(3), 203–214.

Iannella, M., Fiasca, B., Di Lorenzo, T., Biondi, M., Di Cicco, M., Galassi, D.M.P. (2020). Jumping into the grids: Mapping biodiversity hotspots in groundwater habitat types across Europe. *Ecography*, 43, 1825–1841.

Ikeda, H., Nishikawa, M., Sota, T. (2012). Loss of flight promotes beetle diversification. *Nat. Commun.*, 3, 648.

Jaume, B. (2008). Global diversity of spelaeogriphaceans and thermosbaenaceans (Crustacea; Spelaeogriphacea and Thermosbaenacea) in freshwater. *Hydrobiologia*, 595, 219–224.

Jaume, D. and Vonk, R. (2012). Discovery of *Metacrangonyx* in inland groundwaters of Oman (Amphipoda: Gammaridea: Metacrangonyctidae). *Zootaxa*, 3335, 54–68.

Jeannel, R. (1911). Révision des Bathysciinae (Coléoptères Silphides). Morphologie, Distribution géographique, Systématique. *Arch. Zool. Exp. Gén.*, 47, 1–641.

Jeannel, R. (1920). *Étude sur le* Trechus fulvus *Dej. [Col. Carab.], sa phylogénie, son intérêt biogéographique.* Serie Zoológica 41. Museo Nacional de Ciencias Naturales, Madrid.

Jeannel, R. (1926). *Faune cavernicole de la France avec une étude des conditions d'existence dans le domaine souterrain.* Lechevalier, Paris.

Jeannel, R. (1928). Monographie des Trechinae. Morphologie comparée et distribution d'un groupe de Coléoptères. Troisième Livraison : Les Trechini cavernicoles. *L'Abeille*, 35, 1–808.

Jeannel, R. (1943). *Les fossiles vivants des cavernes.* Gallimard, Paris.

Jeannel, R. (1959). Situation géographique et peuplement des cavernes. *Annales de Spéléologie*, 14, 333–338.

Juberthie, C., Delay, B., Bouillon, M. (1980). Sur l'existence d'un milieu souterrain superficiel en zone non calcaire. *C. R. Acad. Sci. III*, 290, 49–52.

Jurado-Rivera, J.A., Pons, J., Alvarez, F., Botello, A., Humphreys, W.F., Page, T.J., Iliffe, T.M., Willassen, E., Meland, K., Juan, C., Jaume, D. (2017). Phylogenetic evidence that both ancient vicariance and dispersal have contributed to the biogeographic patterns of anchialine cave shrimps. *Sci. Rep.*, 7, 2852.

Lamarck, J.B. (1809). *Philosophie zoologique.* Dentu, Paris.

Langille, B., Hyde, J., Saint, K., Bradford, T., Stringer, D., Tierney, S., Humphreys, W., Austin, A., Cooper, S. (2020). Evidence for speciation underground in diving beetles (Dytiscidae) from a subterranean archipelago. *Evolution*, 75(1), 166–175.

Lefébure, T., Douady, C.J., Malard, F., Gibert, J. (2007). Testing dispersal and cryptic diversity in a widely distributed groundwater amphipod (*Niphargus rhenorhodanensis*). *Mol. Phylogenet. Evol.*, 42, 676–686.

Leijs, R., van Nes, E.H., Watts, C.H., Cooper, S.J.B., Humphreys, W.F., Hogendoorn, K. (2012). Evolution of blind beetles in isolated aquifers: A test of alternative modes of speciation. *PLoS One*, 7, e34260.

Leys, R., Watts, C.H.S., Cooper, S.J.B., Humphreys, W.F. (2003). Evolution of subterranean diving beetles (Coleoptera: Dytiscidae: Hydroporini, Bidessini) in the arid zone of Australia. *Evolution*, 57, 2819–2834.

Lozano-Fernandez, J., Giacomelli, M., Fleming, J., Chen, A., Vinther, J., Thomsen, P., Glenner, H., Palero, F., Legg, D., Iliffe, T., Pisani, D., Olesen, J. (2019). Pancrustacean evolution illuminated by taxon-rich genomic-scale data sets with an expanded remipede sampling. *Genome Biol. Evol.*, 11, 10.1093.

Ma, L., Zhao, Y., Yang, J.X. (2019). Cavefish of China. In *Encyclopedia of Caves*, 3rd edition, White, W.B., Culver, D.C., Pipan, T. (eds). Academic Press, Waltham.

Maddison, D.R., Kanda, K., Boyd, O.F., Porch, N., Faille, A., Erwin, T.L., Roig-Juñent, S. (2019). Phylogeny of the beetle supertribe Trechitae (Coleoptera: Carabidae): Unexpected clades, isolated lineages, and morphological convergence. *Mol. Phylogenet. Evol.*, 132, 151–176.

Mammola, S. (2019). Finding answers in the dark: Caves as models in ecology fifty years after Poulson and White. *Ecography*, 42, 1331–1351.

Mammola, S., Amorim, I.R., Bichuette, M.E., Borges, P.A.V., Cheeptham, N., Cooper, S.J.B., Culver, D.C., Deharveng, L., Eme, D., Ferreira, R.L., Fišer, C., Fišer, Ž., Fong, D.W., Griebler, C., Jeffery, W.R., Jugovic, J., Kowalko, J.E., Lilley, T.M., Malard, F., Manenti, R., Martínez, A., Meierhofer, M.B., Niemiller, M.L., Northup, D.E., Pellegrini, T.G., Pipan, T., Protas, M., Reboleira, A.S.P.S., Venarsky, M.P., Wynne, J.J., Zagmajster, M., Cardoso, P. (2020). Fundamental research questions in subterranean biology. *Biol. Rev.*, 95, 1855–1872.

Mashimo, Y., Müller, P., Beutel, R.G. (2019). *Zorotypus Pecten*, a new species of Zoraptera (Insecta) from mid-Cretaceous Burmese amber. *Zootaxa*, 4651(3), 565–577.

Mejía-Ortíz, L.M. (2019). Crustacean. In *Encyclopedia of Caves*, 3rd edition, White, W.B., Culver, D.C., Pipan, T. (eds). Academic Press, Waltham.

Niemiller, M.L., McCandless, J.R., Reynolds, R.G., Caddle, J., Tillquist, C.R., Near, T.J., Pearson, W.D., Fitzpatrick, B.M. (2013). Effects of climatic and geological processes during the Pleistocene on the evolutionary history of the northern cavefish, *Amblyopsis spelaea* (Teleostei: Amblyopsidae). *Evolution*, 67(4), 1011–1025.

Niemiller, M.L., Porter, M.L., Keany, J., Gilbert, H., Fong, D.W., Culver, D.C., Hobson, C.S., Kendall, K.D., Davis, M.A., Taylor, S.J. (2018). Evaluation of eDNA for groundwater invertebrate detection and monitoring: A case study with endangered *Stygobromus* (Amphipoda: Crangonyctidae). *Conserv. Genet. Resour.*, 10, 247–257.

Niemiller, M.L., Bichuette, M.E., Chakrabarty, P., Fenolio, D.B., Gluesenkamp, A.G., Soares, D., Zhao, Y. (2019). Cavefishes. In *Encyclopedia of Caves*, 3rd edition, White, W.B., Culver, D.C., Pipan, T. (eds). Academic Press, Waltham.

Njunjić, I., Perrard, A., Hendriks, K., Schilthuizen, M., Perreau, M., Merckx, V., Baylac, M., Deharveng, L. (2018). Comprehensive evolutionary analysis of the *Anthroherpon* radiation (Coleoptera, Leiodidae, Leptodirini). *PLoS One*, 13(6), e0198367.

Ortuño, V.M., Gilgado, J.D., Jiménez-Valverde, A., Sendra, A, Pérez-Suárez, G., Herrero-Borgoñón, J.J. (2013). The "Alluvial Mesovoid Shallow Substratum", a new subterranean habitat. *PLoS One*, 8(10), e76311.

Ortuño, V.M., Ledesma, E., Gilgado, J.D., Veguillas, L., Barranco, P. (2017). On the distribution and autoecology of *Trechus fulvus* Dejean, 1831 (Coleoptera: Carabidae: Trechinae) in the Iberian Península. *Bol. SEA*, 60, 195–206.

Peck, S.B. and Finston, T.L. (1993). Galápagos Islands troglobites: The questions of tropical troglobites, parapatric distributions with eyed-sister-species, and their origin by parapatric speciation. *Mem. Biospeol.*, 20, 19–37.

Pipan, T. and Culver, D.C. (2019). Shallow subterranean habitats. In *Encyclopedia of Caves*, 3rd edition, White, W.B., Culver, D.C., Pipan, T. (eds). Academic Press, Waltham.

Poinar Jr., G.O. (2018). Burmese amber: Evidence of Gondwanan origin and Cretaceous dispersion. *Hist. Biol.*, 31(10), 1304–1309.

Polato, N.R., Gill, B.A., Shah, A.A., Gray, M.M., Casner, K.L., Barthelet, A., Messer, P.W., Simmons, M.P., Guayasamin, J.M., Encalada, A.C., Kondratieff, B.C., Flecker, A.S., Thomas, S.A., Ghalambor, C.K., Poff, N.L., Funk, W.C., Zamudio, K.R. (2018). Narrow thermal tolerance and low dispersal drive higher speciation in tropical mountains. *Proc. Natl Acad. Sci. USA*, 115, 12471–12476.

Pons, J., Jurado-Rivera, J.A., Jaume, D., Vonk, R., Bauzà-Ribot, M.M., Juan, C. (2019). The age and diversification of metacrangonyctid subterranean amphipod crustaceans revisited. *Mol. Phylogenet. Evol.*, 140, 10659.

Racovitza, E.G. (1907). Essai sur les problèmes biospéologiques. *Arch. Zool. Exp. Gén.*, 6, 371–488.

Ribera, I., Mateu, J., Bellés, X. (2005). Phylogenetic relationships of *Dalyat mirabilis* Mateu, 2002, with a revised molecular phylogeny of ground beetles (Coleoptera, Carabidae). *J. Zool. Syst. Evol. Res.*, 43, 284–296.

Ribera, I., Fresneda, J., Bucur, R., Izquierdo, A., Vogler, A.P., Salgado, J.M., Cieslak, A. (2010). Ancient origin of a western Mediterranean radiation of subterranean beetles. *BMC Evol. Biol.*, 10(29), 1–14.

Ribera, I., Cieslak, A., Faille, A., Fresneda, J. (2018). Chapter 10: Historical and ecological factors determining cave diversity. In *Cave Ecology*, Moldovan, O.T., Kováč, L., Halse, S. (eds). Springer, Cham.

von Rintelen K., Page T.J., Cai Y., Roe K., Stelbrink B., Kuhajda B.R., Iliffe T.M., Hughes J., von Rintelen T. (2012). Drawn to the dark side: A molecular phylogeny of freshwater shrimps (Crustacea: Decapoda: Caridea: Atyidae) reveals frequent cave invasions and challenges current taxonomic hypotheses. *Mol. Phylogenet. Evol.*, 63, 82–96.

Rizzo, V., Comas, J., Fadrique, F., Fresneda, J., Ribera, I. (2013). Early Pliocene range expansion of a clade of subterranean Pyrenean beetles. *J. Biogeogr.*, 40(10), 1861–1873.

Sánchez-Fernández, D., Rizzo, V., Cieslak, A., Faille, A., Fresneda, J., Ribera, I. (2016). Thermal niche estimators and the capability of poor dispersal species to cope with climate change. *Sci. Rep.*, 6, 23381.

Sendra, A., Palero, F., Jiménez-Valverde, A., Reboleira, A.S.P.S. (2021). Diplura in caves: Diversity, ecology, evolution and biogeography. *Zool. J. Linn. Soc.*, 192(3), 675–689.

Souza-Silva, M. and Ferreira, R.L. (2016). The first two hotspots of subterranean biodiversity in South America. *Subterr. Biol.*, 19, 1–21.

Stepien, C., Morton, B., Dabrowska, K., Guarnera, R., Radja, T., Radja, B. (2001). Genetic diversity and evolutionary relationships of the troglodytic "living fossil" *Congeria kusceri* (Bivalvia: Dreissenidae). *Mol. Ecol.*, 10(8), 1873–1879.

Stoch, J.H. (1993). Some remarkable distribution patterns in stygobiont Amphipoda. *J. Nat. Hist.*, 27(4), 807–819.

Stoch F. and Galassi, D.M.P. (2010). Stygobiotic crustacean species richness: A question of numbers, a matter of scale. *Hydrobiologia*, 653, 217–234.

Taiti, S. (2004). Crustacea: Isopoda: Oniscidea (woodlice). In *Encyclopedia of Caves and Karst Science*, Gunn, J. (ed.). Fitzroy Dearborn, New York.

Tian, M.Y., Huang, S.B., Wang, X.H., Tang, M.R. (2016). Contributions to the knowledge of subterranean trechine beetles in southern China's karsts: Five new genera (Insecta, Coleoptera, Carabidae, Trechinae). *ZooKeys*, 564, 121–156.

Torres-Dowdall, J., Karagic, N., Plath, M., Riesch, R. (2018). Evolution in caves: Selection from darkness causes spinal deformities in teleost fishes. *Biol. Lett.*, 14, 20180197.

Vandel, A. (1964) *Biospéologie. La Biologie des animaux cavernicoles*. Gauthier-Villars, Paris.

Vandel, A. (1965). Sur l'existence d'Oniscoïdes très primitifs menant une vie aquatique et sur le polyphylétisme des isopodes terrestres. *Annales de Spéléologie*, 20(4), 489–518.

Wallace, A.R. (1880). *Island Life: Or, the Phenomena and Causes of Insular Faunas and Floras, Including a Revision and Attempted Solution of the Problem of Geological Climates*. MacMillan and Co., London.

Watts, C.H.S. and Humphreys, W.F. (2009). Fourteen new Dytiscidae (Coleoptera) of the genera *Limbodessus* Guignot, *Paroster* Sharp, and *Exocelina* Broun from underground waters in Australia. *Trans. R. Soc. S. Aust.*, 133, 62–107.

Wessel, A., Hoch, H., Asche, M., von Rintelen, T., Stelbrink, B., Heck, V., Stone, F.D., Howarth, F.G. (2013). Founder effects initiated rapid species radiation in Hawaiian cave planthoppers. *Proc. Natl. Acad. Sci. USA*, 110(23), 9391–9396.

Whittaker, R.J., Fernández-Palacios, J.M., Matthews, T.J., Borregaard, M.K., Triantis, K.A. (2017). Island biogeography: Taking the long view of nature's laboratories. *Science*, 357, eaam8326.

Yager, J. (1981). Remipedia, a new class of Crustacea from a marine cave in the Bahamas. *J. Crustac. Biol.*, 1(3), 328–333.

Zagmajster, M., Malard, F., Eme, D., Culver, D.C. (2019). Subterranean biodiversity patterns from global to regional scales. In *Cave Ecology*, Moldovan, O.T., Kováč, L., Halse, S. (eds). Springer, Cham.

Zakšek, V., Delić, T., Fišer, C., Jalžić, B., Trontelj, P. (2019). Emergence of sympatry in a radiation of subterranean amphipods. *J. Biogeogr.*, 46, 657–669.

Zhang, Y. and Li, S. (2013). Ancient lineage, young troglobites: Recent colonization of caves by *Nesticella* spiders. *BMC Evol. Biol.*, 13, 183.

7

Soil Bacterial Biogeography at the Scale of France

Battle KARIMI and Lionel RANJARD
UMR Agroécologie, INRAE, Dijon, France

7.1. Introduction

Micro-organisms are the most abundant and diversified organisms on Earth, but their large-scale spatial distribution and the related driving factors remain poorly known. This can be explained by special features of micro-organisms such as their small size and low accessibility within environmental matrices, their high density and diversity, and low available sampling or monitoring networks at large spatial scales. This knowledge gap about a matrix such as the soil is all the more harmful as micro-organisms play many functions and provide essential services to human societies. Consequently, this explains why we are unable to precisely predict the impact of environmental disturbances on soil quality and biological functioning.

This chapter first deals with the importance and the role of soil bacterial communities, and then with the large-scale (worldwide or national) sampling networks devoted to soil biogeography. Based on one of the most intensive soil sampling networks known to date – the French Monitoring Network of Soil Quality (Réseau de Mesure de la Qualité des Sols) – we synthesize current soil bacterial biogeography knowledge at the scale of the French territory. More precisely, we present the spatial distribution, driving factors and environmental filter hierarchy that drive the bacterial richness. We will show that the taxa–area relationship is also

Biogeography,
coordinated by Eric GUILBERT.

true for micro-organisms. Based on the atlas of bacterial taxa from France, we present the different types of geographical distributions observed for the major soil phyla, and depict their ecological attributes. We demonstrate that co-occurrence networks are potential keys to the understanding of bacterial community regulation and functioning. Finally, we explore the concept of microbial habitat at the macro-ecological scale. To conclude, we illustrate how the same data provide answers to fundamental research questions on microbial biogeography and also account for a precious resource for developing operational bioindicators of the impact of disturbances on soil microbiological quality.

7.2. Soil bacterial communities

7.2.1. *Abundance, diversity and role*

Soil biodiversity represents 25% of the total biodiversity on Earth, and a biomass of several dozen tons per hectare, more than the aboveground biomass. Bacteria represent the largest part of this biological heritage, whether in terms of abundance (10^9 bacteria per g of soil) or diversity (10^6 bacterial species per g of soil), even though the magnitude of their diversity is not completely established (Torsvik and Ovreas, 2002; Flemming and Wuertz, 2019). This reservoir of genetic resources is the keystone of most of the soil's biological functions, encompassing the dynamics of organic matter, the carbon and nitrogen cycles, nutrient bioavailability, degradation of organic pollutants, retention of metal pollutants, soil structure maintenance, etc. (Maron *et al.* 2011). A 30% decrease in bacterial biodiversity induces (1) ~ 40% decrease in organic matter mineralization, and thereby of the recycling of soil mineral elements (Baumann *et al.* 2012), (2) 50% reduction of soil denitrification (Philippot *et al.* 2013), (3) 50% loss of plant productivity (Prudent *et al.* 2020), (4) 40% reduction of the soil structural stability (Maron, personal communication), (5) a total loss of the capacity to degrade pollutants (Hernandez-Raquet *et al.* 2013) and (6) a 3- to 5-fold increase of the residence time of opportunistic pathogenic bacterial populations in the soil (Vivant *et al.* 2013). Such a decrease in bacterial diversity entails consequences on the soil's biological fertility, physical fertility and sanitary state, and also on its resistance and resilience to environmental disturbances such as global changes. This pivotal role of bacteria confirms that it is important to characterize soil bacterial communities and investigate their large-scale ecology to bridge current knowledge gaps and predict the impact of environmental disturbances on soil quality and biological functioning in a context of global change.

7.2.2. *Molecular tools to characterize bacterial communities*

Characterizing the bacterial diversity of a soil is complex because we have to access numerous different populations within a structured and heterogeneous matrix, and also because we cannot retrieve exhaustive data on the multitude of species present in one gram of soil. Soil bacteria were first studied using microscope observations or cultures on synthetic media. Yet, these approaches are limited because taxonomic identification cannot be done based on morphological criteria, and we now know that less than 1% of the soil bacteria are culturable. Nevertheless, soil microbial ecology underwent a true technological revolution in the 2000s thanks to the advent of molecular biology tools (Bouchez *et al.* 2016). These techniques are based on the extraction of the soil genetic information (DNA) and the characterization of bacterial sequences by genotyping or sequencing directly from this nucleic acid matrix (Figure 7.1). They present the advantage of overcoming the biases linked to the culture of bacteria. Thus, they offer new perspectives for understanding the distribution of diversity in the soil and its role in the biological functioning of ecosystems (Maron *et al.* 2011). Moreover, these molecular tools are easy to standardize, and their cost has declined steadily over the last 10 years. It is now possible to carry out medium-throughput analyses to characterize microbial communities from sets of several hundred to several thousand samples integrating large spatial or temporal scales.

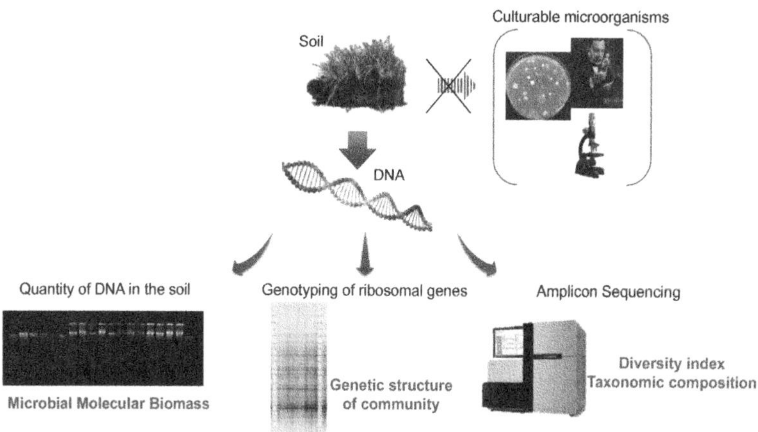

Figure 7.1. *Molecular tools used to characterize the abundance and diversity of soil microbial communities. For a color version of this figure, see www.iste.co.uk/guilbert/biogeography.zip*

Ribosomal genes are used as taxonomic markers for bacterial identification (Figure 7.1). They are amplified from DNA extracted from the soils and sequenced using high-throughput sequencing techniques. Bioinformatic analyses are used to filter high-quality sequences per soil and to compare them to gather those that display more than 95% similarity into operational taxonomic units (also called taxa) (Terrat *et al.* 2017). The notion of species is not robust enough in the microbial world and is not used. The number of bacterial taxa in a soil defines the alpha-diversity in terms of taxonomic richness. This taxonomic richness is determining because it is directly related to the functions supported by the microbial community and its ability to withstand environmental disturbances (e.g. Prudent *et al.* 2020).

7.2.3. *Genesis of microbial biogeography*

Biogeography is the study of the diversity of living organisms at large spatial and temporal scales. It was only applied to the study of micro-organisms at the beginning of the 20th century, based on Beïjerinck's hypothesis: "Everything is everywhere, but the environment selects" (De Witt *et al.* 2006). At the beginning of the 21st century, the simultaneous development of molecular tools and large-scale soil sampling networks allow the emergence of a series of robust and novel works on the spatial distribution of soil microbial diversity (Martiny *et al.* 2006). An increasing number of soil bacterial biogeography studies have indeed been published in the last 10 years. Out of 693 studies addressing this topic since 1994 and referenced in the Web of Science in June 2020, 441 have been published since 2015. A certain number of these studies are focused on one species or one taxon; in that case, the term phylogeography is used (Martiny, 2006). The ecologists who study macro-organisms have long been aware that the study of beta-diversity (the way communities change at the landscape scale) is key to a better understanding of the effects of environmental factors on the amplitude and variability of biodiversity. However, only a small number of works take the whole indigenous microbial community into account. In 2004, the genotyping of soilborne fungal communities on more than 1,500 Australian soils evidenced that diversification of soilborne fungal communities at the regional scale was moderate despite strong local diversity (Green *et al.* 2004). On the American continent, Fierer and Jackson characterized the bacterial communities of around one hundred soils along a north-south transect and showed that bacterial diversity was influenced above all by pedological characteristics such as the pH, very little by climatic factors and geomorphology, and not at all by the geographical distance between communities (Fierer *et al.* 2007). In 2011, Griffiths *et al.* studied the bacterial diversity of more than 1,000 soils in the United Kingdom and established the first microbial diversity atlas at a national scale. They confirmed the important role played by pH, and also by the plant cover and the climate, which was novel compared with previous studies (Griffith *et al.* 2011).

These first studies strongly suggest that micro-organism biogeography inherently differs from macro-organism biogeography. While macro-organisms are more influenced by distal environmental parameters such as the climate and geomorphology, micro-organisms appear to be more dependent on proximal parameters such as the soil type or land use (plant cover, agricultural practices, etc.). Taken together, these results argue in favor of increasing the number of microbial biogeography studies to improve our understanding of microbial diversity, and thereby of its effects on ecosystem services (Martiny *et al.* 2006).

7.3. Soil survey networks around the world

Since 2015, quite a few works at the worldwide, continental, national or regional scales have been aimed to bring generic conclusions on the ecology of microbial communities. Microbial biogeography studies at the worldwide or continental scales are designed to be representative of all terrestrial environments on the planet (Fierer *et al.* 2012; Delgado-Baquerizo *et al.* 2016; Nelson *et al.* 2016; Noronha *et al.* 2017; Bahram *et al.* 2018; Delgado-Baquerizo *et al.* 2018; Bickel *et al.* 2019; Cameron *et al.* 2019; Delgado-Baquerizo *et al.* 2019). Nevertheless, exhaustiveness is limited by the number of sampled sites and their geographical distribution. These studies indeed generally include less than 400 sites, while land surfaces on Earth cover 3.6×10^8 km^2, and a large number of these sites are located on the American continents, especially the United States (Figure 7.2a). Even though these studies highlight global trends, their robustness and their representativeness can be questioned.

Complementary studies at national or regional scales are based on more intensive sampling of smaller surfaces. For instance, in the Netherlands, a sampling strategy displayed a density of 1 point for 140 km^2 and is designed to cover the greatest possible number of soil type × land use combinations (300 points in total) (Rutgers *et al.* 2009 and 2019). In Scotland, 183 points are sampled following a systematic 20 km × 20 km grid, that is, a density of 1 point for 440 km^2 (Yao *et al.* 2013; Powell *et al.* 2015). In Wales, a 436-point sampling randomly distributed across the territory represents the most intensive sampling to date, with 1 point for 47 km^2 (George *et al.* 2019). France also set up a national soil observation and sampling network. Compared with the other existing networks, it represents the largest network in the world, with more than 2,200 sites distributed according to a systematic 16 km × 16 km grid covering a surface of 5.5×10^5 km^2 (density: 1 point for 250 km^2) (Ranjard *et al.* 2010, Figure 7.2b). All these networks present the advantage of being sufficiently intensive, free of *a priori* (e.g. according to a climate gradient or a land use) and based on robust and non-opportunistic sampling strategies (sample availability), so that they provide robust results. However, conclusions may prove insufficiently generic depending on the range of environmental heterogeneity covered by these studies.

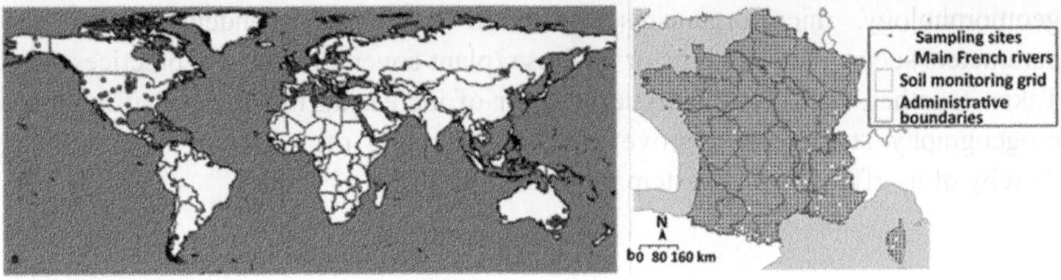

Figure 7.2. *a) Soil sampling strategy at the worldwide scale (Delgado-Baquerizo* et al. *2018). b) Soil sampling strategy at the scale of France, Réseau de Mesure de la Qualité des Sols. Each of the 2,200 cells contained one sampling site located in the center (Ranjard* et al. *2010). For a color version of this figure, see www.iste.co.uk/guilbert/biogeography.zip*

7.3.1. *The French Monitoring Network of Soil Quality*

The great diversity of pedoclimatic conditions encountered in France reinforces the relevance of this territory for studying microbial ecology and biogeography (Figure 7.3). France indeed displays remarkably high pedological, climatic and land use diversity compared with its surface. Mainland France displays the highest soil diversity (10 groups of the WRB soil classification) after Brazil and Argentina (with 11 and 13 groups, respectively) while it only represents 6% of the surface of these countries (Minasny *et al.* 2010). The soils are distributed along an altitudinal gradient ranging between -2 m and 2,500 m, across five bioclimatic regions: marine, continental, mountain, Mediterranean and Mediterranean with influence (Figure 7.3c and 7.3d). The territory is 31% covered by forests (conifers, deciduous trees, mixed forests), 24% by grasslands and natural habitats (old or low-productive grasslands, heathlands, steppes, alpine grasslands, peatlands and marshlands, etc.) and 33% by highly varied agrosystems (field crops, temporary grasslands, market gardens, orchards, vineyards) (Figure 7.3a). Moreover, the gradients of the soil physico-chemical parameters cover a remarkably large range. For instance, soil pH values range between 3 and 9, and soil texture is distributed across 11 of the 12 USDA classes (Figure 7.3b), and both parameters are already known for their implication in soil bacterial diversity regulation.

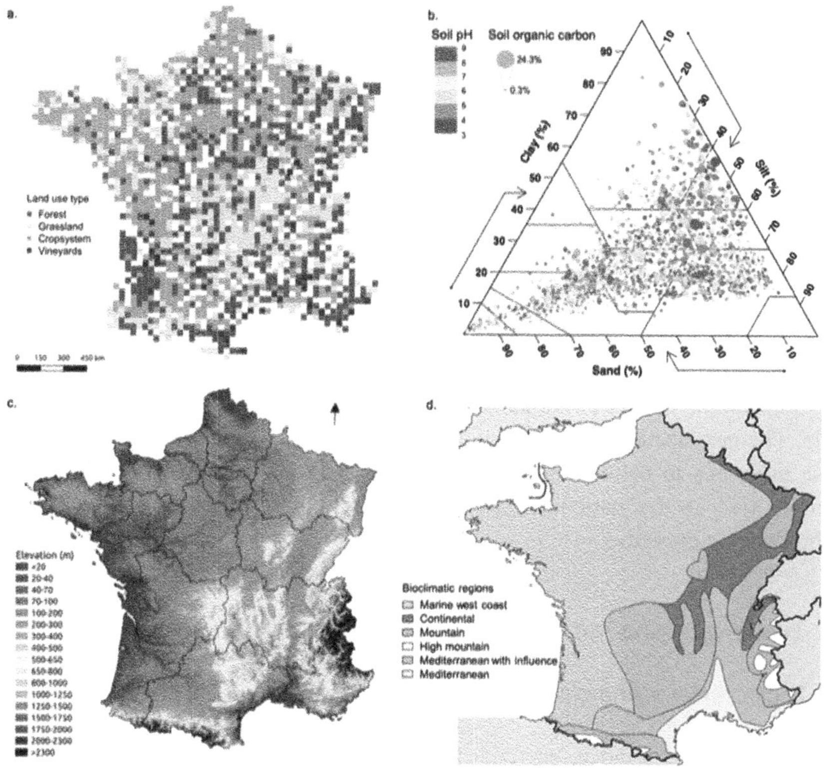

Figure 7.3. *a) Distribution of land uses in France. b) Heterogeneity of the French soil types. c) Altitudinal variation at the scale of the French territory. d) Bioclimatic regions of France. For a color version of this figure, see www.iste.co.uk/guilbert/biogeography.zip*

The sampling network of French soils is the Réseau de Mesure de la Qualité des Sols (RMQS, the Monitoring Network of Soil Quality), set up by the SOL group of scientific interest (GIS) which gathers the ministries in charge of agriculture and the environment, ADEME, Ifen and INRAE. This network constitutes a national framework for monitoring the evolution of soil quality and meets the need for quantitative data on the state of French soils. It aims to detect the emergence and trends of soil degradation in their early phase by measuring their physical, chemical and biological characteristics. The sampling grid is made of 2,200 cells systematically distributed across the whole territory (Figure 7.2a). The sampling sites are located at the center of each cell. They are made of a 20 m × 20 m sampling zone and a pedological pit. During sampling, 25 cores are taken to form composites of two depths: 0–20 cm and 20–40 cm. The site is characterized based on the description of the pedological profile, the land cover (according to Corine Land Cover), and by physico-chemical analyses of soil samples following standardized methods (ISO or NF). These analyses target granulometry, pH, carbon/nitrogen/phosphorus contents, exchangeable cation concentrations, and major and trace elements. Climate parameters (temperature, rainfall and evapotranspiration) and geomorphological parameters (elevation, slope, the presence of natural barriers) are also recorded for each site. Data on the variation of these environmental parameters are used to explain variations in the abundance and diversity of microbial communities at the national scale. Several research programs have answered fundamental microbial biogeography questions and more finalized questions on the impact of human activities on bacterial communities at the national scale. The next section of the present chapter summarizes all the results obtained in these programs to better describe and understand the distribution of soil bacterial communities at the national scale according to environmental parameters and anthropogenic pressures.

7.4. Bacterial alpha- and beta-diversity at the national scale

7.4.1. *Bacterial alpha-diversity*

At the scale of France, bacterial richness strongly varies across soils. It ranges between 870 and 3,074 taxa, with a mean number of 2,079 (Terrat *et al.* 2020). The spatial distribution of richness is heterogeneous and structured in large patches of 110 km in radius (Figure 7.4a) (Terrat *et al.* 2017). Thus, such regions as Brittany in the north-west, the north and the Mediterranean area in the south are hotspots of diversity, whereas the Landes de Gascogne in the south-west, the Centre and Alsace in the east display the lowest levels of bacterial diversity in France. This spatial distribution of bacterial richness cannot be explained by geomorphology (mountains, rivers, coastlines) or by the climate, but rather by the main pedological types and land uses. More precisely, the soil physico-chemical characteristics that most influence bacterial richness rank as follows: pH > texture > C/N ratio. Thus, diversity is particularly enhanced in alkaline soils with a coarse-grained texture (sandy and sandy loam soils) and a low C/N ratio (Terrat *et al.* 2017). Surprisingly, the comparison of land uses indicates the bacterial diversity tended to be higher in agricultural and viticultural soils (more than 2,200 taxa on average) compared with grassland and forest soils (2,140 and 1,930 taxa on average, respectively) (Figure 7.4b).

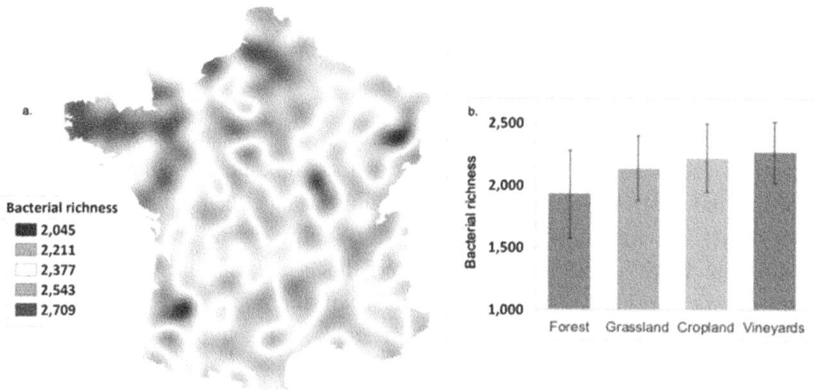

Figure 7.4. *a) Mapping of soil bacterial richness at the French national scale produced by kriging. The regions in red display higher diversity than the regions in gray or black. b) Ranking of land uses as a function of soil bacterial richness. For a color version of this figure, see www.iste.co.uk/guilbert/biogeography.zip*

This greater bacterial diversity in agricultural soils can be partly explained by the higher pH encountered in crop soils compared to forest or grassland soils. In addition, another complementary explanation might be the intermediate disturbance hypothesis, which proposes that an intermediate level of environmental disturbance will lead to the highest species diversity at a local scale (Giller *et al.* 1998). Too low or too strong a disturbance will result in decreased diversity owing to competitive exclusion among species and environmental selection of species, respectively. Thus, forest soils represent natural or semi-natural ecosystems under low disturbance, so their bacterial diversity is lower. Conversely, agricultural and viticultural soils are under intensive management practices, correspond to more disturbed – but not over-disturbed – systems and therefore display higher bacterial diversity. Nevertheless, this greater diversity does not reflect quality, more particularly whether beneficial or deleterious species for its functioning are present or not. If agricultural soils harbor more taxa, some of these taxa can be deleterious to agricultural production, for example, opportunistic pathogenic bacteria, or bacteria implied in too rapid soil organic matter degradation (Lienhard *et al.* 2014).

7.4.2. *The bacterial taxa–area relationship*

While taxonomic richness provides information about local diversity (alpha-diversity), beta-diversity allows comparing diversity among sites and thus assesses it at the territorial scale. The taxa–area relationship is an ecological rule that describes the increase in the number of observed species as a function of the enlargement of the sampling area. It consists of a simple equation:

$$S_A=S_0A^z$$

where S_A is the number of taxa observed in area A, S_0 is the number of taxa initially observed in the smallest sampled area, and z is the rate at which new taxa are sampled as the sampling area is enlarged. This rule has been studied for more than two centuries in macro-organisms, and the value of z has been rigorously determined at large spatial scales. However, this same rule has been very poorly studied in micro-organisms, mainly because of technical limitations in species identification and because of the absence of large-scale sampling for this approach to be applied. The turnover rate (z) of microbial diversity ranges from 0.002 to 0.01 and is generally lower than the turnover rate observed in macro-organisms (usually $0.1 < z < 0.3$) (Terrat *et al.* 2015). In order to evaluate the turnover rate of bacterial diversity ($z_{bacteria}$) at the national scale, the taxa–area relationship was applied to the RMQS using a genotyping tool (ARISA) to characterize the genetic structure of bacterial communities (Figure 7.1). This technique does not provide data about taxon numbers or identification, but about genetic structure similarity among communities. Evaluating the genetic structure similarity of RMQS soil bacterial communities as a function of the distance between them makes it possible to estimate the z-value of the taxa–area relationship. The bacterial taxa–area relationship is significant at the scale of France, with $z_{bacteria}$ estimated at 0.006, in agreement with the literature (Ranjard *et al.* 2013). These results confirm that this ecological rule is true for all living organisms, including the different groups of soil micro-organisms (Horner-Devine *et al.* 2004).

Beyond the taxa–area relationship, the question of the influence of habitat heterogeneity and turnover on soil bacterial diversity turnover is raising. To answer this question, we developed a new measure of habitat characterization based on the soil physico-chemical parameters, climate, land use and geomorphology. All these variables measured at each RMQS point constitute a matrix of data that was analyzed following the taxa–area relationship method. In this case, the habitat–area relationship evaluates habitat turnover depending on the sampled area. $z_{habitat}$ has been estimated to be 0.055 at the scale of France. To reach a robust conclusion about the influence of habitat turnover on bacterial diversity turnover, $z_{bacteria}$ and $z_{habitat}$ were estimated within a circular sliding window of 140 km in radius corresponding to the size of the geographic profiles of the genetic structure variations of bacterial communities (Dequiedt *et al.* 2009). By moving the central point of this sliding window to each RMQS site, 2,100 $z_{bacteria}$ and $z_{habitat}$ values were estimated and compared statistically. Figure 7.5 shows a significant linear regression (r^2=0.65, p<0.001) between $z_{bacteria}$ and $z_{habitat}$, which suggests that the diversification processes of soil bacterial communities at the regional scale (the scale of the sliding window) are mostly influenced by environmental selection processes driven by habitat characteristics. Moreover, the positive regression between the two z values shows that an increase in habitat heterogeneity at the regional scale brought about an increase in the turnover rate of soil bacterial taxa (Ranjard *et al.* 2013). In other words, a diversified landscape is characterized by soils harboring greater cumulated bacterial diversity than a low-diversified landscape. From a fundamental viewpoint, this study is the first demonstration of the influence of landscape environmental variability on soil microbial diversity at the scale of that landscape. This study also demonstrates the need to maintain microbial habitat diversity and spatial structure at a large scale to preserve the soil microbiological heritage. From a more operational viewpoint, maintaining or even increasing agricultural landscape diversity by revising land consolidation rules and diversifying agricultural systems across space and time would be a relevant lever for improving soil biodiversity and the functions and services it can provide.

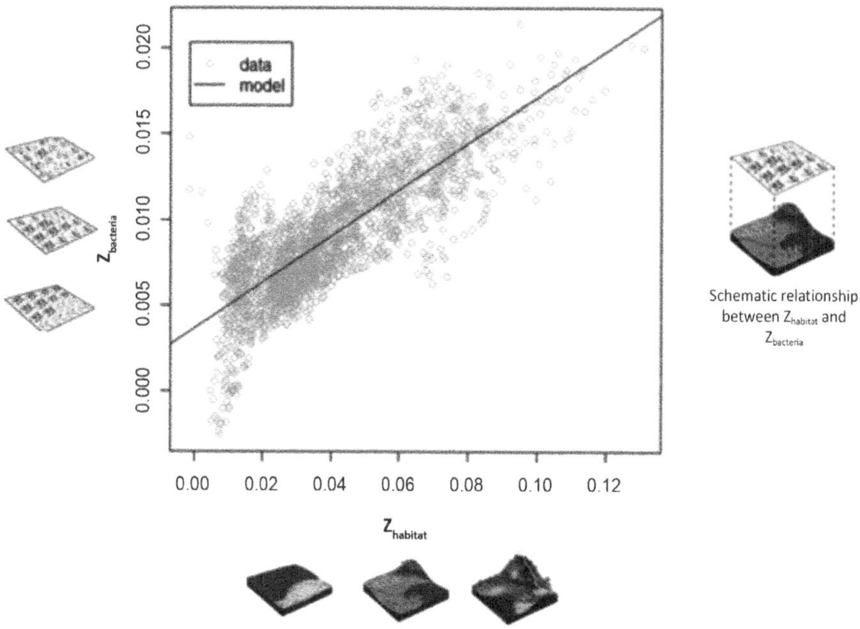

Figure 7.5. *Relation between the turnover rate of the bacterial community structure ($z_{bacteria}$) and the turnover rate of the habitats represented by RMQS soils ($z_{habitat}$). Each dot represents z calculated for each of the 2,100 circular sliding windows of 140 km in radius (modified from Ranjard* et al. *2013). For a color version of this figure, see www.iste.co.uk/guilbert/biogeography.zip*

7.5. Spatial distribution and ecological attributes of bacterial taxa at a large scale

The sequencing of 16S ribosomal RNA genes from all RMQS soils yielded 188,000 bacterial taxa at the scale of France, distributed in 1,355 genera, 267 families, 147 orders, 51 classes and 35 bacterial phyla (32 bacterial phyla and three archaeal phyla) (Karimi *et al.* 2018). Among these phyla, 20 displayed a relative abundance ranging between 14% and 0.01%, and were found in more than 50% of the soils. Therefore, we focused this study on these 20 major and ubiquitous phyla. The most abundant bacterial phyla made up more than 84% of the communities and were also the most ubiquitous ones at the scale of France (Table 7.1) (Karimi *et al.* 2018). These phyla – *Alphaproteobacteria, Actinobacteria, Planctomycetes, Firmicutes, Gammaproteobacteria, Acidobacteria, Betaproteobacteria* and *Deltaproteobacteria* – have also been identified as major phyla in other soil studies (Constancias *et al.* 2015), and also in other environments such as marine sediment (Bienhold *et al.* 2016), oceans (Sunagawa *et al.* 2015) or the digestive tracts of mammals (Donaldson *et al.* 2015). Their strong cosmopolitan character can be explained by their high dispersal capacity and also by their capacity to get adapted to and colonize different environments.

An analysis of their variation at the scale of France allowed us to generate national maps of their spatial distribution (Figure 7.6). The analysis of these maps highlighted different geographic distribution patterns depending on the phyla. For example, *Actinobacteria* displayed a large geographic pattern across areas of more than 200 km in radius, whereas *Bacteroidetes* displayed an intermediate-sized pattern (100–200 km in radius), and *Chloroflexi* were characterized by a small-sized pattern (<100 km) (Table 7.1).

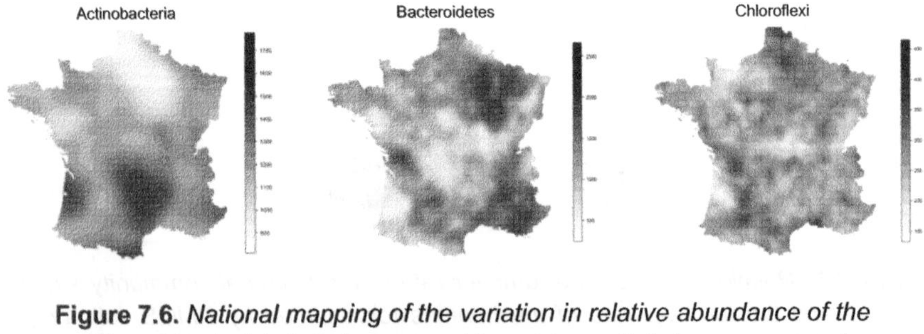

Figure 7.6. *National mapping of the variation in relative abundance of the different bacterial phyla (expressed in numbers of phylum sequences)*

These variations in biogeographic patterns demonstrate that bacterial phyla do not all display the same ecological attributes and depend on different environmental parameters (soils, climates, geomorphology and land use) (Karimi *et al.* 2018). The distributions with the smallest patterns may result from a strong influence of local factors such as the landscape layout or variations in land uses (e.g. *Alphaproteobacteria*, *Chloroflexi* and *Armatimonadetes*), whereas the largest patterns may be explained by the presence of large natural barriers (seas, mountains), by climate type or large-scale soil type variations (e.g. *Actinobacteria*, *Firmicutes* and *Chlorobi*). These different profiles can also be partly explained by differences in dispersal capacities. For instance, *Firmicutes* includes organisms able to disperse over long distances through aerosol processes linked to their capacity to sporulate.

Although the different phyla do not all depend on the same environmental parameters, the most driving factors can be ranked as follows: Soil type > Land use >> Climate > Geomorphology. Regarding the soil physico-chemical parameters, the pH was determining for 17 phyla out of the 20 studied ones, with a stimulating effect for nine of them and an inhibiting effect for the other eight (Table 7.1). Soil texture, and more particularly the clay content (<2μm particles), significantly influenced 15 bacterial phyla, including 10 negatively. In addition, the quality and level of soil resources also played a role in phylum distributions, with nine phyla significantly influenced by the C/N ratio and six phyla influenced by the soil organic matter content. The influence of land use was evaluated by comparing the four main land uses in France (forests, grasslands, croplands and vineyards/orchards). These four uses correspond to increasing levels of farming intensity (tillage, fertilization, pesticides, etc.) and of related disturbances. It is interesting to note that eight phyla (e.g. *Crenarchaeota*) were stimulated by an increasing level of farming intensity, whereas six (e.g. *Alphaproteobacteria*) were repressed (Table 7.1). However, the ecological processes that influence each phylum should be considered altogether as presented in Table 7.1, not independently. For example, the phylum *Deltaproteobacteria* was more abundant in forest and grassland soils, with an alkaline pH, a fine texture and abundant organic matter not very recalcitrant to degradation (a low C/N ratio). Thanks to these data, it is possible not only to present a range of environmental parameters likely to influence the development of each of the bacterial phyla, but also to appraise some of their ecological attributes (tolerance to acidity, oligotrophy, copiotrophy, etc.).

Phyla		Occurrence (% of soils)	Average relative abundance (%)	Size of biogeographical patterns (km)	Effects of soil physico-chemical parameters				Land use effect			
					pH	Clay amount	C:N	C_{org}	Forest	Grassland	Cropland	Vineyards-Orchards
Major phyla	*Bacteroidetes*	100	14.1	124	+++	+	ns	ns	+++	++	++	+
	Alphaproteobacteria	100	13.5	71	–	ns	ns	ns	+++	++	+	+
	Actinobacteria	100	13.1	258	-	+	+	ns	+	+++	+	++
Dominant phyla	*Planctomycetes*	100	8.5	137	-	ns	ns	ns	+++	++	+	+
	Firmicutes	100	8.5	247	–	-	-	ns	+	+++	++	++
	Gammaproteobacteria	100	7.8	206	+	-	+	ns	+++	+	++	++
	Acidobacteria	100	6.9	79	–	ns	+	ns	+++	+	++	++
	Betaproteobacteria	100	6.8	209	ns	-	ns	ns	+	+	++	+++
	Deltaproteobacteria	100	5.5	89	++	+	-	+	+++	+++	++	+
Intermediate phyla	*Thaumarchaeota*	99.4	4.9	105	+	ns	ns	-	+	++	+++	++
	Crenarchaeota	99.8	4.1	119	-	ns	-	ns	+	++	++	+++
	Chloroflexi	100	3.1	43	ns	+	ns	-	+	+	++	+++
	Gemmatimonadetes	99.8	0.8	167	ns	-	-	ns	+++	+++	++	+
	Verrumicrobia	100	0.6	131	+	+	-	ns	+++	++	+	+
	Nitrospirae	99.9	0.5	242	+	-	ns	ns	+	+	++	+++
Minor phyla	*Chlorobi*	99.5	0.2	260	++	-	ns	+	+++	++	++	+
	Armatimonadetes	99.4	0.1	75	-	-	-	ns	+	++	+++	+++
	Elusimicrobia	99.5	0.1	210	+	-	ns	+	+	+	++	+++
	Fibrobacteres	75.1	0.05	45	+	-	ns	ns	+	++	+++	+
	Cyanobacteria	85.1	0.04	60	-	-	ns	+	+	+	++	+++

Table 7.1. *Biogeographic characteristics of the main bacterial phyla, and ranking of the influence of physico-chemical characteristics (from +++, strong positive influence in dark green to ---, strong negative influence in dark orange) and of land uses (from +++, strong influence to +, low influence, following the color gradation) (from Karimi* et al. *2018 Atlas). For a color version of this table, see www.iste.co.uk/guilbert/biogeography.zip*

At a finer taxonomic resolution, only 47 genera displayed relative abundances above 0.5% of the sequences at the scale of France (Karimi *et al.* 2018), and they belonged to 12 different phyla among the most abundant ones. This small part of the community (3.5% of genus diversity) represents an average 62% of the overall taxonomic sequences at the scale of France. These genera are highly cosmopolitan since they were detected in more than 98% of French soils. While *Bacteroidetes* represent the most abundant phylum in French soils, the genus *Holophaga* (*Actinobacteria*) was dominant with 6% of the sequences per soil on average. A comparison of the geographic distributions of the genera with the phyla they belong to evidenced that in some cases the patterns observed at the genus level were conserved at the phylum level. This confirms the ecological coherence of taxa (e.g. *Bacillus* for *Firmicutes*, *Sulfobococcus* for *Crenarchaeota*). In other cases, the distribution of the phylum corresponded to the cumulated distributions of all the genera that formed it (e.g. the eight dominant genera belonging to *Actinobacteria*, the three genera belonging to *Deltaproteobacteria*). Similarly to phylum, environmental filter analysis showed that each genus was influenced by its own parameters. The genera belonging to the same phylum were not necessarily influenced in the same way by identical parameters. For example, while the phylum *Bacteroidetes* was strongly enhanced in acidic soils, the abundance of the genus *Chitinophaga* was much lower in this type of soil. Yet, it was one of the seven dominant genera in this phylum, and the other genera followed the same trend as the global phylum. Whether it was at the phylum or genus level, mapping evidenced pairs of taxa with similar or opposed distribution profiles (e.g. *Gammaproteobacteria* and *Firmicutes*, see Karimi *et al.* 2018). These pairs of taxa are not isolated cases, and correspond to the visible side of bacterial co-occurrence networks.

7.6. Large-scale bacterial co-occurrence networks (also called Bacteriosociology)

Beyond the notions of alpha- or beta-diversity and the ecology of soil taxa, the analysis of microbial interaction networks assesses the complexity level of communities. The underlying hypothesis emanates from works on macro-organisms such as plant-pollinator networks (Schleuning *et al.* 2015), mycorrhizal symbiosis (Van Der Heijden *et al.* 2016) and aquatic or terrestrial foodwebs (Massol and Petit, 2013). It proposes that the complexity and structure of interaction networks are tightly linked to the good functioning of ecosystems. In micro-organisms, a few studies on the gut microbiota have confirmed the link between the microbial network and the patients' health state (Shankar *et al.* 2015; Baldassano and Bassett 2016). These pieces of literature are encouraging and an incentive to investigate microbial networks in soils. The aim is to reach information on the microbiological status of soils, complementary to microbial abundance and richness and the distribution of taxa. Given the technical limitations related to the observation of the multitude of interactions that take place within microbial communities, a widespread practice consists of observing co-occurrences among taxa to reflect interactions (Karimi *et al.* 2017). However, the co-occurrence or co-exclusion of organisms can be explained by direct interactions as well as by the sharing of an ecological niche implying indirect interactions (Röotjers and Faust, 2018). Co-occurrences (or co-abundances) are estimated based on the distribution and relative abundance of taxa following the method detailed in Karimi *et al.* (2019). The synthesis of all statistically identified co-occurrences and co-exclusions forms a network described by a set of metrics.

At the scale of France, bacterial networks have been studied as a function of land use (Karimi *et al.* 2019). The results show bacterial networks with a maximum number of connections in forest soils, followed by grasslands and croplands, and last of all vineyards and orchards. Figure 7.7 shows that in forests, where soils are least disturbed, the network forms a dense, highly connected, tightly closed structure. In grasslands, where the intensity of agricultural practices remains moderate but is more recurrent, the network is not quite as dense as in forest soils, and two poles can be detected, which remain strongly interconnected. In croplands, where tillage and inputs are regular, the network is more degraded, with fewer connections within the network, which displays long chains and also satellites, that is, clusters of taxa strongly connected with one another but isolated from the other clusters.

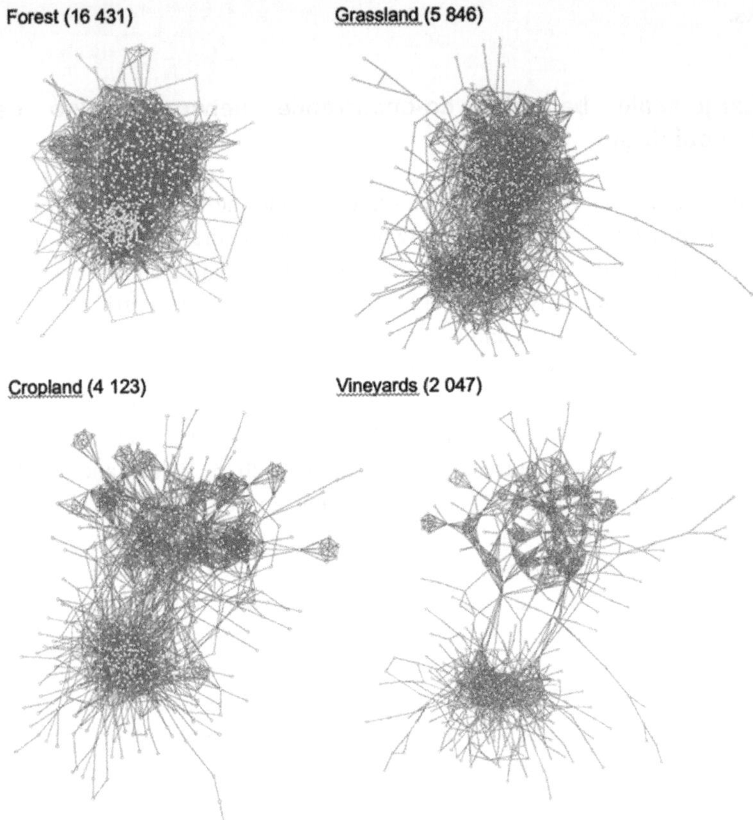

Figure 7.7. *Soil bacterial co-occurrence networks for the four main land uses in France. Green, positive connections; orange, negative connections. Circles represent bacterial genera. The number of network connections is indicated in parentheses (modified from Karimi* et al. *2019). For a color version of this figure, see www.iste.co.uk/guilbert/biogeography.zip*

This transformation of the bacterial network corresponds to an 87% loss of complexity in vineyards compared with forests. Mechanical practices or chemical inputs (fertilizers, pesticides) may have consequences on the lack of connectivity and isolation of microbial populations through three main processes: (1) a reduced microbial biomass, and thereby a reduced number of cells, decreases the probability for organisms to meet and interact (Dequiedt *et al.* 2009; Tecon and Or, 2017); (2) the enhancement of opportunistic or pathogenic bacteria in these soils (Lienhard *et al.* 2014) creates populations metabolically independent from one another; (3) physical soil disturbance destroys spatial connectivity among the different micro-habitats (Cordero *et al.* Datta 2016). The modification of bacterial networks is also characterized by the shifting of hubs – the central, most connected taxa around which the network is organized. In forest and grassland soils, hubs are mainly formed of *Bacteroidetes*, bacteria known to be copiotrophic. In agricultural systems, networks mainly rely on oligotrophic genera belonging to the phyla *Actinobacteria* and *Alphaproteobacteria* (*Bradyrhizobium*) (Karimi *et al.* 2018). These changes are consistent with the different ways soils function depending on land use.

7.7. Do large-scale bacterial habitats exist?

The concept of habitat refers to the set of environmental (biotic and abiotic) conditions that host a particular species or community (Whittaker *et al.* 1973). Although defining macro-organism habitats constitutes a decision-support tool for biodiversity conservation policies, micro-organism habitats have been hardly ever investigated. A few studies address bacterial habitats at the micro-scale, that is, the scale of aggregates, the rhizosphere and the detritusphere (Nunan *et al.* 2017), but these studies are not conclusive about the existence of large-scale bacterial habitats because micro-habitats are found in all soils. The only works that attempt to investigate macro-scale bacterial habitats consider the different biomes as microbial habitats (Fierer *et al.* 2012). This choice relies on the assumption that the plant cover is the main determinant of the soil microbiota; yet, the above-mentioned conclusions provided in the present chapter clearly illustrate that the soil type is the most discriminating factor of the distribution of soil bacterial communities.

In France, the RMQS is a particularly robust tool to define bacterial habitats at the macro-ecological scale in an unbiased manner. Supervised learning methods such as multiple regression trees have been used to identify the environmental variables that best predict community composition. Among more than 50 variables (soil physico-chemical characteristics, land use, climate variables, geomorphology), five are structuring ones, and rank as follows: pH > Land use > C/N ratio > Soil organic carbon > Mean annual temperature. The combination of these five variables defines 16 bacterial habitats that each host a specific community (Karimi *et al.* 2020, Table 7.2, Figure 7.8). These 16 habitats can be grouped in five habitat complexes distributed along a pH gradient, and within each complex, three or four habitats are discriminated based on one or two of the other variables. Land use, which reflects the plant cover, discriminates the habitats of four complexes, while the C/N ratio discriminates the habitats of the 5th complex. Interestingly, the unbiased characterization of bacterial habitats does not rely only on the plant cover, as suggested by the studies that liken microbial habitats to the different terrestrial biomes or phyto-sociological habitats (Fierer 2017). Soil acidity and access to nutrient resources are indeed two of the most significant structuring elements of bacterial habitats.

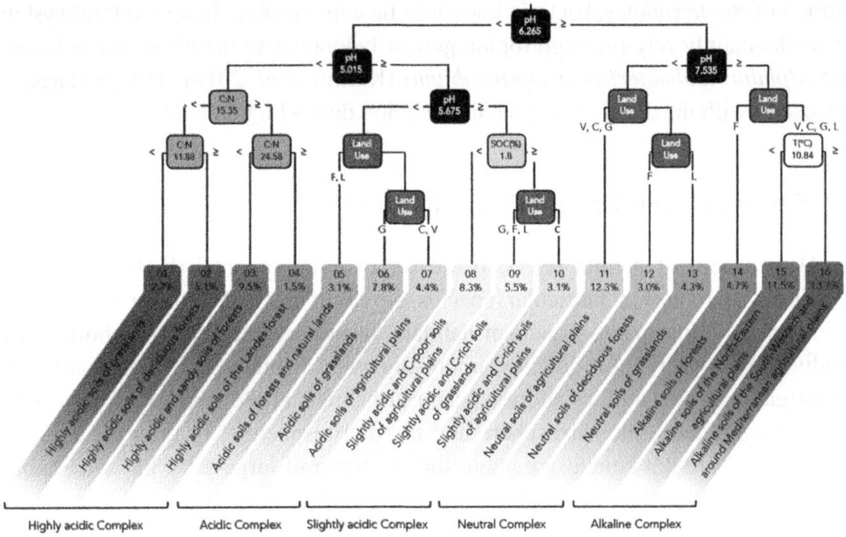

Figure 7.8. *Regression tree defining the 16 bacterial habitats of French soils. Along the tree, nodes represent the splits determined by environmental parameters; values inside rectangles indicate the thresholds for the corresponding parameter. The splits based on land use are indicated in capital letters: F, forest; G, grasslands; C, cropping systems; V, vineyards and orchards; L, low-anthropized environments. The boxes at the end of the branches represent the 16 habitats characterized by a specific bacterial community structure. The values given in the boxes indicate the proportion of sites included in each habitat. SOC: soil organic carbon. For a color version of this figure, see www.iste.co.uk/guilbert/biogeography.zip*

The mapping of these 16 habitats represents a complex mosaic at the scale of the French territory (Figure 7.9). The smallest habitat covers 1.5% of the sampled sites, whereas the largest one represents 13.7% of the sites. Some are located in well-delineated regions, while others are disseminated across the whole territory, in large patches or in thousands of small spots. This suggests that despite high habitat fragmentation, the community is able to build up again if environmental conditions are favorable.

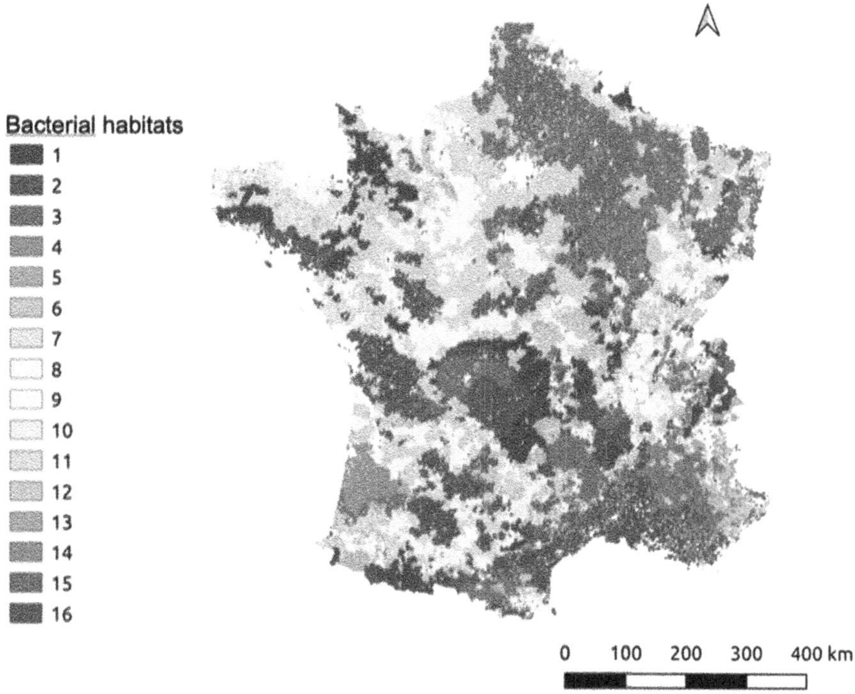

Figure 7.9. *Mapping of the bacterial habitats of French soils (modified from Karimi* et al. *2020). The color code is the same as in Figure 7.8. For a color version of this figure, see www.iste.co.uk/guilbert/biogeography.zip*

The indigenous bacterial communities of the different habitats are differentiated by different diversity levels. While the most acidic habitats display lower bacterial richness, habitats under agricultural land use display higher richness (> 300 genera). This confirms at the habitat scale the previous result found at the territory scale (Terrat *et al.* 2017). According to the positive relationship between community diversity and stability (Loreau *et al.* 2010), low-richness habitats should be more vulnerable to disturbances. Combined with the relationship between biodiversity and ecosystem functioning, these habitats should also be the least fertile, the least suppressive and the least productive ones (Vivant *et al.* 2013; Maron *et al.* 2018). These hypotheses about the vulnerability and low fertility of acidic habitats are consistent with the absence of an agricultural system associated with these habitats.

In terms of composition, despite a pool of 130 highly generalist species shared by all habitats, each habitat displays a different bacterial signature, with its top 10 dominant genera and its own emblematic diversity (taxa detected in only one habitat in more than half of the sites). Moreover, a cosmopolitanism gradient has been evidenced within soil bacterial diversity depending on the demographic performances and adaptive capacities of the taxa (Büchi and Vuilleumier, 2014; Székely and Langenheder, 2014). The habitats with a neutral or alkaline pH harbor the highest numbers of emblematic taxa whatever the land use. This means that habitat conservation policies should not be restricted to natural ecosystems, but should also include anthropized systems. In order to fully assess the bacterial communities of habitats, co-occurrence networks were also investigated. The four most acidic habitats displayed the most complex networks, and the low degradability of the soil organic matter usually observed in these habitats appeared to induce the establishment of connections between organisms, mainly cooperation relationships (habitat 04). Temperature also seemed to enhance microbial interactions: it was the only environmental parameter that distinguished habitat 15 from habitat 16 (with a threshold of 10.8°C), while habitat 16 displayed twice as many connections in its bacterial network as habitat 15 did. Thus, while climatic conditions did not impact richness or hosted taxa, interaction networks appeared to be directly influenced by this global parameter, which suggests consequences on the stability and functioning of soil microbial communities (Montoya *et al.* 2006; Karimi *et al.* 2017).

Habitats	Environment					Distribution			Bacterial signature			
	pH	C:N	Land use	C_{org}	T°	Surface (%)	Localization	Profile size	Genus richness	No. of emblematic genera	Network complexity (%)	Cooperation: Antagonism
1	- - -	-	ns	ns	ns	2.2	localized	patch	295	0	0.15	25
2	- - -	=	ns	ns	ns	5.1	localized	patch	269	0	0.2	21
3	- - -	+	ns	ns	ns	9.5	localized	patch	244	1	0.27	31
4	- - -	+ +	ns	ns	ns	1.5	localized	patch	222	3	0.37	68
5	- -	ns	F,L	ns	ns	3.1	large	spot	300	1	0.12	25
6	- -	ns	G	ns	ns	7.8	large	spot	321	0	0.1	21
7	- -	ns	C,V	ns	ns	4.4	large	spot	346	3	0.11	19
8	-	ns	ns	-	ns	8.3	large	spot	352	0	0.09	29
9	-	ns	G,F,L	+	ns	5.5	large	spot	330	0	0.09	25
10	-	ns	C	+	ns	3.1	large	spot	355	2	0.09	24
11	=	ns	G,C, V	ns	ns	12.3	large	spot	365	0	0.1	25
12	=	ns	F	ns	ns	3	large	spot	326	5	0.09	17
13	=	ns	L	ns	ns	4.3	large	spot	337	1	0.1	19
14	+	ns	F	ns	ns	4.7	large	patch	337	6	0.08	20
15	+	ns	L,G,C,V	ns	-	11.5	large	patch	358	5	0.09	20
16	+	ns	L,G,C,V	ns	+	13.7	large	patch	352	3	0.21	36

Table 7.2. *Characteristics of the 16 bacterial habitats defined at the scale of France. The color code is the same as in Figure 7.8 (from Karimi* et al. *2018, Atlas). For a color version of this table, see www.iste.co.uk/guilbert/biogeography.zip*

To conclude, the dependence of microbial habitats on land uses and climate conditions highlights the direct and indirect impacts of human activities on these habitats. The current environmental crisis questions the critical stake of biodiversity conservation, and large-scale habitat preservation has already proved efficient for macro-organisms. The link between habitat and biodiversity for micro-organisms and the survey of the main soil microbial habitats can lead to the integration of microbial habitats to improve biodiversity protection and conservation policies, especially in the present global change context.

7.8. Biogeography at the service of environmental diagnosis

The bacterial diversity data acquired from the 2,200 RMQS soils were saved in a database (MicroSol database ©, Zenodo files: 10.5281/zenodo.1063503 and 10.5281/zenodo.1065438) that constitutes the first national-scale referential (Morin *et al.* 2013). Thanks to the methodological standardization of molecular biology tools and their application to the high RMQS variability, "bacterial diversity" (in terms of numbers of taxa) was validated as a "national soil indicator" by the Observatoire National de la Biodiversité (Ministry of Environment). Moreover, based on these datasets, we set up a predictive statistical model of bacterial diversity to estimate reference values according to soil pedoclimatic conditions (Terrat *et al.* 2017):

$$\textit{Bacterial diversity} = 1{,}044 + 3.305 \times pH^4 - 0.0457 \times Clay^2 + 0.0597 + 0.00298 \times Clay^2 \times C/N - 1.54 \times 10^{-6} \times Clay^3 \, x \, C/N + 2.336 \, x \, 10^{-5} \times C{:}N^2 \times Longitude.$$

This reference value is the theoretical value of a soil as a function of its physico-chemical characteristics and geographical localization, independently of land use. This value makes it possible to leave the pedoclimatic context aside to carry out a diagnostic of the impact of land uses. For example, the impact of an agricultural practice is positive if the value measured in the field is superior to the reference value, and negative in the opposite case. The sensitivity, repeatability and robustness tests and the results from different experimental systems identified ~ 20% methodological and temporal variability of these indicators, which allowed us to calibrate the standard variation range of the indicator. This bioindicator and the associated diagnostic were applied to different agricultural situations. Figure 7.10 summarizes the results from more than 200 soils from a cropland landscape of around 12 km^2 located in Bourgogne (France). Most of the plots were of a good microbiological quality, with bacterial diversity values superior to the reference value given by the model. Yet, some plots were also strongly affected by agricultural practices, since their values were below the reference value.

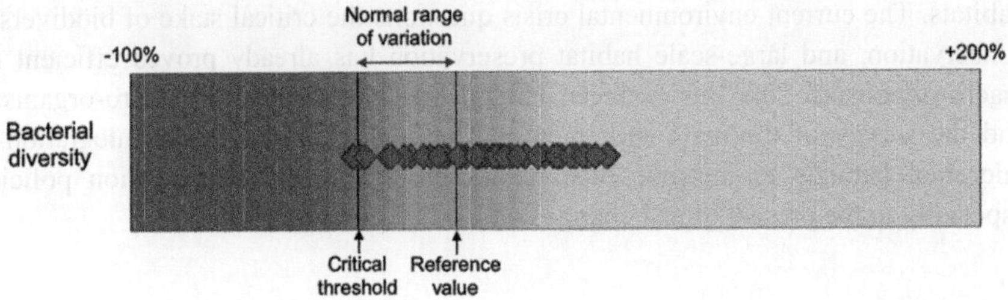

Figure 7.10. *Diagnostic of the microbiological quality of the soils of an agricultural landscape. The squares represent the positions of the values of the plots according to the deviation from the reference value given by the model. For a color version of this figure, see www.iste.co.uk/guilbert/biogeography.zip*

The same tools were applied to a network of more than 250 vineyards and crop systems at a national scale (CASDAR AgrInnov project, 2011–2015), and similar results were recorded. Only 10% of the plots were diagnosed as being of poor microbiological quality with values below the critical threshold. Even though 40% of the plots displayed a very good microbiological quality, the remaining 50% displayed values within the standard range of variation of the reference value and corresponded to a non-critical situation, but to be checked in the future. Moreover, a recent study based on a network of about 100 French vineyards plots demonstrated that organic farming improves significantly (about 20%) the microbiological quality of soil and the connectivity of the microbial network (data not shown).

7.9. Conclusion perspectives

The studies led at the scale of France on the spatial distribution of soil bacteria altogether illustrate that studying these organisms at large spatial scales is relevant. Then, the hypothesis that microbial communities are a black box deprived of a spatial structure and with a homogeneous local and territorial distribution can be rejected. From a fundamental viewpoint, the spatial distribution of micro-organisms is significantly driven by the same processes as those that drive the spatial distribution of macro-organisms, that is, (1) environmental heterogeneity (the selection process) and (2) their dispersal/colonization capacities (neutral processes linked with dispersal). These two processes are not mutually exclusive. This way, we identified and ranked the environmental parameters that had the greatest influence on community selection and also on the different taxa taken individually. The new knowledge accumulated on the ecological attributes of the different bacterial phyla has been compiled in the atlas of French soil bacteria (*Atlas français des bactéries du sol*) published in 2018. This kind of knowledge is all the more important when it contributes to improving the understanding and the prediction of the ecosystem services supported by this biodiversity in a context of sustainability of land uses at a large scale, and of defining a targeted public policy. From a finalized viewpoint, our approach allows for a better evaluation of the impact of land uses on micro-organisms. Furthermore, the large amounts of data generated about microbial community abundance and diversity contributed to set up the first reference bases of soil microbial communities, while validating the molecular tools used to produce these "national indicators" of the biological quality of French soils. Today, these indicators are used to set up a diagnosis of the microbiological quality of agricultural soils directly with the stakeholders of agricultural development. The aim is to lead to a new type of agronomic advice that fits in with the agro-ecological transition.

For more information about soil bacterial biogeography and its operational applications:

– The French atlas of soil bacteria: https://leclub-biotope.com/fr/librairie-naturaliste/1076-atlas-francais-des-bacteries-du-sol.

– Soil ecological quality: https://www.techniques-ingenieur.fr/base-documentaire/environnement-securite-th5/genie-ecologique-42683210/qualite-ecologique-des-sols-ge1051/.

– Bacterial diversity as a bioindicator: http://indicateurs-biodiversite.naturefrance.fr/indicateurs/evolution-de-la-biodiversite-bacterienne-des-sols.

– Molecular microbiology at the service of environmental diagnosis: https://www.ademe.fr/microbiologie-moleculaire-service-diagnostic-environnemental.

– Meta-omics techniques to diagnose soil microbiological quality: https://www.techniques-ingenieur.fr/base-documentaire/procedes-chimie-bio-agro-th2/bioprocedes-dans-les-domaines-de-l-energie-et-de-l-environnement-42161210/nouvelles-techniques-de-meta-omiques-pour-le-diagnostic-de-la-qualite-microbiologique-des-sols-ge1052/.

7.10. References

Bahram, M., Hildebrand, F., Forslund, S.K., Anderson, J.L., Soudzilovskaia, N.A., Bodegom, P.M., Bork, P. (2018). Structure and function of the global topsoil microbiome. *Nature*, 560(7717), 233–237.

Baldassano, S.N. and Bassett, D.S. (2016). Topological distortion and reorganized modular structure of gut microbial co-occurrence networks in inflammatory bowel disease. *Scientific Reports*, 6(May), 1–14.

Baumann, K., Dignac, M.F., Rumpel, C., Bardoux, G., Sarr, A., Steffens, M., Maron, P.A. (2012). Soil microbial diversity affects soil organic matter decomposition in a silty grassland soil. *Biogeochemistry*, 114(1–3), 1–12.

Bickel, S., Chen, X., Papritz, A., Or, D. (2019). A hierarchy of environmental covariates control the global biogeography of soil bacterial richness. *Scientific Reports*, 9(1), 1–10.

Bienhold, C., Zinger, L., Boetius, A.R. (2016). Diversity and biogeography of bathyal and abyssal seafloor bacteria. *PLoS ONE*, 11, e0148016.

Bouchez, T., Blieux, A.L., Dequiedt, S., Domaizon, I., Dufresne, A., Ferreira, S., Ranjard, L. (2016). Molecular microbiology methods for environmental diagnosis. *Environmental Chemistry Letters*, 14(423–441), 1–19.

Büchi, L. and Vuilleumier, S. (2014). Coexistence of specialist and generalist species is shaped by dispersal and environmental factors. *The American Naturalist*, 183(5), 612–624.

Cameron, E.K., Martins, I.S., Lavelle, P., Mathieu, J., Tedersoo, L., Bahram, M., Eisenhauer, N. (2019). Global mismatches in aboveground and belowground biodiversity. *Conservation Biology*, 33(5), 1187–1192.

Constancias, F., Saby, N.P.A., Terrat, S., Dequiedt, S., Horrigue, W., Nowak, V., Ranjard, L. (2015). Contrasting spatial patterns and ecological attributes of soil bacterial and archaeal taxa across a landscape. *MicrobiologyOpen*, (April), 518–531.

Cordero, O.X. and Datta, M.S. (2016). Microbial interactions and community assembly at microscales. *Current Opinion in Microbiology*, 31(Figure 1), 227–234.

De Wit, R. and Bouvier, T. (2006). Everything is everywhere, but, the environment selects: What did Baas Becking and Beijerinck really say? *Environmental Microbiology*, 8(4), 755–758.

Delgado-Baquerizo, M. and Eldridge, D.J. (2019). Cross-biome drivers of soil bacterial alpha diversity on a worldwide scale. *Ecosystems*, 22(6), 1220–1231.

Delgado-Baquerizo, M., Maestre, F.T., Reich, P.B., Trivedi, P., Osanai, Y., Liu, Y.-R., Singh, B.K. (2016). Carbon content and climate variability drive global soil bacterial diversity patterns. *Ecological Monographs*, 86(3), 373–390.

Delgado-Baquerizo, M., Oliverio, A.M., Brewer, T.E., Benavent-gonzález, A., Eldridge, D.J., Bardgett, R.D., Fierer, N. (2018). A global atlas of the dominant bacteria found in soil. *Science*, 325(February), 320–325.

Dequiedt, S., Thioulouse, J., Jolivet, C., Saby, N.P.A., Lelievre, M., Maron, P.A., Ranjard, L. (2009). Biogeographical patterns of soil bacterial communities. *Environmental Microbiology Reports*, 1(4), 251–255.

Donaldson, G.P., Lee, S.M., Mazmanian, S.K. (2015). Gut biogeography of the bacterial microbiota. *Nature Reviews Microbiology*, 14, 20–32.

Fierer, N. (2017). Embracing the unknown: Disentangling the complexities of the soil microbiome. *Nature Reviews Microbiology*, 15(10), 579–590.

Fierer, N., Bradford, M.A., Jackson, R.B. (2007). Toward an ecological classification of soil bacteria. *Ecology*, 88(6), 1354–1364.

Fierer, N., Leff, J.W., Adams, B.J., Nielsen, U.N., Thomas, S., Lauber, C.L., Caporaso, J.G. (2012). Cross-biome metagenomic analyses of soil microbial communities and their functional attributes. *Proceedings of the National Academy of Sciences of the United States of America*, 109, 21390–21395 [Online]. Available at: https://doi.org/10.1073/pnas.1215210110/-/DCSupplemental.www.pnas.org/cgi/doi/10.1073/pnas.1215210110.

George, P.B.L., Lallias, D., Creer, S., Seaton, F.M., Kenny, J.G., Eccles, R.M., Jones, D.L. (2019). Divergent national-scale trends of microbial and animal biodiversity revealed across diverse temperate soil ecosystems. *Nature Communications*, 10(1), 1–11.

Giller, K., Witter, E., McGrath, S. (1998). Toxicity of heavy metals to microorganisms and microbial processes in agricultural soils: A review. *Soil Biology and Biochemistry*, 30, 1389–1414.

Green, J.L., Holmes, A.J., Westoby, M., Oliver, I., Briscoe, D., Dangerfield, M., Beattie, A.J. (2004). Spatial scaling of microbial eukaryote diversity. *Nature*, 432(7018), 747–750.

Griffiths, R.I., Thomson, B.C., James, P., Bell, T., Bailey, M., Whiteley, A.S. (2011). The bacterial biogeography of British soils. *Environmental Microbiology*, 13(6), 1642–1654.

Hernandez-Raquet, G., Durand, E., Braun, F., Cravo-Laureau, C., Godon, J. (2013). Impact of microbial diversity depletion on xenobiotic degradation by sewage-activated sludge. *Environemental Microbiology Reports*, 5(4), 588–594.

Karimi, B., Maron, P.A., Chemidlin Prévost-Bouré, N., Bernard, N., Gilbert, D., Ranjard, L. (2017). Microbial diversity and ecological networks as indicators of environmental quality. *Environmental Chemistry Letters*, 15, 265–281.

Karimi, B., Prévost-Bouré, N.C., Terrat, S., Dequiedt, S., Ranjard, L. (2018a). *Atlas français des bactéries du sol*. Biotope Editions.

Karimi, B., Terrat, S., Dequiedt, S., Saby, N.P.A., Horrigue, W., Lelièvre, M., Ranjard, L. (2018b). Biogeography of soil bacteria and archaea across France. *Science Advances*, 4(eaat1808), 1–14.

Karimi, B., Dequiedt, S., Terrat, S., Jolivet, C., Arrouays, D., Wincker, P., Ranjard, L. (2019). Biogeography of soil bacterial networks along a gradient of cropping intensity. *Scientific Reports*, 9(1), 1–10.

Karimi, B., Villerd, J., Dequiedt, S., Terrat, S., Chemidlin-Prévost Bouré, N., Djemiel, C., Ranjard, L. (2020). Biogeography of soil microbial habitats across France. *Global Ecology and Biogeography*, (December 2019), 1–13.

Lienhard, P., Terrat, S., Prévost-Bouré, N.C., Nowak, V., Régnier, T., Sayphoummie, S., Ranjard, L. (2014). Pyrosequencing evidences the impact of cropping on soil bacterial and fungal diversity in Laos tropical grassland. *Agronomy for Sustainable Development*, 34(2), 525–533.

Loreau, M. (2010). Linking biodiversity and ecosystems: Towards a unifying ecological theory. *Philosophical Transactions of the Royal Society of London. Series B, Biological Sciences*, 365, 49–60.

Maron, P.-A., Mougel, C., Ranjard, L. (2011). Soil microbial diversity: Methodological strategy, spatial overview and functional interest. *Comptes rendus biologies*, 334(5–6), 403–411.

Maron, P.-A., Sarr, A., Kaisermann, A., Lévêque, J., Mathieu, O., Guigue, J., Ranjard, L. (2018). High microbial diversity promotes soil ecosystem functioning. *Applied and Environmental Microbiology*, 84(9), e02738–17.

Martiny, J.B.H., Bohannan, B.J.M., Brown, J.H., Colwell, R.K., Fuhrman, J.A, Green, J.L., Staley, J.T. (2006). Microbial biogeography: Putting microorganisms on the map. *Nature Reviews Microbiology*, 4(February), 102–112.

Massol, F. and Petit, S. (2013). Interaction networks in agricultural landscape mosaics. *Advances in Ecological Research*, 49, 291–338.

Minasny, B., McBratney, A.B., Hartemink, A.E. (2010). Global pedodiversity, taxonomic distance, and the world reference base. *Geoderma*, 155(3–4), 132–139.

Montoya, J.M., Pimm, S.L., Solé, R.V. (2006). Ecological networks and their fragility. *Nature*, 442(7100), 259–264.

Morin, F.E.R., Dequiedt, S., Koyao-Darinest, V., Toutain, B., Terrat, S., Lelièvre, M., Ranjard, L. (2013). MicroSol database©, le Premier Système d'Information Environnemental sur la Microbiologie des Sols. *Etude et gestion des sols*, 20(1), 27–38.

Nelson, M.B., Martiny, A.C., Martiny, J.B.H. (2016). Global biogeography of microbial nitrogen-cycling traits in soil. *Proceedings of the National Academy of Sciences of the United States of America*, 113(29), 8033–8040.

Noronha, M.F., Lacerda Júnior, G.V., Gilbert, J.A., de Oliveira, V.M. (2017). Taxonomic and functional patterns across soil microbial communities of global biomes. *Science of the Total Environment*, 609, 1064–1074.

Nunan, N. (2017). The microbial habitat in soil: Scale, heterogeneity and functional consequences. *Journal of Plant Nutrition and Soil Science*, 180(4), 425–429.

Powell, J.R., Karunaratne, S., Campbell, C.D., Yao, H., Robinson, L., Singh, B.K. (2015). Deterministic processes vary during community assembly for ecologically dissimilar taxa. *Nature Communications*, 6, 1–10.

Prudent, M., Dequiedt, S., Sorin, C., Girodet, S., Nowak, V., Duc, G., Maron, P.A. (2020). The diversity of soil microbial communities matters when legumes face drought. *Plant Cell and Environment*, 43(4), 1023–1035.

Ranjard, L., Dequiedt, S., Jolivet, C., Saby, N.P.A., Thioulouse, J., Harmand, J., Lemanceau, P. (2010). Biogeography of soil microbial communities: A review and a description of the ongoing French national initiative. *Agronomy for Sustainable Environment*, 30(2), 359–365.

Ranjard, L., Dequiedt, S., Chemidlin Prévost-Bouré, N., Thioulouse, J., Saby, N.P.A., Lelievre, M., Lemanceau, P. (2013). Turnover of soil bacterial diversity driven by wide-scale environmental heterogeneity. *Nature Communications*, 4, 1434.

Röttjers, L. and Faust, K. (2018). From hairballs to hypotheses – Biological insights from microbial networks. *FEMS Microbiology Reviews*, (August).

Rutgers, M., Schouten, A.J., Bloem, J., Van Eekeren, N., De Goede, R.G.M., Jagers Op Akkerhuis, G.A.J.M., Breure, A.M. (2009). Biological measurements in a nationwide soil monitoring network. *European Journal of Soil Science*, 60(5), 820–832.

Rutgers, M., van Leeuwen, J.P., Vrebos, D., van Wijnen, H.J., Schouten, T., de Goede, R.G. M. (2019). Mapping soil biodiversity in Europe and the Netherlands. *Soil Systems*, 3(2), 39.

Schleuning, M., Fründ, J., García, D. (2015). Predicting ecosystem functions from biodiversity and mutualistic networks: An extension of trait-based concepts to plant-animal interactions. *Ecography*, 38(4), 380–392.

Shankar, V., Homer, D., Rigsbee, L., Khamis, H.J., Michail, S., Raymer, M., Paliy, O. (2015). The networks of human gut microbe – Metabolite associations are different between health and irritable bowel syndrome. *The ISME Journal*, 9(8), 1899–1903.

Sunagawa, S., Coelho, L.P., Chaffron, S., Kultima, J.R., Labadie, K., Salazar, G., Bork, P. (2015). Structure and function of the global ocean microbiome. *Science*, 384, 1261359.

Székely, A.J. and Langenheder, S. (2014). The importance of species sorting differs between habitat generalists and specialists in bacterial communities. *FEMS Microbiology Ecology*, 87(1), 102–112.

Tecon, R. and Or, D. (2017). Biophysical processes supporting the diversity of microbial life in soil. *FEMS Microbiology Reviews*, 41(5), 599–623.

Terrat, S., Horrigue, W., Dequietd, S., Saby, N.P.A., Lelièvre, M., Nowak, V., Ranjard, L. (2017). Mapping and predictive variations of soil bacterial richness across France. *PLoS ONE*, 12(10).

Terrat, S., Djemiel, C., Journay, C., Karimi, B., Dequiedt, S., Horrigue, W., Ranjard, L. (2020). ReClustOR: A re-clustering tool using an open-reference method that improves operational taxonomic unit definition. *Methods in Ecology and Evolution*, 11(1), 168–180.

Torsvik, V. and Øvreås, L. (2002). Microbial diversity and function in soil: From genes to ecosystems. *Current Opinion in Microbiology*, 5(3), 240–245.

Van Der Heijden, M.G.A., de Bruin, S., Luckerhoff, L., Van Logtestijn, R.S.P., Schlaeppi, K. (2016). A widespread plant-fungal-bacterial symbiosis promotes plant biodiversity, plant nutrition and seedling recruitment. *ISME Journal*, 10(2), 389–399.

Vivant, A.L., Garmyn, D., Maron, P.A., Nowak, V., Piveteau, P. (2013). Microbial diversity and structure are drivers of the biological barrier effect against listeria monocytogenes in soil. *PLoS ONE*, 8(10), 1–11.

Whittaker, R.H., Levin, S.A., Root, R.B. (1975). On the reasons for distinguishing "Niche, habitat, and ecotope". *The American Naturalist*, 109(968), 479–482.

Yao, H., Campbell, C.D., Chapman, S.J., Freitag, T.E., Nicol, G.W., Singh, B.K. (2013). Multi-factorial drivers of ammonia oxidizer communities: Evidence from a national soil survey. *Environmental Microbiology*, 15(9), 2545–2556.

8

Fungal Biogeography

Tarquin NETHERWAY[1] and Mohammad BAHRAM[1,2]
[1] *Swedish University of Agricultural Science, Uppsala, Sweden*
[2] *University of Tartu, Estonia*

8.1. Introduction

Fungi are ubiquitous and extremely diverse heterotrophic eukaryotic organisms existing across terrestrial and aquatic ecosystems, where they bridge the divide between micro- and macro-organisms, existing as single celled yeasts, to multicellular filamentous molds, and complex macro fungi producing structures rivaling plants and animals in terms of size. Thus, they simultaneously defy and conform to historical views of macro and micro-biogeography. These organisms play essential roles in fundamental ecosystem processes, especially carbon and nutrient cycling, by acting as saprotrophs, parasites/pathogens and mutualists (biotrophs) of plants and animals, and antagonists of and mutualists with bacteria (Tedersoo *et al.* 2014). Fungi are considered key decomposers of dead organisms, especially plant material. For example, within the Basidiomycota, one of the largest fungal phyla, several efficient strategies have evolved for breaking down lignin (one of the most abundant recalcitrant biopolymers on Earth) using various exoenzymes. Other non-basidiomycetous fungi decompose chitin and less recalcitrant polymers such as cellulose (the most abundant biopolymer on Earth) and hemicellulose and lignin degradation by-products. The mycorrhizal symbiosis is the most widespread terrestrial biotrophic fungal lifestyle, in which mycorrhizal fungi take up nutrients and provide these to plant partners in exchange for carbon, playing important roles in carbon and nutrient cycling, and plant population and community dynamics (Tedersoo *et al.* 2020). Mycorrhizal fungi have also been implicated as key

Biogeography,
coordinated by Eric GUILBERT.

mediators of biological mineral weathering through their intimate associations with plants and interactions with rhizosphere bacteria (Finlay *et al.* 2020). Fungi represent some of the most devastating plant and animal pathogens, which play a key role in population and community dynamics of the organisms they infect, leading to a loss of host biodiversity in some ecosystems while maintaining host biodiversity in others.

Traditionally, studies on fungi relied on culturable species or on observable reproductive structures, which led to different life cycle stages (i.e. anamorph and teleomorph) of the same species being designated as taxonomically different entities (Liu *et al.* 2001). Furthermore, these structures do not accurately reflect environmental fungal communities (e.g. Gardes and Bruns, 1996; Dahlberg *et al.* 1997; Horton and Bruns, 2001), because some groups do not produce (e.g. *Cenococcum*) or form conspicuous reproductive structures (e.g. Thelephorales and Sebacinales), and most fungi are not easily culturable. The early investigations of mycorrhizal fungi were also based on the morphology and anatomy of mycorrhizal structures (Agerer, 1987–2002). Aside from the potential to neglect morphologically similar species, this method is time-consuming.

The rapid development of high-throughput sequencing techniques has recently facilitated pioneering studies of fungi that have shed light on the biodiversity of fungi and the evolution of genes involved in symbiosis, decomposition and pathogenicity (e.g. Martin *et al.* 2010; Floudas *et al.* 2012; Kohler *et al.* 2015; Nilsson *et al.* 2019). These technological and methodological advancements have also facilitated pioneering biogeographic studies of fungi that were otherwise limited due to the cryptic nature of fungi, and have provided evidence for the importance of both stochastic (i.e. dispersal limitation) and deterministic (i.e. environmental filtering) factors in shaping the biogeographic patterns of fungi (Tedersoo *et al.* 2014; Bahram *et al.* 2015). This contrasts with the long-held paradigm for microbes, including fungi, of "everything is everywhere, but the environment selects" (Baas-Becking, 1934). Yet, due to the infancy of our recent and increasing ability to overcome the technological and methodological limitations of studying fungal ecology, we still know relatively little about their biogeography compared to plants and animals, and whether or not macroecological theories can directly be applied to fungi. Thus, fungal biogeography is an exciting and rapidly improving field that is essential for understanding ecosystem processes and the effects of global environmental change.

While the unique characteristics of fungi (e.g. unique growth habits and reproductive strategies) partially explain their success in nearly all habitats on earth, it is the leveraging of these characteristics through their associations with other organisms and their resources (e.g. ecological strategies as saprotrophs, and plant and animal mutualists and pathogens) that has truly enabled them to become central agents in ecosystem processes. Here, we provide an up-to-date account of fungi and the factors driving their biogeographic patterns, focusing mainly on terrestrial fungi and in particular mycorrhizal fungi due to their more established and studied biogeographic patterns, but we also give some insights into the biogeography of aquatic and animal-associated fungi.

8.2. Fungal evolutionary history

Fungi, historically considered as part of the plant kingdom, form one of the major eukaryotic lineages and belong to their own monophyletic kingdom (Whittaker, 1969). Fungi are more closely related to animals than plants, sharing a common ancestor within the supergroup Opisthokonta, who then share a deeper common ancestor with plants (Nagy *et al.* 2017). The last common ancestor of plants, animals and fungi is thought to have occurred close to around 1,600 million years ago (Ma) (Wang *et al.* 1999), and the split of fungi and animals likely occurred around 800–1,000 Ma (Doolittle *et al.* 1996; Berbee and Taylor, 2010). The common ancestor of fungi is thought to be a unicellular aquatic flagellated (zoosporic) organism, a form still evident in members of the early diverging lineages of Rozellomyceta, Aphelidiomyceta, Blastocladiomyceta, Chytridiomyceta and Olpidiomyceta, which together form the zoosporic fungi (James *et al.* 2006; Spatafora *et al.* 2016; Tedersoo *et al.* 2018). These early diverging aquatic fungi appeared to be endoparasites in other eukaryotic hosts and relied heavily on their hosts for primary metabolism, and some extant members of these groups represent modern plant and animal parasites and pathogens (Berbee *et al.* 2017; Strullu-Derrien *et al.* 2018). Splitting off from the zoosporic fungi was the early diverging (> 700 Ma) non-flagellated fungi from the lineages Basidiobolomyceta, Zoopagomyceta, Mucoromyceta, followed by the more recently (some 642 Ma) diverging Dikarya, which is by far the most diverse fungal lineage (Tedersoo *et al.* 2018). These later waves of fungal diversification were associated with land colonization events, the development of the multicellular hyphal form, which is now the most common fungal form, and the diversification of feeding strategies such as saprotrophy and symbiotrophy, which also coincided with diversification and land colonization events in other eukaryotic groups such as plants and animals (Berbee *et al.* 2017; Minter *et al.* 2017; Lutzoni *et al.* 2018). It is important to understand the evolutionary history of fungi because this could explain some of their biogeographical patterns, especially with reference to symbioses.

8.3. Biogeographic patterns

Understanding biogeographic patterns can inform about the relative role of different processes shaping the distribution of organisms (Vellend, 2010). Under the niche-neutral continuum hypothesis – which is a well-recognized and widely accepted framework in the field of biogeography – such processes can be categorized into niche-based and neutral models (Hubbell, 2001; Chase and Leibold, 2003). Niche-based processes include environmental filtering and niche partitioning – resulting from contrasting habitat preferences and fitness between – whereas stochastic drift and limited dispersal are constituents of neutral processes. Understanding the relative role of these different processes is a central challenge in fungal biogeography.

Regardless of whether they employ sexual or asexual reproduction, most fungi produce spores on which they rely greatly for dispersal via a number of dispersal vectors. One of the principal dispersal vectors of fungal spores are air currents, and fungal spores can make up a large proportion of biogenic aerosols in the atmosphere. In addition to dispersing, fungi may play an important role in physical and chemical processes in the atmosphere such as acting as condensation nuclei and thus directly influencing precipitation (Hassett *et al.* 2015). Furthermore, airborne fungal communities also appear to display distinct vertical biogeographic patterns (Els *et al.* 2019). Most dispersal of fungal spores happens over short time scales (< 1 year) and short spatial scales (< 1 m), yet it is the dispersal over much larger time and spatial scales that drives distinct large-scale biogeographical patterns and processes (Peay *et al.* 2016). Although fungal endemism is low at the level of the operational taxonomic unit (OTU) (Tedersoo *et al.* 2014; Davison *et al.* 2015), at the population level, fungi are well adapted to their local environment as well as their interactive taxa, as can be reflected in their genomic diversity (Ellison *et al.* 2011). Recently, a number of studies have advocated strain and sequence variant level analysis to gain insights into biogeography and underlying drivers of diversity in microbes, including fungi (e.g. Branco *et al.* 2015; Wilson *et al.* 2017; Hartmann *et al.* 2018). Taken together, growing evidence suggests that indeed biogeographical signals on large scales are more pronounced at lower taxonomic levels.

Although much has been reported on the distribution patterns and underlying processes of plants and animals, less has been reported concerning fungi. In the following sections, we discuss some of the more studied fungal biogeographic patterns that are commonly studied for plants and animals.

8.3.1. *Distance-decay of similarity and species area relationship*

A central relationship in the study of biogeography is the distance–decay relationship, which refers to the spatial turnover of species that leads to a decay in the similarity of communities with increasing geographic distance (Morlon *et al.* 2008), as well as the species area relationship that predicts that species richness increases with increasing area (Lomolino, 2000). Most of the scant studies on distance-decay of fungal communities have been on soil fungi. It has been shown that soil fungi experience significant distance-decay over distances as small as meters (Schmidt *et al.* 2013). However, neutral processes dominate in homogeneous environments and in the absence of strong environmental gradients and dispersal limitation (Bahram *et al.* 2016). With a lack of dispersal barriers, a wide distribution range may lead to the homogenization of fungal communities at the fine scale (Bahram *et al.* 2016), similarly to what is expected for plants and animals (Vellend 2010). This could overwhelm the effect of deterministic factors and, thus, lead to weak distance–decay relationships (Mouquet and Loreau, 2003). For ectomycorrhizal (EcM) fungi, distance-decay processes are enhanced in tropical regions, and they decrease with increasing distance from the equator, as well as with increasing host density, hinting at limitations to dispersal and establishment (Bahram *et al.* 2013).

Dispersal limitation and competition have also been used to explain species area relationships for EcM fungi that are equivalent to those observed for macro-organisms, that is, decreasing species richness with decreasing habitat island size (Peay *et al.* 2010). Similar processes have been hinted at for wood inhabiting fungi, where species richness increases with the amount and heterogeneity in sizes of dead wood, with specialist species suffering more from habitat fragmentation compared to generalist species (Nordén *et al.* 2013). For marine fungi, while much less explored than terrestrial fungi, environmental factors – that is, deterministic processes – appear to explain community differences more than geographic distance does; thus, they appear to experience less dispersal limitation compared to most terrestrial fungi (Tisthammer *et al.* 2016). While distance-decay and species area relationships are often viewed from the perspective two-dimensional geographic space, these processes also occur across three-dimensional space, that is, across depth gradients, which for fungi can include different soil depths and layers, water depths and sediment layers, different heights of the phyllosphere (surfaces of above ground vegetation), and different depths of decaying wood. Fungi can exhibit significant community structuring across depth (Bahram *et al.* 2015), with vertical segregation of different fungal guilds (McGuire *et al.* 2013).

At local scales, the balance between species formation, competitive exclusion and adaptation structures fungal communities, whereas history, climate, evolutionary processes and extinction as well as dispersal limitation play important roles in structuring metacommunities at larger (i.e. regional to global) scales (Ricklefs, 1987). Therefore, the underlying processes and the relative role of spatial related processes may differ over various geographical scales (Wiens, 1989; Levin, 1992; Vellend, 2010). The importance of spatial scale and proximity in fungal studies is increasingly recognized (e.g. Bahram *et al.* 2016; Beck *et al.* 2019).

8.3.2. *Latitudinal diversity patterns*

The latitudinal gradient of diversity (LDG) has long been well established for many if not most marine and terrestrial organisms, that is, tropical regions harbor the greatest diversity of species, which then decreases towards the poles, and this remains one of the most fundamental patterns of ecology (Willig *et al.* 2003). The explosion of molecular studies of fungal diversity over the last decade have revealed some interesting and counterintuitive patterns of fungal diversity as it relates to the LDG. Soils are thought to harbor the highest diversity of fungi, and global studies on soil inhabiting fungi have revealed that distance to equator and mean annual precipitation have the strongest effect on fungal species richness, and most fungal classes, including most saprotrophic fungi, fungal endophytes, and fungal plant and animal pathogens follow the patterns of other organisms with diversity peaking in tropical regions and decreasing towards the poles (Tedersoo *et al.* 2014). In contradiction to the general LDG pattern, EcM fungi show higher diversity in temperate and boreal forests compared to tropical forests (Tedersoo and Nara, 2010; Tedersoo *et al.* 2012). Yet, this finding may not be surprising given that the diversity of EcM host tree species is also lower in the tropics compared to temperate regions, and EcM tree basal area increases from tropical to temperate and boreal forests. Several factors have been suggested to drive the temperate peak in diversity for EcM, such as higher species diversification rates, greater soil heterogeneity, higher diversity of host plant lineages, higher density of hosts and greater habitat area (Kennedy *et al.* 2012; Tedersoo *et al.* 2012). These observations about EcM and saprotrophic fungal richness and the LDG from molecular studies have also been observed in a large spatial and temporal study of fruit body collection records (Andrew *et al.* 2019). A non-soil inhabiting fungal group to show a similar pattern to EcM fungi in regard to the LDG is aquatic leaf litter fungi, who also appear to show a hump-shaped LDG distribution (Jabiol *et al.* 2013; Seena *et al.* 2019). Indoor fungi – dominated mostly by fast growing molds that rely on airborne dispersal – also show a similar LDG trend peaking in diversity in the temperate zone and are range limited and unrelated to building type, which hints at the reduced importance of deterministic factors in structuring certain fungal communities (Amend *et al.* 2010).

8.3.3. *Altitudinal diversity patterns*

Next to latitudinal gradients, altitudinal gradients are well-known systems in biogeographical studies, which provide complementary insights into the relative effects of contemporary (i.e. climatic, edaphic and biotic) factors on the underlying mechanisms of diversity (Rahbek, 2005). Large-scale studies of biodiversity along these gradients can further inform about the response of communities to future climate change (Parmesan and Yohe, 2003; Harley, 2011; Nogués-Bravo and Rahbek 2011). Across altitudinal gradients, temperature changes abruptly and therefore montane ecosystems provide a suitable model to address the LDG across a smaller scale. While earlier studies on fungi provided evidence that the fungal altitudinal diversity gradient mirrors the LDG, that is, it decreases towards higher altitudes, which indicates that climate is the main underlying factor (Bahram *et al.* 2012), some other studies show that it may also reflect mid-domain effects (Miyamoto *et al.* 2014). Some studies have shown that fungal diversity tends to show a hump shaped pattern, where species diversity reaches its highest levels at mid-altitudes (Sanders and Rahbek, 2012; Miyamoto *et al.* 2014). This pattern mirrors the pattern observed for EcM fungal diversity across latitudes (Tedersoo and Nara 2010). Long-term fruit body studies have also shed light on the response of certain macrofungal assemblages to climate change across altitudinal gradients. These studies show that both EcM and saprotrophic fungi show similar upwards expansions in fruiting patterns, and that these expansions are most pronounced in communities at the higher edge of the altitudinal range compared to mid- and low-altitude communities (Andrew *et al.* 2018a; Diez *et al.* 2020). Other fungal guilds may show various trends in relation to altitude depending on the ecosystem studied as well as the range of altitude. For example, the diversity of aquatic fungi has been shown to follow no significant trend across altitude (Yeh *et al.* 2018).

8.4. Functional and interactional biogeography of fungi

Taken together, studies based on fungal taxonomic/phylogenetic diversity and community structure suggest that biogeographic patterns appear to be not as clear-cut for fungi as those for animals and plants. In addition, they seem to largely depend on fungal functional guilds, their interactions with other microorganisms as well as their host animal or plant assemblages, within which there is also large variation. In the following section, we discuss how fungal functions and their interactions with other organisms can affect fungal biogeography.

8.4.1. *Functional biogeography of fungi*

Most of our understanding about fungal biogeography is based on species alpha and beta diversity, and there remains a long-standing debate about whether or not species diversity is a useful predictor of ecosystem functioning. Whereas, functional diversity, that is, the diversity of functional traits across different levels of organization, has remained little studied. A functional trait can be considered a feature relevant to the morphological, physiological or phenological behavior of an individual fungus that can influence its ecological performance/fitness (Dawson *et al.* 2019). Functional biogeography is a promising approach that bridges species distribution, their traits, environment and ecosystem properties and can improve our understanding of ecosystem functioning and conservation programs (Violle *et al.* 2014). Recent improvement in databases such as FunGuild (www.funguild.org), FunFun (www.funfun.io) and FungalTraits (Põlme *et al.* 2021) hold great promises for facilitating the functional annotation of fungal communities, that is, assigning taxa to a functional guild such as saprotrophic, pathogenic/parasitic and mutualistic. In addition, metagenomic and metatranscriptomic approaches (to study all potential functional genes and expressed genes in environmental samples, respectively) have become accessible more than ever in the last decade, with great implications for studying fungal functional biogeography at an unprecedented scale and resolution. By leveraging these tools, several studies have already reported great matches as well as mismatches between the biogeography of fungal functions and taxa (Fernandez *et al.* 2016; Bahram *et al.* 2018, 2020). Functional metagenomics can further our understanding of fitness, which in turn helps predict organism abundances according to the maximum entropy approach (Shipley *et al.* 2006; Laughlin *et al.* 2012). Including traits also greatly improves the modeling of biogeographic patterns under global environmental change (Sunday *et al.* 2015).

In addition, an increasing number of genetic, transcriptomic and genomic studies on fungi provides comprehensive characterization of functional gene families and activity centers of enzymes, transporters, signaling molecules, transposable elements and so on. The metabolite pathways and secondary metabolites reflect evolutionary adaptation and vary between obligate biotrophs and saprotrophs, with variation in genome size and reductions in certain gene families, particularly transporters and plant cell wall degradation enzymes (Martin *et al.* 2010; Spanu *et al.* 2010). By examining the metagenomes and metatranscriptomes of fungal communities, and the functional information contained within, across different dimensions of geographical space and time, then we can begin to unpack the biogeography of fungi, their interactions with other organism groups, and their central roles in carbon and nutrient cycling.

8.4.2. *Interactional biogeography of fungi and plants*

Both early and contemporary associations and interactions, including antagonisms, mutualisms and commensalisms, between land plants and fungi represent defining relationships that contributed greatly to extinct and also extant fungal and plant biogeography (Berbee *et al.* 2017; Lutzoni *et al.* 2018; Delavaux *et al.* 2019). Interactions between plants and fungi have been a strong evolutionary pressure that drives contemporary terrestrial biogeochemical cycles as well as plant and fungal community dynamics.

Multiple nutritional modes have evolved several times in fungi, which enables replicating the evolutionary trait shifts from presumably saprotrophic ancestors to both symbiotic and pathogenic forms. For example, plant-parasitic and saprotrophic fungi show a higher number of plant cell wall degrading enzymes compared to beneficial mutualistic fungi. It is thus plausible that mechanisms for evading plant defense systems are common between saprotrophs and parasites, and the division between saprotrophic and parasitic behavior is not so clear (Martin *et al.* 2011). The highly complex genome organization provides the ability of some fungi to exhibit multiple modes of trophic behavior during their life history, such as pathogens switching to saprotrophs upon death of a host plant. Genome-based studies have shown that EcM fungi and pathogens may require specific combinations of genes for establishing relationships with their host plants. EcM lineages have a reduced set of genes involved in the degradation of lignin and cellulose, including class II peroxidases (PODs) and carbohydrate-active enzymes (CAZymes) (e.g. Martin *et al.* 2010; Kohler *et al.* 2015). Instead, EcM fungi have gained a suite of genes involved in inducing the mycorrhizal symbiosis (i.e. mycorrhiza-induced small secreted proteins, MiSSP7) to suppress plant immune response by blocking mycorrhiza-formation compounds such as jasmonic acid, produced to prevent fungal growth in plant cells (Martin *et al.* 2010; Plett *et al.* 2014).

Mycorrhizal symbioses are associated with enhanced plant nutrient acquisition and biotic and abiotic stress tolerance, thus greatly contributing to plant health, fitness and survival (Smith and Read, 2010). One leading hypothesis is that symbiosis with endosymbiotic arbuscular mycorrhizal (AM) fungi from the phylum Glomeromycota, that evolved some 450 Ma, was a key mechanism through which plants were able to colonize land in the absence of an extensive root system. In addition, the AM symbiosis has implications for fungal fitness, as they never developed the ability to photosynthesize and access atmospherically derived carbon (C) (Brundrett, 2002; Bonfante and Genre, 2008; Lutzoni *et al.* 2018). Thus, early land colonization events by plants and fungi that happened rapidly were facilitated by symbiotic relationships (Berbee *et al.* 2017; Lutzoni *et al.* 2018). As early terrestrial habitats were extremely hostile to life, only through coevolutionary symbioses between unrelated organisms could different organism groups overcome such a hostile environment without relying on otherwise slow and inefficient random mutations (Pirozynski and Malloch, 1975). Today, mycorrhizal associations are geographically widespread with multiple distinct types that contribute greatly to plant nutrient dynamics and may play a central role in plant community dynamics and plant biogeography (Tedersoo *et al.* 2020).

There is also evidence that contrasting processes may underlie the biogeographic patterns of fungi from different functional guilds. Perhaps one of the most obvious and interesting comparisons of fungal dispersal and biogeography is between AM and EcM fungi; both are obligate symbionts requiring colonization of a host plant for C acquisition (Figure 8.1). AM fungi are a species poor group with limited evidence of sexual reproduction, producing large asexual resting spores thought to be not well adapted for long-distance dispersal via wind, yet they associate with the majority of land plants. By contrast, EcM fungi are a phylogenetically diverse and species rich group, in which many members produce sexual spores that are relatively small in size and thought to be well adapted for long-distance dispersal by wind. Yet, EcM fungi associate with a small number of plants, mostly woody and across a limited phylogeny. These contrasts seem to have interesting consequences for long-distance dispersal, which requires the production and release of viable propagules, long-distance transport, maintenance of propagule viability, landing in a suitable habitat, establishment in said habitat and completion of the life cycle to reproduction (Golan and Pringle, 2017). Yet, many EcM fungi have been shown to be dispersal limited, with around 96% of their spores potentially landing within one meter of where they are released (Galante *et al.* 2011). In addition, the richness and quantity of EcM spores decreases with increasing distance from habitats dominated by EcM host vegetation (Peay *et al.* 2012). Nevertheless, in other instances, EcM fungi have been found to be capable of cross continental dispersal (Geml *et al.* 2012; Tedersoo *et al.* 2014), suggesting that a combination of dispersal limitation due to distance and host availability, together with environmental filtering, contribute to EcM community assembly (Geml *et al.* 2008). Compared to EcM and saprotrophic fungi, AM fungi with large spores – that should be relatively more dispersal limited – appear to experience much less spatial turnover and endemism than other fungal guilds and their host plant taxa, suggesting efficient dispersal mechanisms with their cosmopolitan distributions, a lack of competition and barriers to establishment or a failure to properly capture their true taxonomic diversity (Davison *et al.* 2015). One such dispersal mechanism may be airborne spore dispersal, which has likely been underestimated for AM fungi. The exact mechanism driving the liberation of AM fungal spores into the air remains unknown (Chaudhary *et al.* 2020), although one such dispersal vector is thought to be migratory birds, including the joint dispersal of plant host seeds and AM fungal spores (Nielsen *et al.* 2016; Correia *et al.* 2019).

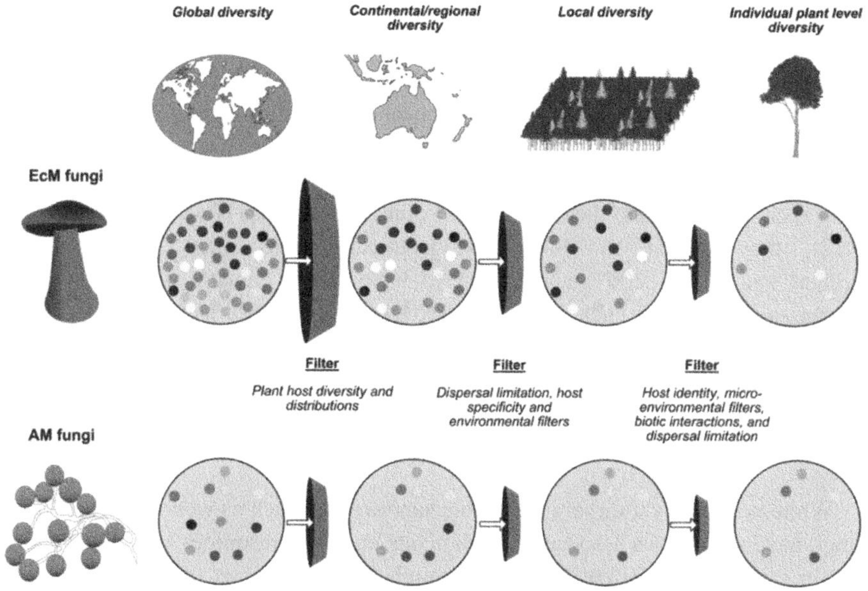

Figure 8.1. *Schematic diagram exploring the potential filters driving the contrasting biogeographical patterns of species diversity of ectomycorrhizal (EcM) and arbuscular mycorrhizal (AM) fungi across different geographic scales. Colored dots represent fungal species diversity and the size of the filter is relative to its proposed role in explaining changes in mycorrhizal diversity patterns across scales. On a global scale, EcM fungi are a diverse group of species with multiple evolutionary origins, while the diversity of their plant hosts is limited to certain lineages of mostly woody plants. By contrast, AM fungi are a species poor group that evolved once coinciding with the colonization of land by plants, and associate with the majority of land plants from all major lineages. On a continental/regional scale, biogeographic patterns are less pronounced for AM fungi due to their low host specificity, high host diversity and global distributions of hosts, whereas distinct biogeographic patterns are expected for EcM fungi with higher host specificities, lower host diversity and distinct biogeographic patterns of their hosts. Moving down to the level of local biomes/ecosystems, we begin to see further biogeographic structuring of EcM and to a lesser extent AM fungi as dispersal limitation, host specificity and abundance along with environmental filters such as climate begin to play a stronger role. On the individual plant level, plant identity, microclimatic and edaphic factors act as strong environmental filters and dispersal limitations become evident for both EcM and AM fungi. For a color version of this figure, see www.iste.co.uk/guilbert/biogeography.zip*

Similarly to mycorrhizal fungi, plant-pathogenic fungi rely on host plants for resources and have been suggested to play key roles in plant population and community dynamics, patchiness, succession and evolutionary segregation (Zadoks, 1987). Plant-pathogenic fungi and plants have been locked in an evolutionary arms race, where fungal virulence and plant resistance genes are highly variable due to constant and strong selection pressures, and co-evolutionary relationships between single fungal pathogen and plant taxa appear to be much stronger in agricultural systems compared to complex interaction networks in natural systems (Karasov *et al.* 2014). Plant-pathogenic fungi tend to have large and flexible genomes that allow them to respond to the development of plant resistance (Raffaele and Kamoun, 2012); they appear to show phylogenetically constrained host specificity (Gilbert and Webb, 2007) and appear to have larger climatic niches than other plant-associated guilds such as EcM (Větrovský *et al.* 2019), with precipitation and land-use as potentially key factors driving their diversity, abundances and distributions. Yet, there remain significant barriers to studying biogeography of plant-pathogenic fungi related largely to the difficulty of determining their pathogenicity, which can be highly variable between fungal strains.

While not strictly a direct interaction, saprotrophs, such as wood and litter decay fungi, have an obvious association with plants and their resources and play a central role in carbon and nutrient cycling. Through their unique set of exoenzymes they can dismantle organic polymers more effectively than other organism groups. The evolution of wood decay mechanisms in fungi fundamentally altered and maintained global carbon and nutrient cycling following the evolution of secondary growth in plants in the form of thickened lignocellulolytic cell walls during the Carboniferous period (Eastwood, 2014). Through their affinity for plant-derived carbon, saprotrophs are thus sensitive to factors affecting plant biogeography. In particular, saprotrophs are influenced by plant-traits that influence the nutritional quality of wood, litter and soil organic matter. Strong differences occur in these resources between gymnosperms and angiosperms, as well as between EcM and AM forming plants, which lead these plant-traits to show distinct biogeographic distributions (Wang and Ran, 2014; Zanne *et al.* 2018; Soudzilovskaia *et al.* 2019). This can be exemplified by the two major modes of wood decay: white rot, the degradation of all plant cell wall components, and brown rot, the preferential decay of cellulose and hemicellulose with minor modification of lignin. These groups show contrasting affinities for gymnosperm and angiosperm wood, where brown rot fungi can be generalists or gymnosperm specialists, while white rot fungi are angiosperm specialists (Krah *et al.* 2018). Furthermore, many wood-decay fungi are considered threatened species, requiring a diversity of large woody debris in terms of size class distribution and tree species. As a result, wood-decay fungi have been impacted heavily by intensive forest management, and many wood-decay fungi are used as surrogates or indicator species of forest habitat quality (Halme *et al.* 2017; Nordén *et al.* 2020).

On a global scale, it appears that the distribution and diversity of fungi is largely linked to plants and their organic residues, with fungal diversity peaking in plant-dominated habitats such as soils, and this is driven by their tightly intertwined coevolutionary history (Lutzoni *et al.* 2018; Bahram *et al.* 2021). Thus, linking vegetation and fungal co-distributions and ranges with alterations to land-use, temperature and precipitation patterns will greatly improve our knowledge of how both of these major organism groups will respond under continued and accelerated global change (Andrew *et al.* 2018a), with obvious implications for ecosystem functioning.

8.4.3. *Interactional biogeography of fungi and animals*

Fungi are a key component of animal microbiomes, animal diets and animal diseases; on the contrary, animals serve as a major dispersal agent for certain fungi (Elliott *et al.* 2019). Fungi are a key food resource both above and belowground for various animals. In soils, fungal biomass and necromass can make up a large proportion of total living biomass and litter, key resources for soil fauna such as mites, collembola and nematodes. These fungivores then place a grazing pressure on fungal mycelium and also consume fungal spores, altering the growth, development and activity on fungi, which in turn can alter the structure and function of fungal communities (Ruess and Lussenhop, 2005). Through consumption and deposition as well as passive transport, soil animals can be a major dispersal agent of fungi through and across soil profiles (Bray and Wickings, 2019; Anslan *et al.* 2016). Aboveground and across larger geographic scales, along with wind, animals are major active and passive long- and short-distance dispersal agents for many fungi, especially saprotrophic and EcM ascomycetes and basidiomycetes that produce fruit bodies, which are an attractive food source for a wide variety of animals of different sizes and movement ranges (Elliot *et al.* 2019). The morphological and physiological traits of fungal fruiting bodies of many fungi seem to be shaped by a trade-off between protection against and attraction for mycophagy by animals (Halbwachs *et al.* 2016). In addition, there are many well-documented fungal–animal mutualisms such as ambrosia beetle–fungus symbioses, and other fungal–farming insect symbioses, which can be a major mechanism by which saprotrophic and plant pathogenic fungi disperse to new substrates and hosts (Hulcr and Stelinski, 2017). Another example is the human cultivation of fungi and plants, which has increased the geographical distribution of certain fungal species, albeit mainly in human land-use associated habitats (Bazzicalupo *et al.* 2019).

On the other end of the spectrum is fungal parasitism of animals, with the most striking example being zombie ant fungi from the species complex *Ophiocordyceps unilateralis*, that manipulate the behavior of their hosts to facilitate reproduction and dispersal. Yet, in general fungi are a common component of animal microbiomes, acting as mutualists through to commensal and parasites, and existing in complex communities. Taking humans as an example, there exists distinct biogeographical patterns of fungal communities inhabiting the human body depending on body sites/organs and host health status (Cui *et al.* 2013), which appear to be driven by microenvironmental filtering (Oh *et al.* 2014). Although, some dispersal limitations may exist, where externally exposed body sites share more fungal taxa with environmental habitats, such as soil, compared to internal body sites (Bahram *et al.* 2021). The effect of land-use change on the composition and diversity of soil fungal communities is potentially evident in skin-associated mycobiomes, although large variations can occur between individuals due to both stochastic and deterministic processes (Barnes *et al.* 2020). Ultimately, animal-associated fungal communities appear to be a subset of associated environmental fungal communities, with the greatest reservoir of diversity in soils (Bahram *et al.* 2021). It is tempting to speculate that factors affecting the biogeography of soil fungi such as climate and land-use patterns are likely to also explain the large-scale biogeographic patterns of animal-associated fungi.

8.4.4. *Interactional biogeography of fungi and bacteria*

Fungi are involved in intimate interactions with bacteria in every habitat they inhabit, often competing for resources, as well as engaging in symbioses ranging from mutualistic to parasitic (Frey-Klett *et al.* 2011). While fungi and bacteria appear to display opposing biogeographic patterns and community assembly processes, the direct interaction between the two organism groups may also be a major driver of these patterns and processes. For example, bacterial community assembly appears to be driven more by environmental filtering, and the greatest environmental filter appears to be pH, while fungi appear to be less sensitive to pH than bacteria, although certain fungal guilds may be more sensitive than others (Bahram *et al.* 2018).

In highly heterogeneous habitats where water, carbon resources and nutrients are unevenly distributed, the ability of fungi to form hyphae and connect resource rich patches, translocate resources and create microsites around their hyphae may facilitate certain bacterial communities in these microsites and serve as dispersal bridges between habitats for bacteria (Worrich *et al.* 2017). On the contrary, bacteria can play a key role in the development and functioning of certain fungi, as exemplified by the important role of mycorrhizal helper bacteria in stimulating formation of the EcM symbiosis and contributing to its functioning (Frey-Klett *et al.* 2007). Another interesting case is in the human microbiome, where *Candida albicans,* a dimorphic fungus and opportunistic pathogen that can switch from a yeast phase to a hyphal phase, where the hyphal phase is associated with pathogenesis. This dimorphic switching of *C. albicans* can be induced by the opportunistic pathogenic bacterium *Pseudomonas aeruginosa* through the release of quorum-sensing molecules, leading to a co-infection, where *P. aeruginosa* likely takes advantage of the hyphal forming ability of *C. albicans* (Hogan *et al.* 2004).

In addition, recent evidence suggests that competitive and antagonistic interactions between fungi and bacteria contribute to their broad-scale biogeographical patterns. For example, the abundance of antibiotic resistance genes, which play important roles in bacterial, ecological and evolutionary processes, have been directly linked to fungi in global soil and ocean metagenomes, suggesting cross-kingdom communication and co-evolution of bacterial and fungal communities (Bahram *et al.* 2018). Whereas, the biomass ratio of bacteria to fungi appears to be negatively correlated to the abundance of fungal carbohydrate active enzyme genes, especially those related to the decomposition of lignin, suggesting that fungi outcompete bacteria for carbon resources in lignin rich habitats (Bahram *et al.* 2021).

8.5. Fungal biogeography under global environmental change

Global environmental change driven by human disturbance, climate change and natural hazards can fundamentally alter the distribution and activity of fungi in time and space, while fungi can buffer or exacerbate the effects of global change on other organism groups. Global warming and alterations to global precipitation patterns directly influence ecological processes that drive biogeographic patterns. For example, since the year 1950 in the UK, climate change has led to an early onset of and increased fruiting periods of the wood-decay fungus *Auricularia auricula-judae.* In turn, this species simultaneously increased its host-range, presumably through an increased combative ability relative to other wood decay species by altering germination and growth rates (Gange *et al.* 2011). This situation is likely to extend to many other macro fungi over large spatial scales, whose fruiting phenology is driven largely by climatic parameters, and in particular, changes in temperature (Andrew *et al.* 2018b). Furthermore, the dismantling of geographic barriers by an interplay of land-use change, climate change and human trade over the last 200 years has facilitated the spread of invasive forest pathogenic fungi throughout Europe (Santini *et al.* 2013). And growing evidence suggests that AM fungi, the most widely distributed plant mutualists, may become less mutualistic and more parasitic with multiple coinciding change factors, in the process also altering the taxonomic composition of AM fungal communities (Weber *et al.* 2019). These are just some examples of the response of fungi to global change. Just as the evolutionary histories of fungi and other organism groups, particularly plants, have been tightly intertwined as have their patterns of biogeography, the response of fungi and other organisms to future global environmental change will also be tightly intertwined and novel biogeographic patterns will arise.

8.6. The role of citizen science in the study of fungal biogeography

We have mainly focused on the utilization of the constantly improving molecular methods available to researchers in answering questions relevant to fungal ecology and biogeography. However, this does not negate the importance of traditional morphological studies of macrofungal biogeography, in which citizen science, facilitated by online and open-source data depositories, has played a central role in collecting some of the largest geographic and temporal datasets in fungal biogeography to date (Andrew *et al.* 2018a, 2019). Citizen science can also play a key role in informing the emerging field of conservation mycology (May *et al.* 2018), overall fungal biodiversity patterns and the discovery of new species (Heilmann-Clausen *et al.* 2019). In addition, molecular citizen science holds great promises for large-scale mycological and microbiological spatial and temporal studies of fungi. This can involve enlisting the public for regional and global scale sampling efforts such as FunHome (www.sisu.ut.ee/funhome) for fungi in house dust and FunLeaf (www.sisu.ut.ee/funleaf) for leaf-associated fungi, which can be further molecularly analyzed. While the continued dismantling of cost and knowledge barriers to undertaking amateur molecular mycological studies will lead to unprecedented data collection, the active engagement of professional researchers will be needed to best use the data to address some of the ongoing questions in fungal biogeography.

8.7. Future directions

There remains plenty of work to do in order to bring the understanding of fungal biogeography to levels of that accumulated for organisms such as plants and animals. This will be facilitated by mycologists and microbial ecologists testing explicit ecological and biogeographical questions and moving beyond mere descriptive studies towards microbial macroecology (Xu *et al.* 2020). Perhaps the most interesting aspect of fungal biogeography is not in studying species themselves, but rather exploring research questions around the factors driving the distribution of fungal functions and their activity in time and space. Such functional fungal biogeography is now beginning to be possible with omics techniques such as genomics, metagenomics, metatranscriptomics, metaproteomics and metabolomics, where we can begin to explore questions such as: What is the functional potential of different fungal guilds across different spatial and temporal scales? Which functional genes are actively being expressed, and which proteins and metabolites are being produced? Addressing such questions will undoubtedly improve our understanding of the role of fungi in biogeochemical cycles and small- and large-scale ecological processes. Furthermore, the interconnected nature of fungi ecologically and evolutionarily with other organisms implores us to consider fungi through an interactional lens, and to consider that ecological interactions are a strong driving force of biogeographical patterns. For example, a recent large-scale metagenomic analysis of fungi together with bacteria across diverse habitats such as soil, to marine and human organs, hinted at the diverse set of carbohydrate-degrading enzymes as the mechanism by which fungi have been able to form stable and complex communities in plant-associated habitats such as soils (Bahram *et al.* 2021). This study further suggested that fungi may be able to outcompete bacteria for carbon resources in soils, whereas in habitats with less direct plant-derived carbon, bacteria may outcompete fungi, and the ratio of bacteria to fungi may be a distinct habitat feature beyond soils, with functional implications. And last but not least, due to the insurmountable task of collecting sufficient data to ask large-scale biogeographical questions with enough statistical power, there remains a strong role for citizen science in collecting such data and answering such questions, which fungal biogeography as a field has a strong history of doing through online fungal atlases.

8.8. References

Agerer, R. (1987–2002). *Colour Atlas of Ectomycorrhizae*. Einhorn-Verlag, Schwäbisch Gmünd.

Amend, A.S., Seifert, K.A., Samson, R., Bruns, T.D. (2010). Indoor fungal composition is geographically patterned and more diverse in temperate zones than in the tropics. *Proceedings of the National Academy of Sciences*, 107(31), 13748–13753.

Andrew, C., Halvorsen, R., Heegaard, E., Kuyper, T.W., Heilmann-Clausen, J., Krisai-Greilhuber, I., Kauserud, H. (2018a). Continental-scale macrofungal assemblage patterns correlate with climate, soil carbon and nitrogen deposition. *Journal of Biogeography*, 45(8), 1942–1953.

Andrew, C., Heegaard, E., Høiland, K., Senn-Irlet, B., Kuyper, T.W., Krisai-Greilhuber, I., Kauserud, H. (2018b). Explaining European fungal fruiting phenology with climate variability. *Ecology*, 99(6), 1306–1315.

Andrew, C., Büntgen, U., Egli, S., Senn-Irlet, B., Grytnes, J.A., Heilmann-Clausen, J., Kauserud, H. (2019). Open-source data reveal how collections-based fungal diversity is sensitive to global change. *Applications in Plant Sciences*, 7(3), e1227.

Anslan, S., Bahram, M., Tedersoo, L. (2016). Temporal changes in fungal communities associated with guts and appendages of Collembola as based on culturing and high-throughput sequencing. *Soil Biology and Biochemistry*, 96, 152–159.

Baas-Becking, L.G.M. (1934). *Geobiologie; of inleiding tot de milieukunde.* WP Van Stockum and Zoon NV.

Bahram, M., Koljalg, U., Courty, P.E., Diedhiou, A.G., Kjøller, R., Polme, S., Tedersoo, L. (2013). The distance decay of similarity in communities of ectomycorrhizal fungi in different ecosystems and scales. *Journal of Ecology*, 101(5), 1335–1344.

Bahram, M., Peay, K.G., Tedersoo, L. (2015). Local-scale biogeography and spatiotemporal variability in communities of mycorrhizal fungi. *New Phytologist*, 205(4), 1454–1463.

Bahram, M., Kohout, P., Anslan, S., Harend, H., Abarenkov, K., Tedersoo, L. (2016). Stochastic distribution of small soil eukaryotes resulting from high dispersal and drift in a local environment. *The ISME Journal*, 10(4), 885–896.

Bahram, M., Hildebrand, F., Forslund, S.K., Anderson, J.L., Soudzilovskaia, N.A., Bodegom, P.M., Huerta-Cepas, J. (2018). Structure and function of the global topsoil microbiome. *Nature*, 560(7717), 233–237.

Bahram, M., Netherway, T., Hildebrand, F., Pritsch, K., Drenkhan, R., Loit, K., Tedersoo, L. (2020). Plant nutrient-acquisition strategies drive topsoil microbiome structure and function. *New Phytologist*, 227, 1189–1199.

Bahram, M., Netherway, T., Frioux, C., Ferretti, P., Coelho, L.P., Geisen, S., Bork, P., Hildebrand, F. (2021). Metagenomic assessment of the global diversity and distribution of bacteria and fungi. *Environmental Microbiology*, 23(1), 316–326.

Barnes, E.M., Kutos, S., Naghshineh, N., Mesko, M., You, Q., Lewis, J.D. (2020). Assembly of the amphibian microbiome is influenced by the effects of land-use change on environmental reservoirs. bioRxiv [Online]. Available at: https://doi.org/10.1101/2020.11.30.405050.

Bazzicalupo, A.L., Whitton, J., Berbee, M.L. (2019). Over the hills, but how far away? Estimates of mushroom geographic range extents. *Journal of Biogeography*, 46(7), 1547–1557.

Beck, S., Anderson, I.C., Drigo, B., Powell, J.R. (2019). A soil fungal metacommunity perspective reveals stronger and more localised interactions above the tree line of an alpine/subalpine ecotone. *Soil Biology and Biochemistry*, 135, 1–9.

Berbee, M.L. and Taylor, J.W. (2010). Dating the molecular clock in fungi – How close are we? *Fungal Biology Reviews*, 24(1–2), 1–16.

Berbee, M.L., James, T.Y., Strullu-Derrien, C. (2017). Early diverging fungi: Diversity and impact at the dawn of terrestrial life. *Annual Review of Microbiology*, 71, 41–60.

Bonfante, P. and Genre, A. (2008). Plants and arbuscular mycorrhizal fungi: An evolutionary-developmental perspective. *Trends in Plant Science*, 13(9), 492–498.

Branco, S., Gladieux, P., Ellison, C.E., Kuo, A., LaButti, K., Lipzen, A., Taylor, J.W. (2015). Genetic isolation between two recently diverged populations of a symbiotic fungus. *Molecular Ecology*, 24(11), 2747–2758.

Bray, N. and Wickings, K. (2019). The roles of invertebrates in the urban soil microbiome. *Frontiers in Ecology and Evolution*, 7, 359.

Brundrett, M.C. (2002). Coevolution of roots and mycorrhizas of land plants. *New Phytologist*, 154(2), 275–304.

Chase, J.M. and Leibold, M.A. (2003). *Ecological Niches: Linking Classical and Contemporary Approaches*. University of Chicago Press, Chicago, IL.

Chaudhary, V.B., Nolimal, S., Sosa-Hernández, M.A., Egan, C., Kastens, J. (2020). Trait-based aerial dispersal of arbuscular mycorrhizal fungi. *New Phytologist*, 228(1), 238–252.

Correia, M., Heleno, R., da Silva, L.P., Costa, J.M., Rodríguez-Echeverría, S. (2019). First evidence for the joint dispersal of mycorrhizal fungi and plant diaspores by birds. *New Phytologist*, 222(2), 1054–1060.

Cui, L., Morris, A., Ghedin, E. (2013). The human mycobiome in health and disease. *Genome Medicine*, 5(7), 63.

Dahlberg, A., Jonsson, L., Nylund, J.E. (1997). Species diversity and distribution of biomass above and below ground among ectomycorrhizal fungi in an old-growth Norway spruce forest in south Sweden. *Canadian Journal of Botany*, 75(8), 1323–1335.

Davison, J., Moora, M., Öpik, M., Adholeya, A., Ainsaar, L., Bâ, A., Johnson, N.C. (2015). Global assessment of arbuscular mycorrhizal fungus diversity reveals very low endemism. *Science*, 349(6251), 970–973.

Dawson, S.K., Boddy, L., Halbwachs, H., Bässler, C., Andrew, C., Crowther, T.W., Jönsson, M. (2019). Handbook for the measurement of macrofungal functional traits: A start with basidiomycete wood fungi. *Functional Ecology*, 33(3), 372–387.

Delavaux, C.S., Weigelt, P., Dawson, W., Duchicela, J., Essl, F., van Kleunen, M., Winter, M. (2019). Mycorrhizal fungi influence global plant biogeography. *Nature Ecology and Evolution*, 3(3), 424–429.

Diez, J., Kauserud, H., Andrew, C., Heegaard, E., Krisai-Greilhuber, I., Senn-Irlet, B., Büntgen, U. (2020). Altitudinal upwards shifts in fungal fruiting in the Alps. *Proceedings of the Royal Society B*, 287(1919), 20192348.

Doolittle, R.F., Feng, D.F., Tsang, S., Cho, G., Little, E. (1996). Determining divergence times of the major kingdoms of living organisms with a protein clock. *Science*, 271(5248), 470–477.

Eastwood, D.C. (2014). Evolution of fungal wood decay. *Deterioration and Protection of Sustainable Biomaterials*, 1158, 93–112.

Ellison, C.E., Hall, C., Kowbel, D., Welch, J., Brem, R.B., Glass, N.L., Taylor, J.W. (2011). Population genomics and local adaptation in wild isolates of a model microbial eukaryote. *Proceedings of the National Academy of Sciences*, 108(7), 2831–2836.

Elliott, T.F., Jusino, M.A., Trappe, J.M., Lepp, H., Ballard, G.A., Bruhl, J.J., Vernes, K. (2019). A global review of the ecological significance of symbiotic associations between birds and fungi. *Fungal Diversity*, 98(1), 161–194.

Els, N., Baumann-Stanzer, K., Larose, C., Vogel, T.M., Sattler, B. (2019). Beyond the planetary boundary layer: Bacterial and fungal vertical biogeography at Mount Sonnblick, Austria. *Geo: Geography and Environment*, 6(1), e00069.

Fernandez, C.W., Langley, J.A., Chapman, S., McCormack, M.L., Koide, R.T. (2016). The decomposition of ectomycorrhizal fungal necromass. *Soil Biology and Biochemistry*, 93, 38–49.

Finlay, R.D., Mahmood, S., Rosenstock, N., Bolou-Bi, E.B., Kohler, S.J., Fahad, Z., Lian, B. (2020). Reviews and syntheses: Biological weathering and its consequences at different spatial levels – From nanoscale to global scale. *Biogeosciences*, 17(6), 1507–1533.

Floudas, D., Binder, M., Riley, R., Barry, K., Blanchette, R.A., Henrissat, B., Aerts, A. (2012). The Paleozoic origin of enzymatic lignin decomposition reconstructed from 31 fungal genomes. *Science*, 336(6089), 1715–1719.

Frey-Klett, P., Garbaye, J.A., Tarkka, M. (2007). The mycorrhiza helper bacteria revisited. *New Phytologist*, 176(1), 22–36.

Frey-Klett, P., Burlinson, P., Deveau, A., Barret, M., Tarkka, M., Sarniguet, A. (2011). Bacterial-fungal interactions: Hyphens between agricultural, clinical, environmental, and food microbiologists. *Microbiology and Molecular Biology Reviews*, 75(4), 583–609.

Galante, T.E., Horton, T.R., Swaney, D.P. (2011). 95% of basidiospores fall within 1 m of the cap: A field-and modeling-based study. *Mycologia*, 103(6), 1175–1183.

Gange, A.C., Gange, E.G., Mohammad, A.B., Boddy, L. (2011). Host shifts in fungi caused by climate change? *Fungal Ecology*, 4(2), 184–190.

Gardes, M. and Bruns, T.D. (1996). Community structure of ectomycorrhizal fungi in a *Pinus muricata* forest: Above- and below-ground views. *Canadian Journal of Botany*, 74(10), 1572–1583.

Geml, J., Tulloss, R.E., Laursen, G.A., Sazanova, N.A., Taylor, D.L. (2008). Evidence for strong inter- and intracontinental phylogeographic structure in Amanita muscaria, a wind-dispersed ectomycorrhizal basidiomycete. *Molecular Phylogenetics and Evolution*, 48(2), 694–701.

Geml, J., Timling, I., Robinson, C.H., Lennon, N., Nusbaum, H.C., Brochmann, C., Taylor, D.L. (2012). An arctic community of symbiotic fungi assembled by long-distance dispersers: Phylogenetic diversity of ectomycorrhizal basidiomycetes in Svalbard based on soil and sporocarp DNA. *Journal of Biogeography*, 39(1), 74–88.

Gilbert, G.S. and Webb, C.O. (2007). Phylogenetic signal in plant pathogen–host range. *Proceedings of the National Academy of Sciences*, 104(12), 4979–4983.

Golan, J.J. and Pringle, A. (2017). Long-distance dispersal of fungi [Online]. Available at: https://doi.org/10.1128/microbiolspec.FUNK-0047-2016.

Halbwachs, H., Simmel, J., Bässler, C. (2016). Tales and mysteries of fungal fruiting: How morphological and physiological traits affect a pileate lifestyle. *Fungal Biology Reviews*, 30(2), 36–61.

Halme, P., Holec, J., Heilmann-Clausen, J. (2017). The history and future of fungi as biodiversity surrogates in forests. *Fungal Ecology*, 27, 193–201.

Harley, C.D. (2011). Climate change, keystone predation, and biodiversity loss. *Science*, 334(6059), 1124–1127.

Hartmann, F.E., McDonald, B.A., Croll, D. (2018). Genome-wide evidence for divergent selection between populations of a major agricultural pathogen. *Molecular Ecology*, 27(12), 2725–2741.

Hassett, M.O., Fischer, M.W., Money, N.P. (2015). Mushrooms as rainmakers: How spores act as nuclei for raindrops. *PLoS One*, 10(10), e0140407.

Heilmann-Clausen, J., Bruun, H.H., Ejrnæs, R., Frøslev, T.G., Læssøe, T., Petersen, J.H. (2019). How citizen science boosted primary knowledge on fungal biodiversity in Denmark. *Biological Conservation*, 237, 366–372.

Hogan, D.A., Vik, Å., Kolter, R. (2004). A Pseudomonas aeruginosa quorum-sensing molecule influences *Candida albicans* morphology. *Molecular Microbiology*, 54(5), 1212–1223.

Horton, T.R. and Bruns, T.D. (2001). The molecular revolution in ectomycorrhizal ecology: Peeking into the black-box. *Molecular Ecology*, 10(8), 1855–1871.

Hubbell, S.P. (2001). *The Unified Neutral Theory of Biodiversity and Biogeography*. Princeton University Press, Princeton, NJ.

Hulcr, J. and Stelinski, L.L. (2017). The ambrosia symbiosis: From evolutionary ecology to practical management. *Annual Review of Entomology*, 62(1), 285–303.

Jabiol, J., Bruder, A., Gessner, M.O., Makkonen, M., Mckie, B.G., Peeters, E.T., Chauvet, E. (2013). Diversity patterns of leaf-associated aquatic hyphomycetes along a broad latitudinal gradient. *Fungal Ecology*, 6(5), 439–448.

James, T.Y., Kauff, F., Schoch, C.L., Matheny, P.B., Hofstetter, V., Cox, C.J., Lumbsch, H.T. (2006). Reconstructing the early evolution of fungi using a six-gene phylogeny. *Nature*, 443(7113), 818–822.

Karasov, T.L., Horton, M.W., Bergelson, J. (2014). Genomic variability as a driver of plant–pathogen coevolution? *Current Opinion in Plant Biology*, 18, 24–30.

Kennedy, P.G., Matheny, P.B., Ryberg, K.M., Henkel, T.W., Uehling, J.K., Smith, M.E. (2012). Scaling up: Examining the macroecology of ectomycorrhizal fungi. *Molecular Ecology*, 21(17), 4151–4154.

Kohler, A., Kuo, A., Nagy, L.G., Morin, E., Barry, K.W., Buscot, F., Colpaert, J. (2015). Convergent losses of decay mechanisms and rapid turnover of symbiosis genes in mycorrhizal mutualists. *Nature Genetics*, 47(4), 410–415.

Krah, F.S., Bässler, C., Heibl, C., Soghigian, J., Schaefer, H., Hibbett, D.S. (2018). Evolutionary dynamics of host specialization in wood-decay fungi. *BMC Evolutionary Biology*, 18(1), 1–13.

Laughlin, D.C., Joshi, C., Bodegom, P.M. (2012). A predictive model of community assembly that incorporates intraspecific trait variation. *Ecology Letters*, 15, 1291–1299.

Levin, S.A. (1992). The problem of pattern and scale in ecology: The Robert H. MacArthur Award Lecture. *Ecology*, 73(6), 1943–1967.

Liu, Z.Y., Yao, Y.J., Liang, Z.Q., Liu, A.Y., Pegler, D.N., Chase, M.W. (2001). Molecular evidence for the anamorph–teleomorph connection in *Cordyceps sinensis*. *Mycological Research*, 105(7), 827–832.

Lomolino, M.V. (2000). Ecology's most general, yet protean 1 pattern: The species-area relationship. *Journal of Biogeography*, 27(1), 17–26.

Lutzoni, F., Nowak, M.D., Alfaro, M.E., Reeb, V., Miadlikowska, J., Krug, M., Hilu, K. (2018). Contemporaneous radiations of fungi and plants linked to symbiosis. *Nature Communications*, 9(1), 1–11.

Martin, F., Kohler, A., Murat, C., Balestrini, R., Coutinho, P.M., Jaillon, O., Porcel, B. (2010). Périgord black truffle genome uncovers evolutionary origins and mechanisms of symbiosis. *Nature*, 464(7291), 1033–1038.

Martin, F., Cullen, D., Hibbett, D., Pisabarro, A., Spatafora, J.W., Baker, S.E., Grigoriev, I.V. (2011). Sequencing the fungal tree of life. *New Phytologist*, 190(4), 818–821.

May, T.W., Cooper, J.A., Dahlberg, A., Furci, G., Minter, D.W., Mueller, G.M., Yang, Z. (2018). Recognition of the discipline of conservation mycology. *Conservation Biology*, 33, 733–736.

McGuire, K.L., Allison, S.D., Fierer, N., Treseder, K.K. (2013). Ectomycorrhizal-dominated boreal and tropical forests have distinct fungal communities, but analogous spatial patterns across soil horizons. *PLoS One*, 8(7), e68278.

Minter, N.J., Buatois, L.A., Mángano, M.G., Davies, N.S., Gibling, M.R., MacNaughton, R.B., Labandeira, C.C. (2017). Early bursts of diversification defined the faunal colonization of land. *Nature Ecology and Evolution*, 1(7), 0175.

Miyamoto, Y., Nakano, T., Hattori, M., Nara, K. (2014). The mid-domain effect in ectomycorrhizal fungi: Range overlap along an elevation gradient on Mount Fuji, Japan. *The ISME Journal*, 8(8), 1739–1746.

Morlon, H., Chuyong, G., Condit, R., Hubbell, S., Kenfack, D., Thomas, D., Green, J.L. (2008). A general framework for the distance–decay of similarity in ecological communities. *Ecology Letters*, 11(9), 904–917.

Mouquet, N. and Loreau, M. (2003). Community patterns in source-sink metacommunities. *The American Naturalist*, 162, 544–557.

Nagy, L.G., Tóth, R., Kiss, E., Slot, J., Gácser, A., Kovacs, G.M. (2017). Six key traits of fungi: Their evolutionary origins and genetic bases. In *The Fungal Kingdom*, Heitman, J., Howlett, B.J., Crous, P.W., Stukenbrock, E.H., James, T.Y., Gow, N.A.R. (eds). ASM Press, Washington, DC.

Nielsen, K.B., Kjøller, R., Bruun, H.H., Schnoor, T.K., Rosendahl, S. (2016). Colonization of new land by arbuscular mycorrhizal fungi. *Fungal Ecology*, 20, 22–29.

Nilsson, R.H., Larsson, K.H., Taylor, A.F.S., Bengtsson-Palme, J., Jeppesen, T.S., Schigel, D., Saar, I. (2019). The UNITE database for molecular identification of fungi: Handling dark taxa and parallel taxonomic classifications. *Nucleic Acids Research*, 47(D1), D259–D264.

Nogués-Bravo, D. and Rahbek, C. (2011). Communities under climate change. *Science*, 334(6059), 1070–1071.

Nordén, J., Penttilä, R., Siitonen, J., Tomppo, E., Ovaskainen, O. (2013). Specialist species of wood-inhabiting fungi struggle while generalists thrive in fragmented boreal forests. *Journal of Ecology*, 101(3), 701–712.

Nordén, J., Abrego, N., Boddy, L., Bässler, C., Dahlberg, A., Halme, P., Junninen, K. (2020). Ten principles for conservation translocations of threatened wood-inhabiting fungi. *Fungal Ecology*, 44, 100919.

Oh, J., Byrd, A.L., Deming, C., Conlan, S., Barnabas, B., Blakesley, R., Gregory, M. (2014). Biogeography and individuality shape function in the human skin metagenome. *Nature*, 514(7520), 59–64.

Parmesan, C. and Yohe, G. (2003). A globally coherent fingerprint of climate change impacts across natural systems. *Nature*, 421(6918), 37–42.

Peay, K.G., Garbelotto, M., Bruns, T.D. (2010). Evidence of dispersal limitation in soil microorganisms: Isolation reduces species richness on mycorrhizal tree islands. *Ecology*, 91(12), 3631–3640.

Peay, K.G., Schubert, M.G., Nguyen, N.H., Bruns, T.D. (2012). Measuring ectomycorrhizal fungal dispersal: Macroecological patterns driven by microscopic propagules. *Molecular Ecology*, 21(16), 4122–4136.

Peay, K.G., Kennedy, P.G., Talbot, J.M. (2016). Dimensions of biodiversity in the Earth mycobiome. *Nature Reviews Microbiology*, 14(7), 434–447.

Pirozynski, K.A. and Malloch, D.W. (1975). The origin of land plants: A matter of mycotrophism. *Biosystems*, 6(3), 153–164.

Plett, J.M., Daguerre, Y., Wittulsky, S., Vayssières, A., Deveau, A., Melton, S.J., Martin, F. (2014). Effector MiSSP7 of the mutualistic fungus Laccaria bicolor stabilizes the Populus JAZ6 protein and represses jasmonic acid (JA) responsive genes. *Proceedings of the National Academy of Sciences*, 111(22), 8299–8304.

Põlme, S., Abarenkov, K., Nilsson, R.H., Lindahl, B.D., Clemmensen, K.E., Kauserud, H., Tedersoo, L. (2021). Fungal traits: A user-friendly traits database of fungi and fungus-like stramenopiles. *Fungal Diversity*, 105(1), 1–16.

Raffaele, S. and Kamoun, S. (2012). Genome evolution in filamentous plant pathogens: Why bigger can be better. *Nature Reviews Microbiology*, 10(6), 417–430.

Rahbek, C. (2005). The role of spatial scale and the perception of large-scale species-richness patterns. *Ecology Letters*, 8(2), 224–239.

Ricklefs, R.E. (1987). Community diversity: Relative roles of local and regional processes. *Science*, 235(4785), 167–171.

Ruess, L. and Lussenhop, J. (2005). Trophic interactions of fungi and animals. *Mycology Series*, 23, 581.

Sanders, N.J. and Rahbek, C. (2012). The patterns and causes of elevational diversity gradients. *Ecography*, 35(1), 1–3.

Santini, A., Ghelardini, L., De Pace, C., Desprez-Loustau, M.L., Capretti, P., Chandelier, A., Hantula, J. (2013). Biogeographical patterns and determinants of invasion by forest pathogens in Europe. *New Phytologist*, 197(1), 238–250.

Schmidt, P.A., Bálint, M., Greshake, B., Bandow, C., Römbke, J., Schmitt, I. (2013). Illumina metabarcoding of a soil fungal community. *Soil Biology and Biochemistry*, 65, 128–132.

Seena, S., Bärlocher, F., Sobral, O., Gessner, M.O., Dudgeon, D., McKie, B.G., Bruder, A. (2019). Biodiversity of leaf litter fungi in streams along a latitudinal gradient. *Science of the Total Environment*, 661, 306–315.

Shipley, B., Vile, D., Garnier, É. (2006). From plant traits to plant communities: A statistical mechanistic approach to biodiversity. *Science*, 314(5800), 812–814.

Smith, S.E. and Read, D.J. (2010). *Mycorrhizal Symbiosis*. Academic Press, Cambridge, MA.

Soudzilovskaia, N.A., van Bodegom, P.M., Terrer, C., van't Zelfde, M., McCallum, I., McCormack, M.L., Tedersoo, L. (2019). Global mycorrhizal plant distribution linked to terrestrial carbon stocks. *Nature Communications*, 10(1), 1–10.

Spanu, P.D., Abbott, J.C., Amselem, J., Burgis, T.A., Soanes, D.M., Stüber, K., Lebrun, M.H. (2010). Genome expansion and gene loss in powdery mildew fungi reveal tradeoffs in extreme parasitism. *Science*, 330(6010), 1543–1546.

Spatafora, J.W., Chang, Y., Benny, G.L., Lazarus, K., Smith, M.E., Berbee, M.L., James, T.Y. (2016). A phylum-level phylogenetic classification of zygomycete fungi based on genome-scale data. *Mycologia*, 108(5), 1028–1046.

Strullu-Derrien, C., Spencer, A.R., Goral, T., Dee, J., Honegger, R., Kenrick, P., Berbee, M.L. (2018). New insights into the evolutionary history of fungi from a 407 Ma Blastocladiomycota fossil showing a complex hyphal thallus. *Philosophical Transactions of the Royal Society B: Biological Sciences*, 373(1739), 20160502.

Sunday, J.M., Pecl, G.T., Frusher, S., Hobday, A.J., Hill, N., Holbrook, N.J., Watson, R.A. (2015). Species traits and climate velocity explain geographic range shifts in an ocean-warming hotspot. *Ecology Letters*, 18(9), 944–953.

Tedersoo, L. and Nara, K. (2010). General latitudinal gradient of biodiversity is reversed in ectomycorrhizal fungi. *New Phytologist*, 185(2), 351–354.

Tedersoo, L., Bahram, M., Toots, M., Diedhiou, A.G., Henkel, T.W., Kjøller, R., Polme, S. (2012). Towards global patterns in the diversity and community structure of ectomycorrhizal fungi. *Molecular Ecology*, 21(17), 4160–4170.

Tedersoo, L., Bahram, M., Põlme, S., Kõljalg, U., Yorou, N.S., Wijesundera, R., Smith, M.E. (2014). Global diversity and geography of soil fungi. *Science*, 346(6213), 1256688.

Tedersoo, L., Sánchez-Ramírez, S., Koljalg, U., Bahram, M., Döring, M., Schigel, D., Abarenkov, K. (2018). High-level classification of the fungi and a tool for evolutionary ecological analyses. *Fungal Diversity*, 90(1), 135–159.

Tedersoo, L., Bahram, M., Zobel, M. (2020). How mycorrhizal associations drive plant population and community biology. *Science*, 367(6480), eaba1223.

Tisthammer, K.H., Cobian, G.M., Amend, A.S. (2016). Global biogeography of marine fungi is shaped by the environment. *Fungal Ecology*, 19, 39–46.

Vellend, M., Srivastava, D.S., Anderson, K.M., Brown, C.D., Jankowski, J.E., Kleynhans, E.J., Kraft, N.J., Letaw, A.D., Macdonald, A.A.M., Maclean, J.E., Myers-Smith, I.H. (2014). Assessing the relative importance of neutral stochasticity in ecological communities. *Oikos*, 123, 1420–1430.

Větrovský, T., Kohout, P., Kopecký, M., Machac, A., Man, M., Bahnmann, B.D., Baldrian, P. (2019). A meta-analysis of global fungal distribution reveals climate-driven patterns. *Nature Communications*, 10(1), 1–9.

Violle, C., Reich, P.B., Pacala, S.W., Enquist, B.J., Kattge, J. (2014). The emergence and promise of functional biogeography. *Proceedings of the National Academy of Sciences*, 111(38), 13690–13696.

Wang, X.Q. and Ran, J.H. (2014). Evolution and biogeography of gymnosperms. *Molecular Phylogenetics and Evolution*, 75, 24–40.

Wang, D.Y.C., Kumar, S., Hedges, S.B. (1999). Divergence time estimates for the early history of animal phyla and the origin of plants, animals and fungi. *Proceedings of the Royal Society of London. Series B: Biological Sciences*, 266(1415), 163–171.

Weber, S.E., Diez, J.M., Andrews, L.V., Goulden, M.L., Aronson, E.L., Allen, M.F. (2019). Responses of arbuscular mycorrhizal fungi to multiple coinciding global change drivers. *Fungal Ecology*, 40, 62–71.

Whittaker, R.H. (1969). New concepts of kingdoms of organisms. *Science*, 163(3863), 150–160.

Wiens, J.A. (1989). Spatial scaling in ecology. *Functional Ecology*, 3(4), 385–397.

Willig, M.R., Kaufman, D.M., Stevens, R.D. (2003). Latitudinal gradients of biodiversity: Pattern, process, scale, and synthesis. *Annual Review of Ecology, Evolution and Systematics*, 34(1), 273–309.

Wilson, A.W., Hosaka, K., Mueller, G.M. (2017). Evolution of ectomycorrhizas as a driver of diversification and biogeographic patterns in the model mycorrhizal mushroom genus *Laccaria*. *New Phytologist*, 213(4), 1862–1873.

Worrich, A., Stryhanyuk, H., Musat, N., König, S., Banitz, T., Centler, F., Miltner, A. (2017). Mycelium-mediated transfer of water and nutrients stimulates bacterial activity in dry and oligotrophic environments. *Nature Communications*, 8(1), 1–9.

Xu, X., Wang, N., Lipson, D., Sinsabaugh, R., Schimel, J., He, L., Tedersoo, L. (2020). Microbial macroecology: In search of mechanisms governing microbial biogeographic patterns. *Global Ecology and Biogeography*, 29(11), 1870–1886.

Yeh, C.F., Soininen, J., Teittinen, A., Wang, J. (2019). Elevational patterns and hierarchical determinants of biodiversity across microbial taxonomic scales. *Molecular Ecology*, 28(1), 86–99.

Zadoks, J.C. (1987). The function of plant pathogenic fungi in natural communities. *Disturbance in Grasslands*. Springer, Dordrecht.

Zanne, A.E., Pearse, W.D., Cornwell, W.K., McGlinn, D.J., Wright, I.J., Uyeda, J.C. (2018). Functional biogeography of angiosperms: Life at the extremes. *New Phytologist*, 218(4), 1697–1709.

9

Freshwater Biogeography in a Nutshell

Anthi OIKONOMOU
Institute of Marine Biological Resources and Inland Waters, Hellenic Centre for Marine Research, Anavyssos, Greece

9.1. Introduction

Freshwater ecosystems occupy only 2.3% of Earth's surface (Reid *et al.* 2019), yet they support an excessive portion of the world's most speciose and endemic taxa. They are estimated to harbor 12% of the world's fauna and one third (18,000 species) of the global vertebrate species richness (Balian *et al.* 2008). These numbers are certainly much larger, given the fact that on a yearly basis many new species are described, even in the case of the most well-studied groups (fish and amphibians). In addition to high species richness, freshwater ecosystems host a large percentage of endemic species (Tisseuil *et al.* 2013). Nonetheless, there is strong evidence that freshwater species are at greater risk than their terrestrial counterparts (Collen *et al.* 2009; IUCN, 2020), and the prognosis for freshwater biodiversity has worsened, with freshwater species exhibiting steeper population declines (Reid *et al.* 2019).

As lakes and river networks are separated from each other by barriers of land, freshwater organisms live in "aquatic biogeographical islands" within the "sea" of terrestrial landscapes and can disperse through the same agent of transport, water (Hugueny, 1989). Because of their well-defined boundaries and isolation, freshwaters are ideal "grains" for understanding patterns, exploring ecological theories and promoting biogeographical research (Hugueny *et al.* 2010, Heino

Biogeography,
coordinated by Eric GUILBERT.

2011). Despite their island-like nature, a few attempts have been made to investigate diversity patterns in the freshwater realm; however, recent studies that have been performed with freshwater fishes have the potential to greatly advance freshwater biogeography (Dias *et al.* 2014; Tedesco *et al.* 2017; Carvajal-Quintero *et al.* 2019). Furthermore, freshwater conservation continues to be under-represented in the literature; it has been canalized to follow strategies applied on terrestrial habitats (Olden *et al.* 2010) and has relied on charismatic taxa (birds and mammals), which proved to be poor surrogates of richness and endemism (Darwall *et al.* 2011) in freshwaters. Moreover, protected area networks have been historically established for terrestrial conservation (Herbert *et al.* 2010; Leal *et al.* 2020). In the EU, a large proportion of threatened freshwater biodiversity is not adequately covered by the Natura 2000 directive, since only 14% of European freshwater fish, 3% of non-marine molluscs, and 19% of dragonflies listed as threatened in the IUCN Red List are designated under the Habitats Directive (Spiliopoulou *et al.* 2021; van Rees *et al.* 2021).

9.2. Freshwater hotspots and patterns in species richness

9.2.1. *Latitudinal gradient in species richness*

Alexander von Humboldt was the first scientist to suggest that the tropics harbor many more species than the colder environments [latitudinal gradient in species richness (von Humboldt and Bonpland, 1805)], and since then, it has been recognized as one of the most commonly described patterns in biogeography (Whittaker *et al.* 2001), being observed in organisms ranging from bacteria to mammals. No consensus has been reached about its underlying causes, associated with hundreds of papers, dozens of hypotheses and disagreements (Pontarp *et al.* 2019). Many potential explanatory factors have been proposed for the grand cline in diversity from low at the poles to high at the equator, including temperature, climate stability, biotic interactions and energy availability that characterize tropical latitudes (Rohde, 1992; Mittelbach *et al.* 2007; Pontarp *et al.* 2019).

Relating to freshwaters, it appears that the European lotic animal diversity richness also declines with increasing latitude (Dehling *et al.* 2010). Fish species richness peaks in equatorial regions and decreases sharply towards the poles (Oberdorff *et al.* 1995; Brucet *et al.* 2013; Griffiths, 2015). Centers of species richness and endemism are concentrated in tropical and subtropical drainage basins for crayfish, amphibians, aquatic birds and aquatic mammals (Tisseuil *et al.* 2013). Neotropics and Afrotropics are also among the global hotspots of freshwater fish endemism (Sayer *et al.* 2018; Leroy *et al.* 2019; Jézéquel *et al.* 2020). Examples of species richness that increase with decreasing latitude have been found also for amphipods (France, 1992), Odonata and Anura (Pearson *et al.* 2009; Heino, 2011).

Exceptions to this rule have been found in many highly diverse and less-studied organism groups that occur in freshwater ecosystems. For instance, Murphy *et al.* (2019) found evidence of nonlinear relationships between latitude and macrophyte α- and γ-diversity (Alahuhta *et al.* 2020; Murphy *et al.* 2020). Macrophyte diversity is the highest in the Neotropics (sub-tropical to low tropical latitudes), intermediate in the Oriental, Nearctic and Afrotropics, and the lowest in Antarctica; the highest rates of endemism are found in the Neotropics and Afrotropics (Chambers *et al.* 2008; Murphy *et al.* 2020). Recently, a reverse species richness-latitudinal gradient has also been reported, which was likely due to high nutrient concentrations in southern latitude lakes of Minnesota (Alahuhta, 2015). Diatoms also show an exception to the general pattern, that is, a U-shaped latitudinal distribution of diatom richness, equally high in subtropical and temperate regions (Passy, 2010). Moreover, Schiaffino *et al.* (2016) reported a non-significant relationship between latitude and autotrophic and heterotrophic eukaryotes species richness (latitudinal range 45–63°S) [but see (Stomp *et al.* 2011)]. Underlying reasons for inverse richness patterns are context-dependent and may be caused when a factor other than latitude drives richness (e.g. nutrient supply for diatoms) (Soininen and Teittinen, 2019). Contrary to the European lotic animal species, as it has been stated before, lentic animal species richness showed a hump-shaped relationship with latitude (Dehling *et al.* 2010). Moreover, it has been suggested that Ephemeroptera and Plecoptera diversity is greater at higher latitudes (Pearson *et al.* 2009), probably because they have diversified in mountainous and northerly regions (Heino, 2011, 2009). For Trichoptera and Caudata, latitudinal gradients were also not apparent (Pearson *et al.* 2009). The absent or reversed latitudinal gradients may be also related to the Linnean and Wallacean shortfalls that are mostly evident at low latitudes (Lomolino, 2004).

9.2.2. *Geography, environment and biogeographical history*

Ecological studies suggest that there is presumably no single factor that entirely explains distribution of biodiversity. For species diversity, various historical, environmental and geographical factors have long been recognized as major drivers, with their relative importance being the key approach to understanding their underlying mechanisms (Ricklefs, 1987). Regional factors in freshwaters encompass hydrology (flooding and drying), geomorphology of river basins (drainage density and channel morphology), canopy cover, climate seasonality and gradients such as precipitation and temperature (driven by elevation), chemistry, regional species pool, biotic interactions and differentiation rate (Jackson *et al.* 2001 and references therein; Fine, 2015; Dodds *et al.* 2019). Three possible mechanisms have been proposed for understanding species richness patterns: (1) ecological hypotheses that focus on mechanisms of species coexistence and species diversity maintenance, (2) evolutionary hypotheses that focus on diversification rates and (3) historical biogeography that focuses on the role of historical events in Earth's history (Mittelbach *et al.* 2007). Indeed, area, energy, time, climate stability, temperature and biotic interactions have been reported to contribute to global patterns of species diversity (reviewed in Fine, 2015).

Recent findings in freshwaters underscore the importance of studying simultaneously historical processes, drainage basin characteristics and local environmental conditions to understand variation in species richness (Tisseuil *et al.* 2013). Indeed, freshwater species richness and endemism patterns are the result of climate, productivity and biogeographical history. For instance, the freshwater animal richness gradient is driven by postglacial recolonization and dispersal ability of taxa, with richness differences across realms reflecting differences in speciation, extinction and connectivity (e.g. Pearson *et al.* 2009; Griffiths, 2015). The importance of past and current environmental factors in explaining the present-day patterns of freshwater fish α-diversity was also highlighted by Dias *et al.* (2014). A general pattern of fish endemism is low endemic species richness in areas subjected to Quaternary glaciation. For instance, high levels of endemism were observed for European hotspots such as the Balkan, Iberian and Italian peninsula (Oikonomou *et al.* 2014), since their fish faunas diverged from each other and from that of northern Europe from Late Oligocene to Early Miocene times (Levy *et al.* 2009) and these areas had not been affected to intense glaciation. Indeed, Pleistocene glaciations played an important role in a secondary recolonization from the Danubian refuge determining the current European distribution of freshwater fish (Sommerwerk *et al.* 2009). Thus, the Danubian basin and local river captures in the southern Adriatic Sea and the eastern Balkan Peninsula have promoted the expansion and genetic homogenization of freshwater species across Europe (Perea *et al.* 2010), leading to low endemism.

Previous work on freshwater fish has shown that total species richness was also influenced by factors related to energy availability (i.e. net primary productivity) and climatic parameters (annual rainfall and average annual temperature) (Oberdorff *et al.* 1999, 2011; Tedesco *et al.* 2005; Tisseuil *et al.* 2013; Griffiths *et al.* 2014; Pelayo-Villamil *et al.* 2015; Boll *et al.* 2016). Moreover, water depth, altitude, pH and total phosphorus concentration determine fish richness (Allen *et al.* 1999; Helminen *et al.* 2000; Zhao *et al.* 2006; Brucet *et al.* 2013; Boll *et al.* 2016; Oikonomou and Stefanidis, 2020). Phytoplankton species richness was associated to lake chlorophyll concentration and water temperature, resembling the effects of productivity, habitat area and temperature on diversity patterns (Stomp *et al.* 2011). Conductivity, pH, total phosphorus, elevation and silica ice cover and thermal stratification in lakes, and shading and water depth in streams were among the most important predictors of diatom species richness (Allen *et al.* 1999; Soininen and Teittinen, 2019 and references herein). Landscape aridity, altitude and land area also influenced macrophyte diversity (Murphy *et al.* 2019). pH and altitude were also highlighted as significant descriptors of macrophyte richness in Balkan lakes (Oikonomou and Stefanidis, 2020). Turbidity, lake depth and minimum water temperature were among the most significant predictors of zooplankton species richness (Allen *et al.* 1999).

9.2.3. *Species–area relationship (SAR)*

The general increase in species richness with increasing area is called "species-area relationship (SAR)" and is one of the most studied global biogeographical patterns. The pattern dates back to at least the mid-19th century (Watson, 1835). Arrhenius (1921) mathematically described the pattern and since then, the positive relationship between the area and species richness encompasses thousands of studies of a wide variety of taxa, realms and scales (e.g. Eadie *et al.* 1986; Triantis *et al.* 2012; Matthews *et al.* 2021). In addition, the relationship has been also confirmed for habitat islands (island-like biological systems), such as mountain peaks or regions with high altitudes and freshwaters (lakes, rivers) (Matthews *et al.* 2016). The first models to be proposed for SARs were the power model (Arrhenius, 1921, 1920) and the exponential model (Gleason, 1922). The power model remains the most frequently preferred model, both for fitting curves to species–area data and for explaining species diversity patterns (Triantis *et al.* 2012).

Area is one of the most significant constraints acting on species richness and endemism of freshwater fishes, crayfish, amphibians, aquatic mammals and birds (Tisseuil *et al.* 2013). Previous work in freshwaters has shown that total species richness was influenced by area in a series of conditions. For instance, species richness macrophytes were significantly associated with lake surface area (Oikonomou and Stefanidis, 2020). The freshwater invertebrate fauna was also associated with area both in streams and lakes (Douglas and Lake, 1994; Carlsson, 2001; Pearson *et al.* 2009). Moreover, riverine and lacustrine fish species richness increases with basin area (Oberdorff *et al.* 1999, 2011; Tedesco *et al.* 2005; Oikonomou and Stefanidis, 2020). Nonetheless, for Ephemeroptera, Plecoptera and Trichoptera, the relationships between regional diversity and area were non-significant on the global scale (Pearson *et al.* 2009).

SAR studies for freshwater microscopic organisms have received much less attention in the literature (Soininen *et al.* 2019), but similar patterns to those found for larger organisms have been observed. For instance, a positive relationship between bacterial richness and lake area has been reported in 11 high-mountain lakes from Sierra Nevada (Spain) (e.g. Reche *et al.* 2005). However, Logue *et al.* (2012) found that bacterioplankton richness decreased with increasing catchment area of 14 Swedish nutrient-poor lakes, highlighting nutrient availability as a major driver of species richness and re-opening the debate about whether SARs can indeed be found for microbial organisms. Stomp *et al.* (2011) reported a significant positive SAR for lake phytoplankton communities in 540 lakes and reservoirs distributed across the continental United States. For diatoms, Teittinen and Soininen (2015) did not find a significant SAR in 50 boreal springs in Finland, whereas Bolgovics *et al.* (2016) clearly demonstrated that the size of the water body (ponds, river oxbows, pools and lakes) is a key variable affecting the richness of benthic diatom communities.

Although, the area *per se* explains most of the variation in the number of species in riverine ecosystems (an increase in the area implies an increase in species richness) (Hugueny, 1989), it is necessary to take into account parameters other than area, such as environmental heterogeneity, energy availability, speciation rate and extinction (Hugueny *et al.* 2010). Indeed, three non-exclusive theories have been put forward to explain the pattern (Hugueny *et al.* 2010): (1) the size-dependent extinction rate, which assumes that probability of extinction of a species increases with a reduction in its population size, which in turn is a function of surface area (MacArthur and Wilson, 1967), (2) the size-dependent speciation rate (Losos and Schluter, 2000) which assumes a positive effect of area on speciation rate, if larger areas support larger species ranges, and if a large range is more likely to expose species to greater ecological heterogeneity and/or geographical barriers (Rosenzweig, 1995) and (3) the increasing habitat heterogeneity with increasing area, thus offering more available niches and consequently favoring the coexistence of a larger number of species with dissimilar ecological requirements (Williamson, 1988).

9.2.4. *Community assembly in freshwater*

A key question in freshwater ecology is what determines the composition and structure of biological communities which vary dramatically in time and space (e.g. Nikolski, 1933). The answer to this question is the subject of ecological research for decades, as the distribution of species is governed by many factors operating at different temporal and spatial scales. The assembly of communities is often viewed in a conceptual model involving the dispersal of species from a regional species pool that is constrained by historical processes and is determined by speciation and extinction processes at very large spatial and temporal scales. A subset of the regional species pool (influenced by chance and dispersal limitation) is available for colonization of a particular site. The species of this subpool pass through an environmental (abiotic) filter and a biotic filter and successfully colonize and interact to form local communities (LeRoy Poff, 1997; Mittelbach and Schemske, 2015; Figure 9.1). The study of community assembly unites disciplines as diverse as evolutionary biology, biogeography and community ecology and the theories of island biogeography (MacArthur and Wilson, 1967), the neutral theory of biodiversity (Hubbell, 2001), metacommunity concept (Leibold *et al.* 2004), and the modern species coexistence theory (HilleRisLambers *et al.* 2012).

It is already stated that patterns in community composition are scale-dependent (spatiotemporal scales), and this wide range of temporal and spatial scales affects the relative importance of ecological versus evolutionary factors on community assembly. On the spatial scale, species interactions and adaptation play an important role inside communities (i.e. local scale) while immigration of new species from outside the community (i.e. regional scale) affects the dynamic nature of the species pool (Hubert *et al.* 2015), being also shaped by metacommunity dynamics in addition to speciation, extinction and dispersal (Mittelbach and Schemske, 2015). On the temporal scale, species pool composition is influenced by metacommunity dynamics over short timescales and by speciation, extinction and dispersal over long timescales (Leibold and Chase, 2017; Mittelbach and Schemske, 2015).

9.2.5. *Local scale*

Local freshwater community composition results from an interplay of local and regional factors, both abiotic factors and biotic interactions (LeRoy Poff, 1997). According to the influential research of Smith and Powell (1971), the local fish community in a small creek in the U.S. was a "snapshot" and the product of selection on the global fish fauna by various environmental, biogeographical and evolutionary factors that were portrayed graphically as a series of filters (see also Figure 9.1). Tonn (1990) also developed a framework in lake fish communities where environmental variables prevailing at different scales, ranging from global to habitat, are understood as filters selecting species from the species pool at larger scales to coexist in local communities.

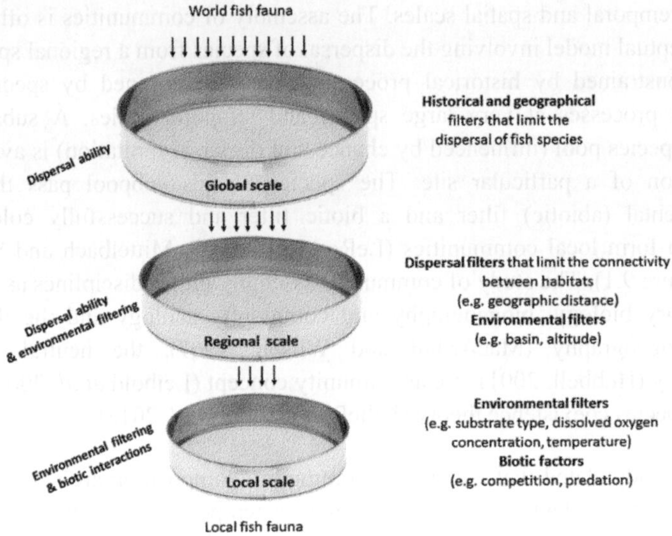

Figure 9.1. *Conceptual diagram representing freshwater fish community assembly mechanisms driving species composition at multiple scales*

At the basin scale, pioneering research emphasized a linear upstream–downstream schema where the community assembly is shaped by many landscape factors that operate at a variety of spatial and temporal scales and change from headwaters to river mouths (Allan and Castillo, 2007). Water temperature, flow regime, water physicochemical properties, channel hydraulics, pool-riffle complexity and bottom composition interactively influence instream habitat and biota. Indicatively for fishes, abiotic parameters such as substrate, width and depth (Schlosser, 1985) and physicochemical factors (e.g. oxygen and temperature) are strongly related to the composition of fish communities (Wang *et al.* 2006). In addition, riparian vegetation affects fish communities and anthropogenic activities change species composition (e.g. dams and introduction of alien species) (Rowe *et al.* 2009). Competition and predation as well as habitat heterogeneity interact and determine the composition of fish communities (Jackson *et al.* 2001).

In order to understand the longitudinal changes in structure and ecological function along an entire river, many theories and frameworks have been integrated on this concept (reviewed by Wang *et al.* 2006). Models of stream ecology attempt to explain changes in the structure and function of stream communities within the context of the landscape, including longitudinal patterns and shifts in energy inputs between forested and open sites. Arguably, the most influential paper on river ecology over the past four decades is the milestone research "The River Continuum Concept (RCC)" of Vannote *et al.* (1980). This study integrated stream physicochemical and geomorphic conditions from headwater streams to river mouths with patterns in habitat characteristics and metabolism dynamics (i.e. depth, channel width, velocity, discharge, temperature and entropy gain), functional traits and corresponding biota (see also Jones, 2011; Doretto *et al.* 2020). RCC is currently the dominant theory employed *de facto* by riverine ecologists and environmental managers. It is also taught in classrooms throughout the world and represents one of the very few acknowledged contributions of stream ecology to general ecology (Thorp *et al.* 2008).

The serial discontinuity concept (SDC) proposed later (Ward and Stanford, 1983) provided another major step forward and recognized that no specific river is a continuum (Poole, 2002); at any spatial scale, the downstream physicochemical and biological continuum are interrupted in response to channel and valley morphological changes. After receiving criticism for its simplistic view, the SDC was expanded to include lateral and vertical connectivity (Ward and Stanford, 1995) and since then, it has been extensively applied in ecohydrological case studies to evaluate the influence of tributaries and dams (Rice *et al.* 2001; Ellis and Jones, 2013; Mellado-Díaz *et al.* 2019). Indeed, SDC is supported widely in the literature, with benthic invertebrates responding differently to river regulation. Much of this variability stems from the degree of flow variability and resource subsidies from the upstream reservoir (Ellis and Jones, 2013). Ward (1989) has also extended the continuum and discontinuum concepts to a holistic perspective of lotic ecosystems and proposed a four-dimensional framework to conceptualize the dynamic and hierarchical nature of river ecosystems. The upstream–downstream gradient constitutes the longitudinal dimension; the channel morphology and the extent of floodplain form the lateral dimension; the vertical dimension refers to interactions between the channel and contiguous groundwater; and the fourth dimension represents the temporal scale. This approach has been followed in studies on longitudinal, lateral and vertical patterns to illustrate complex responses to flow regulation (Davies *et al.* 2000; Fausch *et al.* 2002).

9.2.6. *Metacommunity concept*

Aforementioned patterns identify local processes (environmental filtering and biotic interactions) as the main driver of community composition, although dispersal limitation of organisms is known to be a key driver of assemblages (Gilbert and Lechowicz, 2004). Indeed, communities are no longer considered to be exclusively shaped by local processes; regional processes such as dispersal are now considered of paramount importance to understand biotic assemblages at the local scale. To this end, the metacommunity framework (Leibold *et al.* 2004) is a recently emerged subdiscipline of ecology and biogeography, which explicitly considers the combined roles of local and regional processes in community assembly. Simply put, a metacommunity represents a larger-scale "region" consisting of several smaller-scale "localities" (i.e. communities); these localities are connected by dispersal and may be heterogeneous in abiotic and biotic variables (Leibold and Chase, 2017). In this context, to understand both the patterns and processes of coexistence and species composition, the interactions between scale, dispersal and heterogeneity must be considered along with the milieu of local-scale processes. Metacommunity ecology may inform also about phylogenetic and historical biogeography and the consequences of these effects on patterns of diversity and composition across different spatial and temporal scales (Mittelbach and Schemske, 2015).

Within the metacommunity context, four different paradigms can be distinguished in explaining the importance of local- and regional-scale processes. The patch dynamics (PD) (Levins, 1969) paradigm assumes that local species diversity is maintained by dispersal of individuals between patches irrespective of the quality of patches; species show different colonization and competitive potential for patches that are identical in environmental conditions. In the neutral paradigm, strongly influenced by the Neutral Theory of Biodiversity (NT) (Hubbell, 2001), species are subject to a stochastic balance between speciation and extinction at the regional scale; they are ecologically equivalent, and their presence in local communities are stochastically determined and only dictated by their limited spatial dispersion; in this perspective, environmental drivers do not play a significant role shaping local communities. The mass-effect (ME) paradigm (Shmida and Wilson, 1985) assumes variation in environmental conditions across a set of sites and emigration of individuals from areas of high environmental value (core areas or sources) to sink sites where environmental conditions are suboptimal and where species cannot maintain viable populations. The species sorting (SS) paradigm (Chase and Leibold, 2003) assumes that species differ in their responses to heterogeneous local environmental conditions; each species can persist in any habitat where it can achieve positive population growth and species are sorted in the local community according to their preferences to environmental gradients (species sorting).

The metacommunity framework has been intensively applied to freshwater ecosystems to add on the role of dispersal to environmental heterogeneity (e.g. Heino, 2013; Heino *et al.* 2017a, 2017b; Tonkin *et al.* 2018). As freshwater organisms portray a range from weak (e.g. fish and mussels), intermediate (e.g. aquatic insects) to high (e.g. bacteria, diatoms and aquatic macrophytes) dispersal abilities, organisms are constrained in different ways depending on their mode of dispersal. For instance, the stream corridor is the main "highway" for dispersal of benthic invertebrates (Petersen *et al.* 2004) or fish, but many diatom, macrophyte and macroinvertebrate species are capable of overland dispersal (Bunn and Hughes, 1997; see also Schmera *et al.* 2018).

Of the four paradigms, species sorting and mass effects have a predominant role in structuring communities of freshwater ecosystems (Soininen, 2014). Dispersal limitation and species sorting may be conceptualized extremes along a continuum of processes underlying observed biogeographical patterns (Heino *et al.* 2015b). In rivers, metacommunity patterns depended on a variety of factors, including dispersal mode (aquatic versus aerial vs. terrestrial) and landscape type (arid vs. mesic), location-specific factors, such as network connectivity, land use, topographic heterogeneity and biotic interactions (Tonkin *et al.* 2018).

Freshwater diatom metacommunities are typically structured by species sorting; indeed, this mechanism was demonstrated as the primary role in determining Swedish diatom communities (Keck *et al.* 2018). Additionally, Dong *et al.* (2016) indicated that both environmental filtering and dispersal processes influenced metacommunity structuring, with dispersal contributing more than environmental processes to benthic diatom composition in high mountain streams. Local abiotic factors (nutrients, conductivity and lake morphology) were the most important predictors controlling the compositional variation of diatom assemblages in European shallow lakes (Rodríguez-Alcalá *et al.* 2020). Nonetheless, at smaller spatial scales, diatoms could be driven by mass effects. To this end, Leboucher *et al.* (2020) developed a method for easier detection of mass effects in order to increase the ability to disentangle metacommunity structuring in passive dispersers as benthic diatoms. Reche *et al.* (2005) suggested that microbial biogeography in lakes was driven only by geographical distance, while Van Der Gucht *et al.* (2007) found that it is primarily controlled by environmental factors. At the global scale, elevation range and dispersal limitation were found as the most significant predictors structuring lake macrophytes (Alahuhta *et al.* 2018), which is in accordance with the findings of García-Girón *et al.* (2019) suggesting that both dispersal limitation and species sorting drive community assembly in northwestern Spanish ponds. Havel and Shurin (2004) concluded that zooplankton species show extensive dispersal such that local selection processes were most important for shaping populations and communities. Furthermore, European cladoceran communities were not generally limited by dispersal at the regional scale, with species sorting explaining the high species turnover (Viana *et al.* 2016). From an evaluation comprising >1.23 million invertebrates collected from rivers across nine biogeographic regions on three continents, community assembly modeling indicated that dispersal was the primary mechanism driving invertebrate community response to decreasing glacier cover, closely following the patch dynamics paradigm of metacommunity theory (Brown *et al.* 2018). Li *et al.* (2019) found that both deterministic (e.g. environmental filtering) and stochastic (e.g. dispersal) processes controlled macroinvertebrate assemblage structure depending on dispersal mode; indeed, spatial processes were more important in regulating variation of the assemblage structure of passive dispersers compared to environmental filtering that showed a greater importance for active dispersers. The authors found no effect of river basin identity for either passive or active dispersers, in contrast to Heino, Soininen, *et al.* (2017) who found that basin identity was a slightly better predictor of community structure of diatoms, bryophytes and zooplankton across various drainage basins in Finland. Chironomid metacommunities were controlled by both the environment (i.e. environmental filtering – species sorting) and spatial drivers (i.e. dispersal limitation and mass effect) according to Specziár *et al.* (2018), whereas Datry *et al.* (2016) reported that dispersal in shaping stream metacommunities might have been underestimated in systems prone to disturbance.

Contrary to plankton, fish are largely obligate active dispersers through surface waters and their dispersal abilities are in general higher than for invertebrates; thus, fish communities are a simpler model to explore how species sorting and dispersal shape river communities. Indeed, fish communities were more influenced by environmental filtering than dispersal (Heino, 2013; Datry *et al.* 2016), which is in agreement with findings from the Neotropical ecoregion where environmental filters (i.e. climate and hydrology) shaped fish assemblages (Borges *et al.* 2020). Species sorting was found important in shaping intermittent river fish metacommunities (Faustino and Terra, 2020) and was also reported as the most important process in controlling the variation in fish species composition in Mediterranean lakes (Oikonomou and Stefanidis, 2020) (see also Heino *et al.* 2015b and references therein). By contrast, the absence of niche filtering processes and the predominant role of mass effect structuring fish assemblages in the main stem of a large Neotropical river were reported by Vitorino Júnior *et al.* (2016). Recently, according to Vardakas *et al.* (2020), mass effects model controlled the structure of a fish metacommunity in a Mediterranean intermittent river.

9.2.7. *Beta diversity*

The metacommunity framework places different emphasis on the importance of environmental factors and spatial factors, and as a result, disentangling the relative influence of environment and space on the similarity of species composition among sites plays a key role in unraveling the mechanisms and understanding the possible underlying processes. This issue is the main research frontier of beta diversity studies (β-diversity, broadly, refers to variation in species composition between communities within a region), which is receiving increasing scientific interest in the last decades, with many developments in theoretical and analytical grounds (Anderson *et al.* 2011; Legendre, 2019; Magurran *et al.* 2019).

β-diversity is linked with α-diversity gradients, and both components result from community assembly through local and regional filters. In general, β-diversity indicates the spatial and temporal variation of species composition among communities (Anderson *et al.* 2011; Magurran *et al.* 2019) and is essentially related to two processes (Baselga, 2010): spatial turnover (species replacement), referring to the replacement of some species by others as a consequence of environmental filtering and/or spatial and historical constraints, and nestedness (i.e. species-poor sites being a subset of the richest site in the region) that stems from species loss or gain caused by different processes, such as selective colonization and extinctions. To date, many quantitative frameworks and metrics have been proposed for the estimation of β-diversity (Koleff *et al.* 2003; Baselga, 2010; Anderson *et al.* 2011; Carvalho *et al.* 2012; Legendre, 2019; and also see Leprieur and Oikonomou, 2014).

The variation in the replacement and richness difference components of β-diversity has been rarely studied in freshwater ecosystems. However, it has been shown that freshwater β-diversity patterns were driven by various factors often resulting from a combination of regional and local processes (Dias *et al.* 2014), and therefore must be interpreted within the respective biogeographical and historical context (Ricklefs and Schluter, 1993). The most important predictors encompass environmental filters (latitude, altitude, water temperature, conductivity, pH, lake area, stream width, river depth, water velocity and shading) showing the importance of using multiple sets of predictors to understand the processes structuring biodiversity distribution (e.g. Alahuhta *et al.* 2017; Heino *et al.* 2017b; Soininen *et al.* 2018; Perez Rocha *et al.* 2019; Oikonomou and Stefanidis, 2020). According to Heino *et al.* (2015a), environmental control may not be the sole mechanism affecting community composition dissimilarity among sites, but it is likely to be most important when dispersal rates are intermediate (i.e. among sites within a river basin). In contrast, if dispersal rates are very high (i.e. within a stream) or very low (i.e. within an ecoregion), environmental control is probably masked by high dispersal rates or is prevented from occurring because not all species can reach all localities, respectively.

Recent research on freshwater β-diversity has concluded that turnover is the dominant component of overall β-diversity; typically, turnover is more than five times larger than nestedness (reviewed by Soininen *et al.* 2018). Indeed, high β-diversity of European lentic macrophytes, chironomids and cladoceran was primarily explained by species replacement (turnover) rather than differences in species richness (i.e. nestedness) (Viana *et al.* 2016; Alahuhta *et al.* 2017; Specziár *et al.* 2018). The same pattern has been observed for lacustrine fishes and diatoms (Heino *et al.* 2017; Szabó *et al.* 2019; Oikonomou and Stefanidis, 2020). High β-diversity was also largely attributable to the turnover component for stream bryophytes, diatoms, insects and fishes (Heino *et al.* 2017b; Oikonomou *et al.* 2014). In contrast, the richness difference component was more important in contributing to variation in β-diversity for macroinvertebrates across Arctic Fennoscandia (Brittain *et al.* 2020) and fish faunas of Europe (Leprieur *et al.* 2009).

9.3. Conclusion

Freshwater ecosystems are among the most vulnerable systems in the Anthropocene era. Multiple species have already gone extinct, and many are at the brink of extinction. Climate change is expected to magnify existing pressures, leading to a further degradation of freshwater communities at multiple scales, from local to global. Freshwater biogeography can provide insightful tools and concepts for understanding and predicting potential consequences of current human-induced global changes (e.g. climate change, habitat loss and invasive species), and thus, help for scientifically informed decisions about what and where to protect. Researchers often focus on small spatial and short temporal scales, but these may be only weakly linked to questions that managers must address at larger spatial and longer temporal scales. In this chapter, the author draws together threads of recent theoretical and empirical results and patterns at multiple scales; both may offer a useful roadmap of theoretical background for identifying new paths of investigation and future challenges into the field of freshwater biogeography, which needs to be considered to safeguard the status of aquatic ecosystems.

9.4. Acknowledgments

Anthi Oikonomou's research is co-financed by Greece and the European Union (European Social Fund – ESF) through the Operational Program "Human Resources Development, Education and Lifelong Learning" in the context of the project "Reinforcement of Postdoctoral Researchers – 2nd Cycle" (MIS-5033021), implemented by the State Scholarships Foundation (IKY).

9.5. References

Alahuhta, J. (2015). Geographic patterns of lake macrophyte communities and species richness at regional scale. *J. Veg. Sci.*, 26, 564–575.

Alahuhta, J., Kosten, S., Akasaka, M., Auderset, D., Azzella, M.M., Bolpagni, R., Bove, C.P., Chambers, P.A., Chappuis, E., Clayton, J., de Winton, M., Ecke, F., Gacia, E., Gecheva, G., Grillas, P., Hauxwell, J., Hellsten, S., Hjort, J., Hoyer, M.V., Ilg, C., Kolada, A., Kuoppala, M., Lauridsen, T., Li, E.H., Lukács, B.A., Mjelde, M., Mikulyuk, A., Mormul, R.P., Nishihiro, J., Oertli, B., Rhazi, L., Rhazi, M., Sass, L., Schranz, C., Søndergaard, M., Yamanouchi, T., Yu, Q., Wang, H., Willby, N., Zhang, X.K., Heino, J. (2017). Global variation in the beta diversity of lake macrophytes is driven by environmental heterogeneity rather than latitude. *J. Biogeogr.*, 44, 1758–1769.

Alahuhta, J., Lindholm, M., Bove, C.P., Chappuis, E., Clayton, J., de Winton, M., Feldmann, T., Ecke, F., Gacia, E., Grillas, P., Hoyer, M.V., Johnson, L.B., Kolada, A., Kosten, S., Lauridsen, T., Lukács, B.A., Mjelde, M., Mormul, R.P., Rhazi, L., Rhazi, M., Sass, L., Søndergaard, M., Xu, J., Heino, J. (2018). Global patterns in the metacommunity structuring of lake macrophytes: Regional variations and driving factors. *Oecologia*, 188, 1167–1182.

Alahuhta, J., Antikainen, H., Hjort, J., Helm, A., Heino, J. (2020). Current climate overrides historical effects on species richness and range size of freshwater plants in Europe and North America. *J. Ecol.*, 108, 1262–1275.

Allan, J.D. and Castillo, M.M. (2007). *Stream Ecology: Structure and Function of Running Waters*. Chapman and Hall, London.

Allen, A.P., Whittier, T.R., Larsen, D.P., Kaufmann, P.R., O'Connor, R.J., Hughes, R.M., Stemberger, R.S., Dixit, S.S., Brinkhurst, R.O., Herlihy, A.T., Paulsen, S.G. (1999). Concordance of taxonomic composition patterns across multiple lake assemblages: Effects of scale, body size, and land use. *Can. J. Fish. Aquat. Sci.*, 56, 2029–2040.

Anderson, M.J., Crist, T.O., Chase, J.M., Vellend, M., Inouye, B.D., Freestone, A.L., Sanders, N.J., Cornell, H.V., Comita, L.S., Davies, K.F., Harrison, S.P., Kraft, N.J.B., Stegen, J.C., Swenson, N.G. (2011). Navigating the multiple meanings of β diversity: A roadmap for the practicing ecologist. *Ecol. Lett.*, 14, 19–28.

Arrhenius, O. (1920). Distribution of the species over the area. *Medd. från Vetenskapsakadmiens Nobelinstitut*, 4, 1–6.

Arrhenius, O. (1921). Species and area. *J. Ecol.*, 9, 95–99.

Balian, E.V., Segers, H., Lévèque, C., Martens, K. (2008). The freshwater animal diversity assessment: An overview of the results. *Hydrobiologia*, 595, 627–637.

Baselga, A. (2010). Partitioning the turnover and nestedness components of beta diversity. *Glob. Ecol. Biogeogr.*, 19, 134–143.

Bolgovics, Á., Ács, É., Várbíró, G., Görgényi, J., Borics, G. (2016). Species area relationship (SAR) for benthic diatoms: A study on aquatic islands. *Hydrobiologia*, 764, 91–102.

Boll, T., Levi, E.E., Bezirci, G., Özuluğ, M., Tavşanoğlu, Ü.N., Çakıroğlu, A.İ., Özcan, S., Brucet, S., Jeppesen, E., Beklioğlu, M. (2016). Fish assemblage and diversity in lakes of western and central Turkey: Role of geo-climatic and other environmental variables. *Hydrobiologia*, 771, 31–44.

Borges, P.P., Dias, M.S., Carvalho, F.R., Casatti, L., Pompeu, P.S., Cetra, M., Tejerina-Garro, F.L., Súarez, Y.R., Nabout, J.C., Teresa, F.B. (2020). Stream fish metacommunity organisation across a Neotropical ecoregion: The role of environment, anthropogenic impact and dispersal-based processes. *PLoS One*, 15, 1–18.

Brittain, J.E., Heino, J., Friberg, N., Aroviita, J., Kahlert, M., Karjalainen, S.M., Keck, F., Lento, J., Liljaniemi, P., Mykrä, H., Schneider, S.C., Ylikörkkö, J. (2020). Ecological correlates of riverine diatom and macroinvertebrate alpha and beta diversity across Arctic Fennoscandia. *Freshw. Biol.*, 1–15 [Online]. Available at: https://doi.org/10.1111/fwb.13616.

Brown, L.E., Khamis, K., Wilkes, M., Blaen, P., Brittain, J.E., Carrivick, J.L., Fell, S., Friberg, N., Füreder, L., Gislason, G.M., Hainie, S., Hannah, D.M., James, W.H.M., Lencioni, V., Olafsson, J.S., Robinson, C.T., Saltveit, S.J., Thompson, C., Milner, A.M. (2018). Functional diversity and community assembly of river invertebrates show globally consistent responses to decreasing glacier cover. *Nat. Ecol. Evol.*, 2, 325–333.

Brucet, S., Pédron, S., Mehner, T., Lauridsen, T.L., Argillier, C., Winfield, I.J., Volta, P., Emmrich, M., Hesthagen, T., Holmgren, K., Benejam, L., Kelly, F., Krause, T., Palm, A., Rask, M., Jeppesen, E. (2013). Fish diversity in European lakes: Geographical factors dominate over anthropogenic pressures. *Freshw. Biol.*, 58, 1779–1793.

Bunn, S.E. and Hughes, J.M. (1997). Dispersal and recruitment in streams: Evidence from genetic studies. *J. North Am. Benthol. Soc.*, 16, 338–346.

Carlsson, R. (2001). Species-area relationships, water chemistry and species turnover of freshwater snails on the Åland Islands, Southwestern Finland. *J. Mollusc. Stud.*, 67(1), 17–26.

Carvajal-Quintero, J., Villalobos, F., Oberdorff, T., Grenouillet, G., Brosse, S., Hugueny, B., Jézéquel, C., Tedesco, P.A. (2019). Drainage network position and historical connectivity explain global patterns in freshwater fishes' range size. *Proc. Natl. Acad. Sci. USA*, 116, 13434–13439.

Carvalho, J.C., Cardoso, P., Gomes, P. (2012). Determining the relative roles of species replacement and species richness differences in generating beta-diversity. *Glob. Chang. Biol.*, 21, 760–771.

Chase, J. and Leibold, M. (2003). *Ecological Niches*. University of Chicago Press, Chicago.

Collen, B., Ram, M., Dewhurst, N., Clausnitzer, V., Kalkman, V., Cumberlidge, N., Baillie, J. (2009). Broadening wildlife, the coverage of biodiversity assessments. In *IUCN, in a Changing World: An Analysis of the 2008 Red List of Threatened Species*, Vié, J., Hilton-Taylor, C., Stuart, S.N. (eds). IUCN, Gland.

Darwall, W.R.T., Holland, R.A., Smith, K.G., Allen, D., Brooks, E.G.E., Katarya, V., Pollock, C.M., Shi, Y., Clausnitzer, V., Cumberlidge, N., Cuttelod, A., Dijkstra, K.D.B., Diop, M.D., García, N., Seddon, M.B., Skelton, P.H., Snoeks, J., Tweddle, D., Vié, J.C. (2011). Implications of bias in conservation research and investment for freshwater species. *Conserv. Lett.*, 4, 474–482.

Datry, T., Melo, A.S., Moya, N., Zubieta, J., de la Barra, E., Oberdorff, T. (2016). Metacommunity patterns across three Neotropical catchments with varying environmental harshness. *Freshw. Biol.*, 61, 277–292.

Davies, B.R., Beilfuss, R.D., Thoms, M.C. (2000). Cahora Bassa retrospective, 1974–1997: Effects of flow regulation on the Lower Zambezi River Cahora Bassa retrospective, 1974–1997: Effects of flow regulation on the Lower Zambezi River. *Verhandlungen des Int. Verein Limnol.*, 27, 1–9.

Dehling, D.M., Hof, C., Brändle, M., Brandl, R. (2010). Habitat availability does not explain the species richness patterns of European lentic and lotic freshwater animals. *J. Biogeogr.*, 37, 1919–1926.

Dias, M.S., Oberdorff, T., Hugueny, B., Leprieur, F., Jézéquel, C., Cornu, J.F., Brosse, S., Grenouillet, G., Tedesco, P.A. (2014). Global imprint of historical connectivity on freshwater fish biodiversity. *Ecol. Lett.*, 17, 1130–1140.

Dodds, W.K., Bruckerhoff, L., Batzer, D., Schechner, A., Pennock, C., Renner, E., Tromboni, F., Bigham, K., Grieger, S. (2019). The freshwater biome gradient framework: Predicting macroscale properties based on latitude, altitude, and precipitation. *Ecosphere*, 10, e02786.

Dong, X., Li, B., He, F., Gu, Y., Sun, M., Zhang, H., Tan, L., Xiao, W., Liu, S., Cai, Q. (2016). Flow directionality, mountain barriers and functional traits determine diatom metacommunity structuring of high mountain streams. *Sci. Rep.*, 6, 1–11.

Doretto, A., Piano, E., Larson, C.E. (2020). The river continuum concept: Lessons from the past and perspectives for the future. *Can. J. Fish. Aquat. Sci.*, 12, 1–12.

Douglas, M. and Lake, P. (1994). Species richness of stream stones: An investigation of the mechanisms generating the species-area relationship. *Oikos*, 69, 387–396.

Eadie, J.M.A., Hurly, T.A., Montgomerie, R.D., Teather, K.L. (1986). Lakes and rivers as islands: Species-area relationships in the fish faunas of Ontario. *Environ. Biol. Fishes*, 15, 81–89.

Ellis, L.E. and Jones, N.E. (2013). Longitudinal trends in regulated rivers: A review and synthesis within the context of the serial discontinuity concept. *Environ. Rev.*, 21, 136–148.

Fausch, K.D., Torgersen, C.E., Baxter, C.V., Li, H.W. (2002). Landscapes to riverscapes: Bridging the gap between research and conservation of stream fishes. *Bioscience*, 52, 483–498.

Faustino-Queiroz, A.C. and Terra, B.F. (2020). Ecological drivers of fish metacommunities: Environmental and spatial factors surpass predation in structuring metacommunities of intermittent rivers. *Ecol. Freshw. Fish*, 29, 145–155.

García-Girón, J., Fernández-Aláez, C., Fernández-Aláez, M., Alahuhta, J. (2019). Untangling the assembly of macrophyte metacommunities by means of taxonomic, functional and phylogenetic beta diversity patterns. *Sci. Total Environ.*, 693, 133616.

Gilbert, B. and Lechowicz, M.J. (2004). Neutrality, niches, and dispersal in a temperate forest understory. *Proc. Natl. Acad. Sci. USA*, 101, 7651–7656.

Gleason, H.A. (1922). On the relation between species and area. *Ecology*, 3, 158–162.

Griffiths, D. (2015). Connectivity and vagility determine spatial richness gradients and diversification of freshwater fish in North America and Europe. *Biol. J. Linn. Soc.*, 116, 773–786.

Griffiths, D., McGonigle, C., Quinn, R. (2014). Climate and species richness patterns of freshwater fish in North America and Europe. *J. Biogeogr.*, 41, 452–463.

Havel, J.E. and Shurin, J.B. (2004). Mechanisms, effects, and scales of dispersal in freshwater zooplankton. *Limnol. Oceanogr.*, 49, 1229–1238.

Heino, J. (2009). Biodiversity of aquatic insects: Spatial gradients and environmental correlates of assemblage-level measures at large scales. *Freshw. Rev.*, 2, 1–29.

Heino, J. (2011). A macroecological perspective of diversity patterns in the freshwater realm. *Freshw. Biol.*, 56, 1703–1722.

Heino, J., Melo, A.S., Bini, L.M. (2015a). Reconceptualising the beta diversity-environmental heterogeneity relationship in running water systems. *Freshw. Biol.*, 60, 223–235.

Heino, J., Melo, A.S., Siqueira, T., Soininen, J., Valanko, S., Bini, L.M. (2015b). Metacommunity organisation, spatial extent and dispersal in aquatic systems: Patterns, processes and prospects. *Freshw. Biol.*, 60, 845–869.

Heino, J., Soininen, J., Alahuhta, J., Lappalainen, J., Virtanen, R. (2017). Metacommunity ecology meets biogeography: Effects of geographical region, spatial dynamics and environmental filtering on community structure in aquatic organisms. *Oecologia*, 183, 121–137.

Helminen, H., Karjalainen, J., Kurkilahti, M., Rask, M., Sarvala, J. (2000). Eutrophication and fish biodiversity in Finnish lakes. *Int. Vereinigung für Theor. und Angew. Limnol. Verhandlungen*, 27, 194–199.

Herbert, M.E., Mcintyre, P.B., Doran, P.J., Allan, J.D., Abell, R. (2010). Terrestrial reserve networks do not adequately represent aquatic ecosystems. *Conserv. Biol.*, 24, 1002–1011.

HilleRisLambers, J., Adler, P., Harpole, W., Levine, J., Mayfield, M. (2012). Rethinking community assembly through the lens of coexistence theory. *Annu. Rev. Ecol. Evol. Syst.*, 43, 227–248.

Hubbell, S.P. (2001). *The Unified Neutral Theory of Biodiversity and Biogeography.* Princeton University Press, Princeton, NJ.

Hubert, N., Calcagno, V., Etienne, R., Mouquet, N. (2015). Metacommunity speciation models and their implications for diversification theory. *Ecol. Lett.*, 18, 864–881.

Hugueny, B. (1989). West African rivers as biogeographic islands: Species richness of fish communities. *Oecologia*, 79, 236–243.

Hugueny, B., Oberdorff, T., Tedescco, P.A. (2010). Community ecology of river fishes: A large-scale perspective. *Community Ecol., Stream Fishes Concepts, Approaches, Tech.*, 73, 29–62.

von Humboldt, A. and Bonpland, A. (1805). *Essai sur la géographie des plantes.* Chez Levrault, Schoell et compagnie, Paris.

IUCN (2020). The IUCN Red List of Threatened Species. Version 2020–6. Available at: https://www.iucnredlist.org.

Jackson, D.A., Peres-Neto, P.R., Olden, J.D. (2001). What controls who is where in freshwater fish communities – The roles of biotic, abiotic, and spatial factors. *Can. J. Fish. Aquat. Sci.*, 58, 157–170.

Jézéquel, C., Tedesco, P.A., Darwall, W., Dias, M.S., Frederico, R.G., Hidalgo, M., Hugueny, B., Maldonado-Ocampo, J., Martens, K., Ortega, H., Torrente-Vilara, G., Zuanon, J., Oberdorff, T. (2020). Freshwater fish diversity hotspots for conservation priorities in the Amazon Basin. *Conserv. Biol.*, 34, 956–965.

Jones, N. (2011). Spatial patterns of benthic invertebrates in regulated and natural rivers. *River Res. Appl.*, 29, 343–351.

Keck, F., Franc, A., Kahlert, M. (2018). Disentangling the processes driving the biogeography of freshwater diatoms: A multiscale approach. *J. Biogeogr.*, 45, 1582–1592.

Koleff, P., Gaston, K.J., Lennon, J.J. (2003). Measuring beta diversity for presence – Absence data. *J. Anim. Ecol.*, 72, 367–382.

Leal, C.G., Lennox, G.D., Ferraz, S.F.B., Ferreira, J., Gardner, T.A., Thomson, J.R., Berenguer, E., Lees, A.C., Hughes, R.M., MacNally, R., Aragão, L.E.O.C., De Brito, J.G., Castello, L., Garrett, R.D., Hamada, N., Juen, L., Leitão, R.P., Louzada, J., Morello, T.F., Moura, N.G., Nessimian, J.L., Oliveira-Junior, J.M.B., Oliveira, V.H.F., De Oliveira, V.C., Parry, L., Pompeu, P.S., Solar, R.R.C., Zuanon, J., Barlow, J. (2020). Integrated terrestrial-freshwater planning doubles conservation of tropical aquatic species. *Science*, 80(370), 117–121.

Leboucher, T., Tison-Rosebery, J., Budnick, W.R., Jamoneau, A., Vyverman, W., Soininen, J., Boutry, S., Passy, S.I. (2020). A metacommunity approach for detecting species influenced by mass effect. *J. Appl. Ecol.*, 1–10.

Legendre, P. (2019). A temporal beta-diversity index to identify sites that have changed in exceptional ways in space-time surveys. *Ecol. Evol.*, 9, 3500–3514.

Leibold, M. and Chase, J. (2017). *Metacommunity Ecology*. Princeton University Press, Princeton, NJ.

Leibold, M.A., Holyoak, M., Mouquet, N., Amarasekare, P., Chase, J.M., Hoopes, M.F., Holt, R.D., Shurin, J.B., Law, R., Tilman, D., Loreau, M., Gonzalez, A. (2004). The metacommunity concept: A framework for multi-scale community ecology. *Ecol. Lett.*, 7, 601–613.

Leprieur, F. and Oikonomou, A. (2014). The need for richness-independent measures of turnover when delineating biogeographical regions. *J. Biogeogr.*, 41, 417–420.

Leprieur, F., Olden, J.D., Lek, S., Brosse, S. (2009). Contrasting patterns and mechanisms of spatial turnover for native and exotic freshwater fish in Europe. *J. Biogeogr.*, 36, 1899–1912.

LeRoy Poff, N. (1997). Landscape filters and species traits: Towards mechanistic understanding and prediction in stream ecology. *J. North Am. Benthol. Soc.*, 16, 391–409.

Leroy, B., Dias, M.S., Giraud, E., Hugueny, B., Jézéquel, C., Leprieur, F., Oberdorff, T., Tedesco, P.A. (2019). Global biogeographical regions of freshwater fish species. *J. Biogeogr.*, 46, 2407–2419.

Levins, R. (1969). The effect of random variations on different types of population growth. *Proc. Natl. Acad. Sci. USA*, 1061–1065.

Levy, A., Doadrio, I., Almada, V.C. (2009). Historical biogeography of European leuciscins (Cyprinidae): Evaluating the Lago Mare dispersal hypothesis. *J. Biogeogr.*, 36, 55–65.

Li, Z., Wang, J., Meng, X., Heino, J., Sun, M., Jiang, X., Xie, Z. (2019). Disentangling the effects of dispersal mode on the assembly of macroinvertebrate assemblages in a heterogeneous highland region. *Freshw. Sci.*, 38, 170–182.

Logue, J.B., Langenheder, S., Andersson, A.F., Bertilsson, S., Drakare, S., Lanzén, A., Lindström, E.S. (2012). Freshwater bacterioplankton richness in oligotrophic lakes depends on nutrient availability rather than on species-area relationships. *ISME J.*, 6, 1127–1136.

Lomolino, M.V. (2004). Conservation biogeography. In *Frontiers of Biogeography: New Directions in the Geography of Nature*, Lomolino, M.V. and Heaney, L.R. (eds). Sinauer Associates, Sunderland, MA.

Losos, J.B. and Schluter, D. (2000). Analysis of an evolutionary species–area relationship. *Nature*, 408, 847–850.

MacArthur, R.H. and Wilson, E.O. (1967). *The Theory of Island Biogeography*. Princeton University Press, Princeton, NJ.

Magurran, A.E., Dornelas, M., Moyes, F., Henderson, P.A. (2019). Temporal β diversity – A macroecological perspective. *Glob. Ecol. Biogeogr.*, 28, 1949–1960.

Matthews, T.J., Guilhaumon, F., Triantis, K.A., Borregaard, M.K., Whittaker, R.J. (2016). On the form of species–area relationships in habitat islands and true islands. *Glob. Ecol. Biogeogr.*, 25, 847–858.

Matthews, T.J., Triantis, K.A., Whittaker, R.J. (2021). *The Species–Area Relationship: Theory and Application (Ecology, Biodiversity and Conservation)*. Cambridge University Press, Cambridge.

Mittelbach, G.G. and Schemske, D.W. (2015). Ecological and evolutionary perspectives on community assembly. *Trends Ecol. Evol.*, 30, 241–247.

Mittelbach, G.G., Schemske, D.W., Cornell, H.V., Allen, A.P., Brown, J.M., Bush, M.B., Harrison, S.P., Hurlbert, A.H., Knowlton, N., Lessios, H.A., McCain, C.M., McCune, A.R., McDade, L.A., McPeek, M.A., Near, T.J., Price, T.D., Ricklefs, R.E., Roy, K., Sax, D.F., Schluter, D., Sobel, J.M., Turelli, M. (2007). Evolution and the latitudinal diversity gradient: Speciation, extinction and biogeography. *Ecol. Lett.*, 10, 315–331.

Murphy, K., Efremov, A., Davidson, T.A., Molina-Navarro, E., Fidanza, K., Crivelari Betiol, T.C., Chambers, P., Tapia Grimaldo, J., Varandas Martins, S., Springuel, I., Kennedy, M., Mormul, R.P., Dibble, E., Hofstra, D., Lukács, B.A., Gebler, D., Baastrup-Spohr, L., Urrutia-Estrada, J. (2019). World distribution, diversity and endemism of aquatic macrophytes. *Aquat. Bot.*, 158, 103127.

Murphy, K., Carvalho, P., Efremov, A., Tapia Grimaldo, J., Molina-Navarro, E., Davidson, T.A., Thomaz, S.M. (2020). Latitudinal variation in global range-size of aquatic macrophyte species shows evidence for a Rapoport effect. *Freshw. Biol.*, 65, 1622–1640.

Nikolski, G.V. (1933). On the influence of the rate of flow on the fish fauna of the rivers of central Asia. *J. Anim. Ecol.*, 2, 266.

Oberdorff, T., Lek, S., Guégan, J.F. (1999). Patterns of endemism in riverine fish of the Northern Hemisphere. *Ecol. Lett.*, 2, 75–81.

Oberdorff, T., Tedesco, P.A., Hugueny, B., Leprieur, F., Beauchard, O., Brosse, S., Dürr, H.H. (2011). Global and regional patterns in riverine fish species richness: A review. *Int. J. Ecol.*, 12, Article ID 967631.

Oikonomou, A. and Stefanidis, K. (2020). A- and β-diversity patterns of macrophytes and freshwater fishes are driven by different factors and processes in lakes of the unexplored Southern Balkan biodiversity hotspot. *Water (Switzerland)*, 12(7), 1984.

Oikonomou, A., Leprieur, F., Leonardos, I.D. (2014). Biogeography of freshwater fishes of the Balkan Peninsula. *Hydrobiologia*, 738, 205–220.

Olden, J.D., Kennard, M.J., Leprieur, F., Tedesco, P.A., Winemiller, K.O., García-Berthou, E. (2010). Conservation biogeography of freshwater fishes: Recent progress and future challenges. *Divers. Distrib.*, 16, 496–513.

Passy, S.I. (2010). A distinct latitudinal gradient of diatom diversity is linked to resource supply. *Ecology*, 91, 36–41.

Pearson, R.G., Boyero, L., Journal, S., American, N., Society, B., June, N. (2009). Gradients in regional diversity of freshwater taxa. *J. N. Am. Benthol. Soc.*, 28, 504–514.

Pelayo-Villamil, P., Guisande, C., Vari, R.P., Manjarrés-Hernández, A., García-Roselló, E., González-Dacosta, J., Heine, J., González Vilas, L., Patti, B., Quinci, E.M., Jiménez, L.F., Granado-Lorencio, C., Tedesco, P.A., Lobo, J.M. (2015). Global diversity patterns of freshwater fishes – Potential victims of their own success. *Divers. Distrib.*, 21, 345–356.

Perea, S., Böhme, M., Zupančič, P., Freyhof, J., Šanda, R., Özulu, M., Abdoli, A., Doadrio, I. (2010). Phylogenetic relationships and biogeographical patterns in Circum-Mediterranean subfamily Leuciscinae (Teleostei, Cyprinidae) inferred from both mitochondrial and nuclear data. *BMC Evol. Biol.*, 10, 1–27.

Petersen, I., Masters, Z., Hildrew, A.G., Ormerod, S.J. (2004). Dispersal of adult aquatic insects in catchments of differing land use. *J. Appl. Ecol.*, 41, 934–950.

Pontarp, M., Bunnefeld, L., Cabral, J.S., Etienne, R.S., Fritz, S.A., Gillespie, R., Graham, C.H., Hagen, O., Hartig, F., Huang, S., Jansson, R., Maliet, O., Münkemüller, T., Pellissier, L., Rangel, T.F., Storch, D., Wiegand, T., Hurlbert, A.H. (2019). The latitudinal diversity gradient: Novel understanding through mechanistic eco-evolutionary models. *Trends Ecol. Evol.*, 34, 211–223.

Poole, G.C. (2002). Fluvial landscape ecology: Addressing uniqueness within the river discontinuum. *Freshw. Biol.*, 47, 641–660.

Reche, I., Pulido-Villena, E., Morales-Baquero, R., Casamayor, E.O. (2005). Does ecosystem size determine aquatic bacteria richness? *Ecology*, 86, 1715–1722.

van Rees, C.B., Waylen, K.A., Schmidt-Kloiber, A., Thackeray, S.J., Kalinkat, G., Martens, K., Domisch, S., Lillebø, A.I., Hermoso, V., Grossart, H.P., Schinegger, R., Decleer, K., Adriaens, T., Denys, L., Jarić, I., Janse, J.H., Monaghan, M.T., De Wever, A., Geijzendorffer, I., Adamescu, M.C., Jähnig, S.C. (2021). Safeguarding freshwater life beyond 2020: Recommendations for the new global biodiversity framework from the European experience. *Conserv. Lett.*, 14, 1–17.

Reid, A.J., Carlson, A.K., Creed, I.F., Eliason, E.J., Gell, P.A., Johnson, P.T.J., Kidd, K.A., MacCormack, T.J., Olden, J.D., Ormerod, S.J., Smol, J.P., Taylor, W.W., Tockner, K., Vermaire, J.C., Dudgeon, D., Cooke, S.J. (2019). Emerging threats and persistent conservation challenges for freshwater biodiversity. *Biol. Rev.*, 94, 849–873.

Ricklefs, R.E. (1987). Community diversity: Relative roles of local and regional processes. *Science*, 235(80), 167–171.

Rodríguez-Alcalá, O., Blanco, S., García-Girón, J., Jeppesen, E., Irvine, K., Nõges, P., Nõges, T., Gross, E.M., Bécares, E. (2020). Large-scale geographical and environmental drivers of shallow lake diatom metacommunities across Europe. *Sci. Total Environ.*, 707, 135887.

Rohde, K. (1992). Latitudinal gradients in species diversity: The search for the primary cause. *Oikos*, 65, 514–527.

Rosenzweig, M. (1995). *Species Diversity in Space and Time*. Cambridge University Press, Cambridge.

Rowe, D.C., Pierce, C.L., Wilton, T.F. (2009). North American journal of fisheries physical habitat and fish assemblage relationships with landscape variables at multiple spatial scales in Wadeable Iowa streams. *North Am. J. Fish. Manag.*, 29, 1333–1351.

Sayer, C.A., Maiz-Tome, L., Darwall, W.R.T. (eds) (2018). Freshwater biodiversity in the Lake Victoria Basin: Guidance for species conservation, site protection, climate resilience and sustainable livelihoods. *A Critical Sites Network for Freshwater Biodiversity in the Lake Victoria Catchment*. IUCN, Cambridge, UK and Gland, Switzerland.

Schiaffino, M.R., Lara, E., Fernández, L.D., Balagué, V., Singer, D., Seppey, C.C.W., Massana, R., Izaguirre, I. (2016). Microbial eukaryote communities exhibit robust biogeographical patterns along a gradient of Patagonian and Antarctic lakes. *Environ. Microbiol.*, 18, 5249–5264.

Schlosser, I.J. (1985). Flow regime, juvenile abundance, and the assemblage structure of stream fishes. *Ecology*, 66, 1484–1490.

Schmera, D., Árva, D., Boda, P., Bódis, E., Bolgovics, Á., Borics, G., Csercsa, A., Deák, C., Krasznai, E., Lukács, B.A., Mauchart, P., Móra, A., Sály, P., Specziár, A., Süveges, K., Szivák, I., Takács, P., Tóth, M., Várbíró, G., Vojtkó, A.E., Erős, T. (2018). Does isolation influence the relative role of environmental and dispersal-related processes in stream networks? An empirical test of the network position hypothesis using multiple taxa. *Freshw. Biol.*, 63, 74–85.

Shmida, A. and Wilson, M.V. (1985). Biological determinants of species diversity. *J. Biogeogr.*, 12, 1.

Smith, C. and Powell, C. (1971). The summer fish communities of brier creek, Marshall County, Oklahoma. *Am. Museum Novit.*, 2458, 1–30.

Soininen, J. (2014). A quantitative analysis of species sorting across organisms and ecosystems. *Ecology*, 95, 3284–3292.

Soininen, J. and Teittinen, A. (2019). Fifteen important questions in the spatial ecology of diatoms. *Freshw. Biol.*, 64, 2071–2083.

Soininen, J., Heino, J., Wang, J. (2018). A meta-analysis of nestedness and turnover components of beta diversity across organisms and ecosystems. *Glob. Ecol. Biogeogr.*, 27, 96–109.

Sommerwerk, N., Baumgartner, C., Bloesch, J.D., Hein, T., Ostojić, A., Paunović, M., Schneider-Jacoby, M., Siber, R., Tockner, K. (2009). The Danube River Basin. *Rivers Eur.*, 59–112.

Spiliopoulou, K., Dimitrakopoulos, P.G., Brooks, T.M., Kelaidi, G., Paragamian, K., Kati, V., Oikonomou, A., Vavylis, D., Trigas, P., Lymberakis, P., Darwall, W., Stoumboudi, M.T., Triantis, K.A. (2021). The Natura 2000 network and the ranges of threatened species in Greece. *Biodivers. Conserv.*, 30, 945–961.

Stomp, M., Huisman, J., Mittelbach, G.G., Litchman, E., Klausmeier, C.A. (2011). Large-scale biodiversity patterns in freshwater phytoplankton. *Ecology*, 92, 2096–2107.

Szabó, B., Lengyel, E., Padisák, J., Stenger-Kovács, C. (2019). Benthic diatom metacommunity across small freshwater lakes: Driving mechanisms, β-diversity and ecological uniqueness. *Hydrobiologia*, 828, 183–198.

Tedesco, P.A., Oberdorff, T., Lasso, C.A., Zapata, M., Hugueny, B. (2005). Evidence of history in explaining diversity patterns in tropical riverine fish. *J. Biogeogr.*, 32, 1899–1907.

Tedesco, P.A., Paradis, E., Lévêque, C., Hugueny, B. (2017). Explaining global-scale diversification patterns in actinopterygian fishes. *J. Biogeogr.*, 44, 773–783.

Teittinen, A. and Soininen, J. (2015). Testing the theory of island biogeography for microorganisms patterns for spring diatoms. *Aquat. Microb. Ecol.*, 75, 239–250.

Thorp, J., Thoms, M., Delong, M. (2008). The Riverine ecosystem synthesis. *Toward Conceptual Cohesiveness in River Science*. Academic Press, London.

Tisseuil, C., Cornu, J.F., Beauchard, O., Brosse, S., Darwall, W., Holland, R., Hugueny, B., Tedesco, P.A., Oberdorff, T. (2013). Global diversity patterns and cross-taxa convergence in freshwater systems. *J. Anim. Ecol.*, 82, 365–376.

Tonkin, J.D., Olden, J.D., Altermatt, F., Finn, D.S., Heino, J., Pauls, S.U., Lytle, D.A. (2018). The role of dispersal in river network metacommunities: Patterns, processes, and pathways. *Freshw. Biol.*, 63, 141–163.

Tonn, W.M. (1990). Transactions of the American fisheries society: Climate change and fish communities: A conceptual framework. *Trans. Am. Fish. Soc.*, 119, 37–41.

Triantis, K.A., Guilhaumon, F., Whittaker, R.J. (2012). The island species-area relationship: Biology and statistics. *J. Biogeogr.*, 39, 215–231.

Van Der Gucht, K., Cottenie, K., Muylaert, K., Vloemans, N., Cousin, S., Declerck, S., Jeppesen, E., Conde-Porcuna, J.M., Schwenk, K., Zwart, G., Degans, H., Vyverman, W., De Meester, L. (2007). The power of species sorting: Local factors drive bacterial community composition over a wide range of spatial scales. *Proc. Natl. Acad. Sci. USA*, 104, 20404–20409.

Vardakas, L., Kalogianni, E., Smeti, E., Economou, A.N., Skoulikidis, N.T., Koutsoubas, D., Dimitriadis, C., Datry, T. (2020). Spatial factors control the structure of fish metacommunity in a Mediterranean intermittent river. *Ecohydrol. Hydrobiol.*, 20(3), 346–356.

Viana, D.S., Figuerola, J., Schwenk, K., Manca, M., Hobæk, A., Mjelde, M., Preston, C.D., Gornall, R.J., Croft, J.M., King, R.A., Green, A.J., Santamaría, L. (2016). Assembly mechanisms determining high species turnover in aquatic communities over regional and continental scales. *Ecography (Cop.)*, 39, 281–288.

Vitorino Júnior, O.B., Fernandes, R., Agostinho, C.S., Pelicice, F.M. (2016). Riverine networks constrain β-diversity patterns among fish assemblages in a large Neotropical river. *Freshw. Biol.*, 61, 1733–1745.

Wang, L., Seelbach, P.W., Hughes, R.M. (2006). Introduction to landscape influences on stream habitats and biological assemblages. *Am. Fish. Soc. Symp.*, 48, 1–23.

Ward, J.V. (1989). The four-dimensional nature of lotic ecosystems reviewed work(s): The four-dimensional nature of lotic ecosystems. *J. North Am. Benthol. Soc.*, 8, 2–8.

Ward, J.V. and Stanford, J.A. (1983). The serial discontinuity concept of lotic ecosystems. In *Dynamics of Lotic Ecosystems*, Fontaine, T.D. and Bartell, S.M. (eds). Ann Arbor Science, Ann Arbor, MI.

Watson, H.C. (1835). *Remarks on the Geographical Distribution of British Plants*. Longman, Rees, Orme, Brown, Green, and Longman, London.

Whittaker, R.J., Willis, K.J., Field, R. (2001). Scale and species richness: Towards a general, hierarchical theory of species diversity. *J. Biogeogr.*, 28, 453–470.

Williamson, M. (1988). Relationship of species number to area, distance and other variables. *Anal. Biogeogr.*, 91–115.

Zhao, S., Fang, J., Peng, C., Tang, Z., Piao, S. (2006). Patterns of fish species richness in China's lakes. *Glob. Ecol. Biogeogr.*, 15, 386–394.

10

Marine Biogeography

Jorge GARCÍA MOLINOS and Irene D. ALABIA
Arctic Research Center, Hokkaido University, Sapporo, Japan

10.1. Introduction

Marine biogeography as a scientific subdiscipline emerged amid the 19th century. Transoceanic voyages like those of the *Beagle* (1826–1843), the U.S. Exploring Expedition (1838–1842) or the *Challenger* expedition (1872–1876), and the illustrious names of Charles Darwin (1809–1882), Edward Forbes Jr. (1815–1854), James D. Dana (1813–1895) or John Murray (1841–1914), among others, are associated with the earliest attempts to document the taxonomy and geographic distribution of marine taxa (for a comprehensive account on the subject, see Egerton, 2019) (Figure 10.1). Some of the earliest biogeographical summaries include the remarkable global map of coral distributions by Darwin (1842), Dana's work using isocrymes to delineate broad distribution regions of corals and crustaceans (1853, 1875) or the worldwide map of the distribution of marine life (1856) and Forbes's bioregionalization of European seas (1859).

Advances in biogeography have traditionally been a terrestrial-dominated endeavor. The obvious logistic constraints of accessing and working in remote marine areas, such as the open ocean and the deep sea, account to a large extent for the historical lag of marine biogeography relative to terrestrial biogeography. Marine biogeography has thus largely followed the path marked by terrestrial biogeography in terms of fundamental principles, basic concepts and analytical methods. However, as we will see in this chapter, the unique abiotic and biotic characteristics particular to oceans make knowledge from terrestrial biogeography not always transferable.

Biogeography,
coordinated by Eric GUILBERT.

Figure 10.1. *a) The USS Vincennes during the U.S. Exploring Expedition sailing through Disappointment Bay, Antarctica, around 1840 (unknown artist). James D. Dana, pictured on the right, was among the scientific crew of the expedition. b) Published in "The structure and distribution of coral reefs" (1842), Darwin's map of global coral distribution represented a major advance for marine biogeography (colors in the extract of the map stand for different types of reef formations). For a color version of this figure, see www.iste.co.uk/guilbert/biogeography.zip*

Although division of the discipline into historical and ecological biogeography was first proposed two centuries ago (de Candolle, 1820), in practice (marine) biogeography has largely remained "systematic and historical rather than ecological in its approach" (Hedgpeth, 1957), while exchanges between historical biogeography and ecology have been limited (Wiens and Donoghue, 2004). It was not until the 1970s, with the pioneering work of R. MacArthur (1972) and G. Vermeij (1978), that solid attempts to bridge across both disciplines were first formally made. This allowed us to move from the traditional biogeography focus on the distribution of systematic units (a shift or higher taxa), to that of populations and ecological communities. More recently, this tendency towards an integrative biogeography has consolidated thanks to advancements in data acquisition and accessibility, analytical methods (e.g. Ronquist and Sanmartín, 2011), such as remote sensing, spatial statistics or molecular tools, as well as the emergence of new integrative disciplines such as metacommunity ecology (Leibold *et al.*, 2004) and macroecology (Brown and Maurer, 1989; Brown, 1995). Marine ecology has been particularly receptive to the integration of historical biogeography given the openness and connectedness of marine ecosystems, and the relevance of large-scale temporal and spatial processes, such as larval recruitment, in regulating population dynamics and patterns in biological communities (Valiela, 1995; Witman and Roy, 2009).

10.2. Diversification in the oceans

The oceans present an interesting diversity paradox (Figure 10.2). They comprise 71% of the Earth's surface, making for approximately 665 times more of the habitable volume than terrestrial environments, yet they only host approximately 15% of all described species (May, 1994; Dawson, 2012). Differences between aquatic environments are also obvious. For example, continental waters host about 40% of all fish species although they represent just ~0.01% of available aquatic habitat in contrast to the >99% comprised by marine environments (Lévêque *et al.*, 2008; Eschmeyer *et al.*, 2010). Taking into account the habitable volume, Dawson (2012) estimates marine eukaryotic species density to be three to four orders of magnitude less than that of non-marine environments. Estimates cipher approximately 12–25% of all exiting eukaryotic species (~8.7 million) to be marine, which makes for about 68–91% of them still awaiting for discovery (Mora *et al.*, 2011; Appeltans *et al.*, 2012; Costello *et al.*, 2012). This is counterintuitive given the many properties of marine environments that should make them, at least on first thought, more specious. Other than sheer numbers related to habitable area, which theory predicts should be related to species richness, the oceans are less fragmented and more stable than terrestrial environments, highly productive as a whole; plankton photosynthesis accounts for about half of the biosphere primary production (Field *et al.*, 1998), and, from an evolutionary perspective, supported life hundreds of millions of years before the first species appeared on land during the late Silurian. In contrast, phyletic diversity is far greater in the oceans (Figure 10.2), with numbers of marine species much more spread across phyla than on land, where a single phyla (Arthropoda), for example, accounts for about 56% of all species (May, 1994).

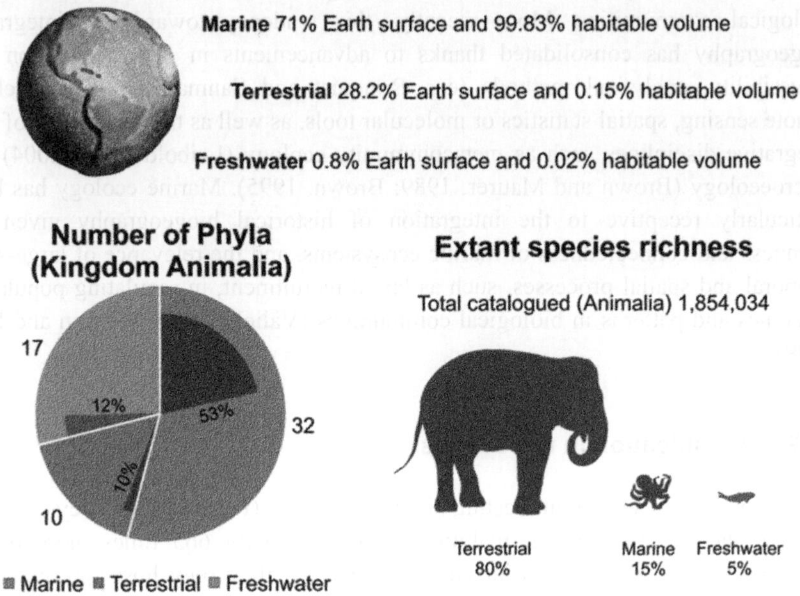

Figure 10.2. *Characteristics of terrestrial, marine and freshwater realms in terms of extent and habitable volume, extant species richness and phyletic diversity. Habitable volume defined from estimates of biome areal coverage and the depth/height of their corresponding habitable environments (Dawson, 2012). Of the 35 extant metazoan phyla comprising the kingdom Animalia, which makes for about 70–80% of all species, the marine realm hosts 32, 53% of them exclusively marine, while only 17 and 10 are present in inland waters and on land, of which 12% and 10% are exclusively freshwater and terrestrial phyla (May, 1994; Zhang* et al., *2013). Contrastingly, the proportion of extant terrestrial species grossly outnumber that of marine species (May, 1994; Grosberg* et al., *2012). For a color version of this figure, see www.iste.co.uk/guilbert/biogeography.zip*

Numerous hypotheses have been proposed to explain the obvious stark differences in patterns of diversification and richness between land and sea. These include differences in air and water as a living medium, the greater endemism, narrower ecological specialization and smaller geographical distributions of terrestrial biota, the more complex and fragmented terrestrial habitats, the larger size of primary producers on land (plankton dominating the ocean), higher relative net-primary productivity on land, lower terrestrial herbivorous pressure, higher diversification rates in terrestrial clades and biased transition rates between lineages over evolutionary time (May, 1994; Briggs, 1995; Benton, 2001; Maddison, 2006; Vermeij and Grosberg, 2010; Grosberg *et al.*, 2012; Wiens, 2015). In aquatic environments, where general differences in productivity (freshwaters as a whole are far less productive) and herbivory pressure (comparable on both environments) cannot be invoked as an explanation, some phylogenetic analyses point towards higher speciation and extinction rates in freshwater linages (Hou *et al.*, 2011; Bloom *et al.*, 2013; Adams *et al.*, 2018). The much higher fragmentation and isolation of freshwater habitats appear as a key driver imposing limits to gene flow and promoting genetic divergence and allopatric (vicariance) speciation while also making freshwater species more exposed to environmental disturbances. Another phylogenetic study comparing patterns of diversity in marine and freshwater actinopterygian (ray-finned) fishes, the most species-rich clade of marine vertebrates representing 96% of fish species, found similar richness and net diversification rates in freshwater and marine habitats (Carrete Vega and Wiens, 2012). Although earliest actinopterygians are marine in origin, the study found evidence suggesting that all contemporary extant marine fishes derive from freshwater ancestors. This result, together with the fact that both marine and freshwater actinopterygians are dominated by highly diverse but relatively recent clades, suggests that ancient marine extinctions may have had a major role in explaining the lower relative richness and diversity patterns of modern marine actinopterygian fish, a pattern that the authors speculate may apply to other ancient marine clades.

Diversification trajectories of marine and terrestrial organisms over geological time show contrasting patterns (Figure 10.3). In contrast to the more sustained, exponential terrestrial diversification, patterns of marine diversification have been traditionally explained using an equilibrium model where recurrent mass extinction events trigger periods of rapid diversification resulting from the intense radiation of life forms as they occupy the new vacant environments (Sepkoski, 1984). These are alternated with equilibrium periods of slower or stalled diversification driven by diversity-dependent processes. Equilibrium is successively attained at higher levels of diversity because evolution makes organisms better adapted or more versatile to exploit the new resources and environments. Although the concept of equilibrium in marine diversification is contested (Benton, 2001), obvious differences in diversification trajectories do reflect fundamental differences in diversification rates between non-marine and marine clades (Wiens, 2015).

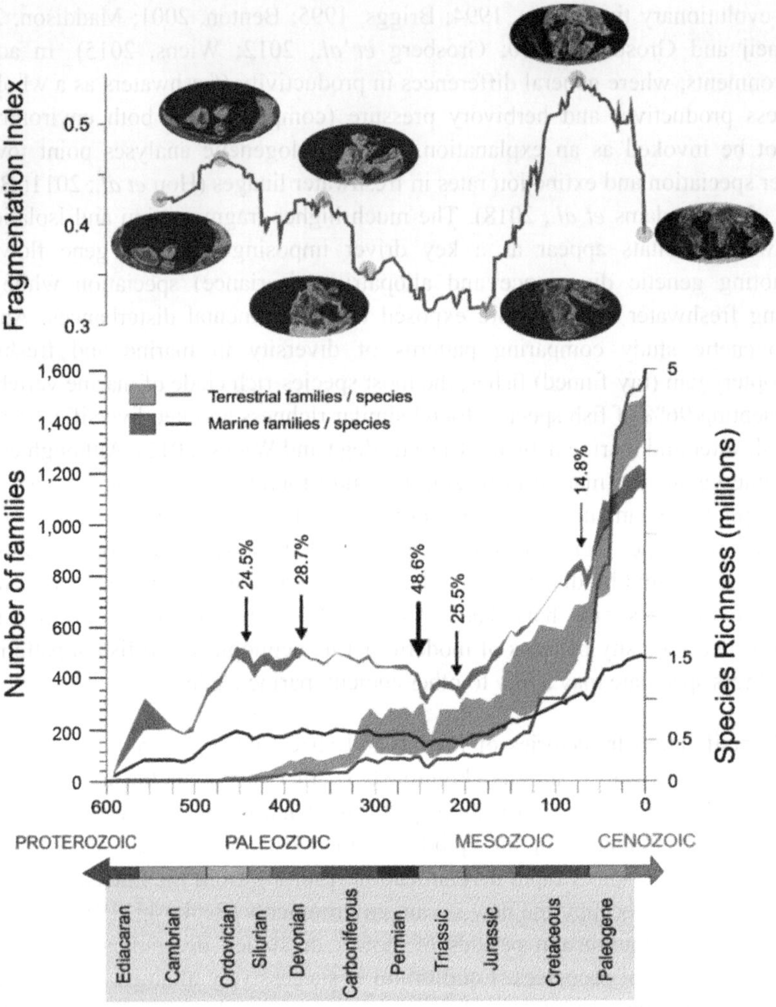

Figure 10.3. *Diversification patterns in marine and terrestrial organisms over geological time with indication of major extinction events, diversification periods and emergence of representative clades. Species- and family-level plots are extracted from Benton (1999, 2016). Family envelopes denote maximum and minimum estimates. The arrows identify major mass extinction events with thickness proportional to their specific extinction rate, as identified in Benton (1999). Note that diversity loss following mass extinction events is masked to different degrees by the subsequent rapid diversification. The upper panel shows the fragmentation index describing the degree of continental fragmentation during the Phanerozoic from Zaffos* et al. *(2017) with accompanying paleogeographic maps for reference periods (Scotese, 2016). For a color version of this figure, see www.iste.co.uk/guilbert/biogeography.zip*

From an evolutionary standpoint, the fossil record indicates that oceans remained richer in life for 3.6 billion years until the great diversification of terrestrial life during the mid-Cretaceous approximately 100–125 Mya ago (Benton, 2016) (Figure 10.3). This diversification explosion has been related to the Cretaceous Terrestrial Revolution signaled by the replacement of ferns and gymnosperms by angiosperms (flowering plants), and the subsequent rapid emergence and diversification of related terrestrial groups, chiefly insects as well as others like birds, lizards and mammals. Although the fossil record indicates the existence of several other marine diversification periods (Vermeij, 2011), most notably the Cambrian explosion when most major animal groups appeared (Briggs, 2015), oceans experienced their own "diversification revolution" during the Mesozoic paving the road towards modern marine biota from approximately 251 to 66 Mya ago (Vermeij, 1977). The Mesozoic Marine Revolution (MMR) saw an enormous diversification of marine life, particularly benthic groups (e.g. echinoids, gastropods, bivalves, crustaceans) and plankton (e.g. foraminifera, coccoliths, dinoflagellates, diatoms), together with a major restructuring of shallow-water benthic communities characterized by a transition from epifaunal (surface-dwelling) and semi-infaunal to infaunal dominated communities (Vermeij, 1977). The MMR involved at least two independent, protracted episodes, the first from the Late Triassic to Early Jurassic and the second during the mid-Cretaceous, overlapped with and related to the intense terrestrial diversification that occurred in parallel (Vermeij, 2011). Vermeij formulated the escalation hypothesis (1987) to explain these changes from the perspective of the role of predation and competition as drivers of natural selection and evolution through adaptation. Essentially, escalation refers to the intense selection resulting from the evolution of more powerful and effective predation and competition on preys and other competitors leading to the rapid development of morphological, physiological and behavioral adaptive and ecological (e.g. colonization of new habitats) responses. In particular, the emergence and proliferation of more powerful and efficient predators, such as durophagous (shell-crushing) predators, and the intensification of grazing during the MMR triggered a number of related important adaptive innovations such as increased metabolism and mobility, the development of shell defenses in molluscs or the colonization of infaunal benthic habitats (Vermeij, 2011).

But what particular conditions concurred at that period that allowed this explosive adaptive breakthrough and the rapid diversification of marine communities? The breakup of Pangaea, which initiated during the mid-Jurassic and culminated during the Late Cretaceous representing the highest rate of continental fragmentation experienced throughout the Phanerozoic (Figure 10.3), has been correlated with the explosive radiation of marine life during the MMR (Zaffos *et al.*, 2017). The rapid continental breakup effectively led to the formation of distinctive ocean basins and biogeographical provinces, while the recursive formation of land bridges and ocean gateways allowed for increased rates of speciation as well as mixing of previously unconnected communities (Owen and Crame, 2002). The Mesozoic was dominated by polytaxic episodes, protracted periods of globally warm, homogeneous climate resulting in higher and more uniform ocean temperatures, increased sea levels that created large areas of shallow seas, reduced oceanic convection, and an expansion and intensification of the oxygen minimum layer (Fischer and Arthur, 1977). Shallow ocean conditions during the tectonically very active Mesozoic were nutrient-rich as a result of large quantities of carbon dioxide and minerals released into the ocean from periodic extreme submarine volcanic eruptions, weathering and erosion of rocks during mountain building events and other tectonic processes (Vermeij, 2011). The Cretaceous diversification of terrestrial angiosperms also meant an important volume of nutrient inputs delivered into the oceans from the increased productivity on land. Together, these conditions of extended warm, nutrient-rich shallow seas would have fueled primary productivity enabling diversification and the metabolically costly novel predatory strategies and related adaptations of the MMR (Bambach, 1999). Increased productivity seems to also have triggered an explosive radiation of fauna developing fertilization via copulation or some other mechanism of reproductive interaction among adults during the Cretaceous–Cenozoic that would have contributed to promote diversification, especially in predators, by facilitating reproductive isolation and smaller population sizes relative to fertilization mechanisms broadcasting sperm into the water (Bush *et al.*, 2016). Major taxa in this nonbroadcasting group includes all marine Vertebrata, Arthropoda, Cephalopoda, Chaetognatha and some Gastropoda (Bush *et al.*, 2016).

The origin of modern deep-sea fauna has proved somehow more controversial. Early explanations formed during the 19th century (Thomson, 1873), reinforced by discoveries of living Paleozoic deep-sea relics like stalked crinoids (sea lilies), conceptualized the deep sea as a stable refuge for ancient clades that had left shallower shelf habitats driven by competition displacement and predation avoidance. This vision is now largely abandoned given evidence for major deep-sea paleoceanographic perturbations with drastic variations in temperature, oxygenation and circulation mode (Kennett and Stott, 1991; Foster *et al.*, 2013; Clarkson *et al.*, 2021). More recent inference-based hypotheses call for a relatively young origin of modern deep-sea fauna resulting from repeated episodes of mass extinction triggered by major perturbation events followed by recolonization from shelf habitats. Particularly, those related to Mesozoic oceanic anoxic events and the posterior development of thermohaline circulation, stratification of the oceans and cooling of deep-water masses during the Cenozoic point to a latest Mesozoic or early Cenozoic origin for modern deep-sea fauna on which biogeographic patterns and molecular clock estimates predominantly converge (Jacobs and Lindberg, 1998; Smith and Stockley, 2005; Strugnell *et al.*, 2008). Arguments have also been made as to whether global deep-sea anoxia effectively prevented colonization of deep-sea environments until the late Cretaceous (Horne 1999). However, recent fossil evidence (Thuy *et al.*, 2012, 2014) demonstrates the existence of macrofaunal groups from the lower Cretaceous and Jurassic for a number of extant deep-sea groups. These findings contest the idea of mass extinction during the Mesozoic and suggest that some groups survived in deep-sea refugia isolated from shallow waters by the formation of anoxic layers (Wilson, 1999). Both immigration from shelf habitats and in situ deep-sea diversification appeared, therefore, to have contributed to shaping modern deep-sea biodiversity. Modern deep-sea fauna comprise a mixture of clades with origins spanning throughout the Phanerozoic. Foraminifera alone, a group with an excellent sediment fossil record, includes genera that range in origin from the early Paleozoic to the Eocene with some 500 million years in between.

10.3. Diversity gradients in the oceans

The search for overarching, global patterns in the geographical distribution of species has been a central focus of biogeography since its origin (Darwin, 1842; Dana, 1853). Early realization of higher accrual of species in the tropics motivated a search for causal drivers and mechanisms, which has become a longstanding goal of biogeography. Recent decades have seen substantial progress on this front thanks to taxonomically and spatially better resolved biological records, increasingly accessible through open-access online data publication infrastructures such as the Ocean Biogeographic Information System (OBIS) and technological and analytical advances such as remote sensing or spatial statistics. Yet a unified, testable theory that coherently explains the mechanisms generating and maintaining patterns of biodiversity remains elusive. This section provides a general overview of the major gradients of diversity distribution in the ocean and possible driving mechanisms at continental to global scales.

10.3.1. *Latitudinal diversity gradients*

The latitudinal diversity gradient (LDG), postulating a decrease of diversity (richness) with latitude from its maximum at the tropics towards a minimum at the poles, is probably the most pervasive and studied global diversity pattern both on land and in the ocean. Traditionally stated as unimodal and symmetric, recent evidence points towards a prevalence in many marine groups of asymmetric bimodal LDGs with tropical or mid-latitude peaks and a marked dip at the equator (Brayard *et al.*, 2005; Chaudhary *et al.*, 2016; Saeedi *et al.*, 2017, 2019a) (Figure 10.4(a)). Nonetheless, others have suggested that bimodality in LDG might result from artifactual absences of species near the equator due to consistent undersampling at low latitudes within the latitudinal range of species, hence generating a false equatorial dip in species richness (Menegotto and Rangel, 2018). Irrespectively, it seems evident that there is no single LDG but rather a gradient of patterns reflecting how different ecological and evolutionary drivers have shaped the distribution of organisms over geological time (Worm and Tittensor, 2018).

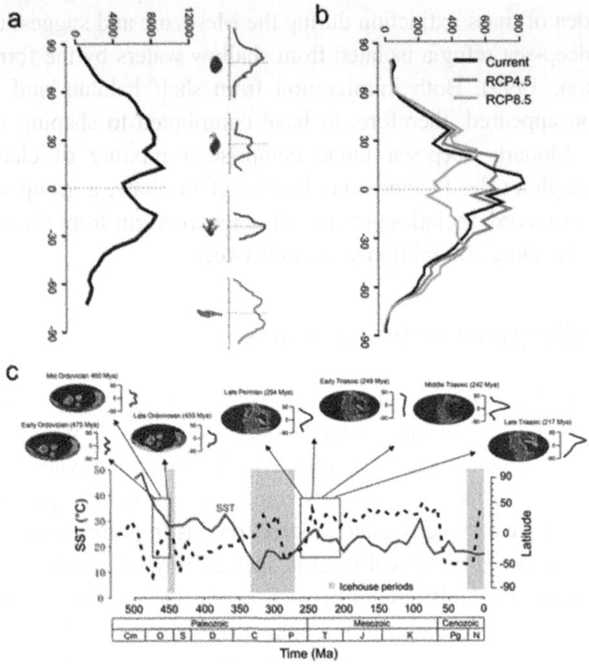

Figure 10.4. *Examples of the variability in historical, current and future patterns of global marine latitudinal diversity gradients (LDGs). (a) Current global LDG derived from occurrence records of 51,670 species (source: Chaudary* et al., *2017), with some simplified LDGs for specific marine groups (bivalves, source: Schumm* et al., *2019; phytoplankton, source: Righetti* et al., *2019; reef (staghorn) corals, source: Muir* et al., *2015; marine mammals, source: Kaschner* et al., *2011). (b) Shifts in the LDGs (12,796 species) projected by year 2100 under the emission stabilization scenario RPC4.5, yielding moderate levels of warming by the end of the century, and the "business as usual" RCP8.5 scenario within the highest warming rates and end points (source: García Molinos* et al., *2016). (c) Global average, low-latitude sea surface temperature (SST) across the Phanerozoic (source: Song* et al., *2019). The dotted line shows the latitudinal position of the peak in brachiopod genus richness reconstructed from the fossil record (source: Powell* et al., *2015). The insets highlight shifts in global richness (4,081 genera) LDGs associated with "greenhouse" and "icehouse" transitions during the Ordovician (source: Kröger, 2018) and Permian–Triassic (source: Song* et al., *2020) (see the main text for details). For a color version of this figure, see www.iste.co.uk/guilbert/biogeography.zip*

Diversity patterns differ among coastal, pelagic and deep-sea habitats (Figure 10.4(a)) (for a comprehensive review on this topic, see Worm and Tittensor, 2018). With exceptions (e.g. cold-adapted groups like pinnipeds), coastal species richness typically exhibit tropical or subtropical latitudinal peaks, driven by the Caribbean and central Indo-Pacific (Coral Triangle) biodiversity hotspots (Worm and Tittensor, 2018). Coastal waters, transitional between terrestrial and marine environments, are characterized by higher nutrient availability and productivity, habitat heterogeneity and complexity, and sharper environmental gradients than off-shore waters. They are also exposed to much higher levels of human impacts. In contrast, the vast pelagic domain is often seen as an open, well-connected and more homogeneous system with low overall productivity. Pelagic species richness displays for many invertebrate and vertebrate groups a more uniform pattern with broader and less pronounced richness peaks at mid- or high latitudes. Biodiversity hotspots are much less pronounced for pelagic species and sparsely concentrated around productive areas associated with frontal systems such as poleward boundaries of subtropical gyres and western boundary currents (Worm and Tittensor, 2018). Finally, despite being far less documented due to limitations with data availability and spatial coverage, deep-sea LDGs appear to also significantly differ from those of other marine domains. Recent work on global diversity patterns of ophiuroid (brittle star) has documented consistent richness peaks in tropical Indo-West Pacific and Caribbean (0–30°) latitudes in coastal to upper slope waters (<2,000 m), a pattern that correlated well with temperature gradients, while richness of deep-sea species follows a bimodal pattern with richness peaking at higher latitudes (30–40° S and 40–50° N) and marked minima near the poles and at northern subtropical latitudes (15–30° N) (Woolley *et al.*, 2016). Deep-sea regions of high richness matched areas of high carbon export flux and were close to continental margins. The authors proposed a species-energy framework to explain these patterns, where kinetic (thermal) energy drives shallow water richness gradients and chemical energy (export productivity) drives deep-sea diversity. The observed relationship with distance from continental margins has been linked to the effect of periodic extinctions of deep-sea abyssal fauna over geological time and the immigration flow and subsequent radiation of deep-sea fauna from bathyal communities (source) into abyssal communities (sink) (Rex *et al.*, 2005). Dissolved oxygen has also been found to be an important predictor of deep-sea richness gradients, particularly in relation to the distribution of oxygen minimum zones (Saeedi *et al.*, 2020).

The underlying ecological and evolutionary drivers that generate and maintain these patterns remain debated, with none of the dozens of hypothesis that have been put forward reaching general consensus (Willig *et al.*, 2003; Pontarp *et al.*, 2019). Hypotheses proposing null models as explanation of the LDG, such as the mid-domain hypothesis postulating highest species richness accrued by chance at the middle (e.g. the tropics) of a geographical domain (the Earth) defined by hard boundaries (the poles) (Colwell and Hurtt, 1994), are useful to generate null expectations independent of ecological and evolutionary drivers against which observed patterns can then be tested (Hillebrand, 2004). For example, Brayard *et al.* (2005) used a simulation model to demonstrate how the coupling of the mid-domain effect with the unimodal sea surface temperature gradient and the geographic origin of randomly generated clades sufficed to realistically reproduce a bimodal LDG with a marked drop of species near the equator observed for Atlantic planktonic foraminifera. Interestingly, thermal gradients imposed a mid-domain effect themselves on the resulting simulated LDG; stronger gradients accentuating the bimodal shape of the LDG that became unimodal only when the applied thermal gradients were very weak. The origin of the clade influenced the latitudinal position of the richness peaks, shifting towards higher latitudes as the ancestral origin moved from the tropics towards mid-latitudes, an effect attributed to the modeled partial inheritance of the thermal tolerance range through speciation.

Deterministic hypotheses for the LDG can be loosely grouped into three categories: (i) ecological, focusing on adaptation to the environment and mechanisms of species coexistence; (ii) evolutionary, with a focus on diversification rates and dispersal; and (iii) historical, related to time for species accumulation and the role of geological (e.g. plate tectonics) and climatic (e.g. glaciations) events through Earth's history in shaping extant patterns of diversity (Mittelbach *et al.*, 2007; Brown, 2014; Pontarp *et al.*, 2019). For example, temperature is often singled out as the strongest driver for extant LDGs at a global scale. Global unimodal temperature gradients have been frequently correlated with observed latitudinal richness gradients across multiple marine groups (Roy *et al.*, 1998; Rutherford *et al.*, 1999; Fuhrman *et al.*, 2008; Tittensor *et al.*, 2010; Yasuhara *et al.*, 2012; Righetti *et al.*, 2019). This positive relationship has been mainly attributed to the effects of temperature on diversity in line with energy-productivity (Clarke and Gaston, 2006) and metabolic (Brown *et al.*, 2004) theory. Higher net energy input into the tropics allows for more productive habitats, hence larger population sizes that may in turn lead to higher diversity of primary producers and consumers through niche specialization. On the other hand, warmer environments can accelerate the rate of evolutionary diversification through the effect of high temperature on increased metabolic rates, faster mutation rates, lower generation times and increased population size (Machac *et al.*, 2012; Zhang *et al.*, 2018). Faster net diversification rates in the tropics can result from faster speciation (tropics seen as a "cradle"), slower extinction rates (tropics seen as a "museum") or a combination of both. The later view leads Jablonski *et al.* (2006) to formulate their "out of the tropics" (OOT) model where fast speciation and low extinction rates lead to higher sustained species richness in the tropics from which some species emigrated and radiated into higher latitudes where higher extinction rates have kept diversity at comparatively lower levels. Evolutionary adaptation of species to the stable warmer and aseasonal tropical climates inhibiting expansion into higher latitudes due to intolerance to more variable and extreme environments (phylogenetic niche conservatism) would explain why only a small portion of species have expanded and radiated outside the tropics. Nevertheless, whereas the fossil record supports the hypothesis of faster tropical speciation for several groups including plankton (Allen and Gillooly, 2006), corals (Kiessling *et al.*, 2010) and molluscs (Jablonski *et al.*, 2006), a recent phylogenetic study found an inverse pattern of highest speciation rates in marine fishes at high latitudes in species-poor regions with high endemism and colder environments (Rabosky *et al.*, 2018).

Temperature adaptation and radiation of species at mid-latitudes overlapping in range with their tropical precursors have also been proposed as a plausible explanation for the bimodality of LDGs observed in multiple marine groups (Brayard *et al.*, 2005; Chaudhary *et al.*, 2016). Temperature-mediated selection imposed by strong thermal gradients, such as those often found at the boundaries of marine bioregions, can also lead to allopatric speciation even in the absence of dispersal barriers (e.g. Teske *et al.*, 2019). Here, niche-related physiological (thermal stress) and ecological constraints are assumed to contribute to the equatorial dip in species richness. The observed bimodality of LDG may also reflect in part a global restructuring of diversity patterns forced by anthropogenic climate change (Chaudhary *et al.*, 2016). Biogeographical shifts are the most common, ubiquitous response to climate warming resulting from a combination of range expansions and contractions due to the spatial reconfiguration of their climatic niches. This is particularly evident in the marine environment (Poloczanska *et al.*, 2013), which is less constrained by dispersal limitation and where species ranges tend to occupy more fully their thermal niche (Sunday *et al.*, 2012) and hence are more sensitive to warming (Pinsky *et al.*, 2019). Under an overarching pattern of shifts towards higher latitudes and deeper waters, mid-latitudes are posed to accrue higher species richness as they receive a large proportion of warm-water climate migrants while retaining most of their local species (García Molinos *et al.*, 2016). In contrast, the tropics, as regions unconnected to warmer environments where species have typically narrower thermal ranges, are projected to experience a strong decline in current diversity levels under future climate warming (Figure 10.4(b)).

The notion of shifting LDGs under opposing global climate states appears to have strong support from the Phanerozoic fossil record depicting alternative scenarios of tropical peaks in diversity during "icehouse" periods (including the last 30 million years), characterized by colder climates and steeper environmental gradients, with temperate peaks or flattened gradients occurring during "greenhouse" and interglacial periods with prevailing warmer and more homogeneous climates (Mannion *et al.*, 2014) (Figure 10.4(c)). These patterns are thought to be due to chronic extinctions of high-latitude, tropical-adapted clades and their reclusion into remnant tropical biomes in equatorial regions during icehouse conditions, whereas the expansion of tropical species and more equivalent diversification rates across latitudes, combined with decreases in tropical diversity, would account for the LDG during greenhouse periods (Meseguer and Condamine, 2020). For example, Kröger (2017) found a clear transition from a bimodal LDG with marked tropical dip following during the Early Ordovician greenhouse period to a modern-type, unimodal LDG during the Mid- to Early Ordovician (Figure 10.4(c)), resulting from a transition from greenhouse to icehouse conditions during the Mid-Ordovician and coincident with the diversification climax of the Great Ordovician Biodiversity Event. Similarly, a flat LDG that lasted for ~5 Ma during the Early Triassic has been documented for marine biota following the Permian–Triassic mass extinction, stronger at tropical latitudes, resulting from dramatic global environmental changes (extreme warming and anoxia) (Song *et al.*, 2020) (Figure 10.4(c)). Surprisingly, the authors document a reversal to the unimodal, modern-like LDG during the still greenhouse conditions of the Mid- to Late Triassic, suggesting that only extreme and variable greenhouse climates produce flat LDGs while more stable periods can still maintain sufficiently strong latitudinal environmental gradients (e.g. solar radiation, day light) to produce "normal" unimodal LDG.

10.3.2. *Bathymetric diversity gradients*

Bathymetric diversity gradients (BDGs) constitute the other main paradigm on geographical gradients of species diversity. Variation of total species richness with depth displays a unimodal gradient following a decreasing trend from shallow to deep waters and maximum richness attained at shallow depths (0–500 m) (Costello and Chaudhary, 2017; Saeedi *et al.*, 2019a). High productivity, habitat complexity and warm temperatures are often invoked as main environmental drivers for these patterns for reasons akin to those explained for the LDGs (Smith and Brown, 2002; Costello and Chaudhary, 2017). As it can be expected, BDGs vary across taxonomic groups. For example, whereas pelagic fishes display maximum diversity at shallow depths (<200 m) (Smith and Brown, 2002), deep-sea benthic groups typically increase from the continental shelf to the bathyal zone, attaining maximum richness at mid-slopes (1,000–2,000 m), and then decrease again into the abyssal zone (Rex, 1973; Danovaro *et al.*, 2008; Saeedi *et al.*, 2019a). Hernández-Ávila *et al.* (2018) reported bimodal BDGs for deep-sea megafauna along continental slopes in the Caribbean. Geographically, variation in BDGs within the same group is also common. This is particularly apparent in deep-sea benthic groups in which spatial distribution is mainly driven by substrate heterogeneity and energy input in the form of labile organic matter supplied from the pelagic compartment (Danovaro *et al.*, 2008; Hernández-Ávila *et al.*, 2018). The observed decline in species richness of deep-sea fauna with distance from continental margins also suggests the existence of source–sink dynamics between bathyal and abyssal communities operating over geological time (Rex *et al.*, 2005).

10.3.3. *Compositional diversity gradients*

Research on global marine diversity patterns has historically focused on the study of local (alpha) or regional (gamma) taxonomic diversity. However, new analytical tools and increased availability of faunal and environmental data, sampled more intensively and over larger scales, have made it possible to extend this focus to other facets of diversity offering important, complementary insight into the same fundamental questions. For example, observed changes in diversity gradients may be caused by regional differences in rates of turnover in species composition (beta diversity) associated with latitude or depth. Different studies have documented high spatial variation in species composition of deep-sea macrobenthos and meiofauna at different depths as well as at similar depths across ecoregions or with longitude/latitude, revealing high biogeographic complexity that is missed by species richness patterns (Danovaro *et al.*, 2008; Hernández-Ávila *et al.*, 2018). Deep-sea benthic communities have been reported to follow a general pattern of increasing similarity (lower turnover) with increasing depth and rapid turnover coincident with abrupt transitions in environmental gradients such as oxygen minimum zones or major topographic features, whereas horizontal changes in community composition are typically more moderate except for the shallower depths influenced by coastal processes (McClain and Rex, 2015). Environmental filtering, dispersal limitation and energy availability represent the main drivers structuring community composition of deep-sea benthos. For example, Brault *et al.* (2013) analyzed patterns of compositional turnover for deep-sea bivalves to find that, whereas abyssal assemblages on the western North Atlantic are nested impoverished subsets of bathyal assemblages, turnover (species replacement) dominates at all depths in the eastern North Atlantic. These contrasting patterns were attributed to differences in available energy: the low food supply into abyssal depths in the western North Atlantic sector, depressing species diversity to an impoverished subset of the bathyal fauna, was maintained to some extent by source–sink dynamics. On the other hand, community composition of deep-sea nematodes in the Mediterranean shows extremely high turnover both across different depths and at constant depths within and among the two Mediterranean basins in line with differences and patterns observed for phylogenetic diversity (PD) (Danovaro *et al.*, 2008).

Groundfish communities along the North American Pacific continental shelf follow an overall decrease in beta diversity (i.e. increased compositional similarity) with increasing depth (Anderson *et al.*, 2013). Species turnover also showed a decreasing gradient with increasing latitude, although weaker than the depth gradient and limited to shallower waters (<200 m). Several peaks in compositional turnover were observed for both bathymetric and latitudinal gradients in agreement with environmental gradients (i.e. oxygen minimum zone and oceanographic boundaries). Interestingly, observed bathymetric beta diversity gradients were stronger at lower latitudes, suggesting latitude-by-depth interactions. Zintzen *et al.* (2017) also report the observed strong gradients of latitudinal variation in beta diversity of fish communities in the New Zealand continental shelf (<1,200 m) becoming exponentially weaker with increasing depth, resulting in higher community similarity across latitudes with increasing depth. In contrast, the magnitude of the decay in beta diversity along the depth gradient remained consistent across latitude.

10.3.4. *Functional and phylogenetic diversity gradients*

Species richness as a metric of diversity does not reflect evolutionary and ecological differences among species. As such, its sole use does not suffice to understand the mechanisms driving spatial and temporal biodiversity patterns. Alternative measures of diversity, mainly functional and PD, have been developed to address this shortcoming. Functional diversity (FD) is a relatively complex, abstract concept that refers to the combination of roles or functions species have in ecological communities (Petchey and Gaston, 2006). Therefore, the study of macroecological FD gradients and its environmental correlates offers a distinct and complementary perspective to that of taxonomic diversity by linking species to ecosystem through mechanisms such as niche conservatism, species coexistence, resource use and functional redundancy. Similarly, by capturing the shared ancestry of species, PD links to evolution and historical contingencies and mechanisms of speciation and extinction (Mouquet *et al.*, 2012).

The few existing studies that have analyzed latitudinal FD gradients at global scales in marine fauna (reef fish and intertidal-to-outer-shelf bivalves) have found a consistent unimodal latitudinal gradient of strong decrease in functional richness (i.e. the number of unique functional groups in a community) from a maximum at the tropics towards minima at the poles (Stuart-Smith *et al.*, 2013; Berke *et al.*, 2014; Schumm *et al.*, 2019). These latitudinal gradients were similar to those of taxonomic richness although with a much wider peak caused by the decoupling between species richness and functional richness as tropical functional groups become saturated in species. Interestingly, functional bivalve evenness (i.e. the distribution of species richness in functional trait space) showed an inverse pattern, reaching its minimum at the tropics and increasing exponentially towards the poles. This pattern, reflecting heavy occupation by tropical bivalve species of specific regions of the functional trait space (functional redundancy), suggests that specific combinations of functional groups promote species coexistence and/or reflect higher origination or lower extinction rates (Schumm *et al.*, 2019). Historically, the higher stability and amenable climatic conditions of the tropics together with its productive and biogenic complex habitats would have favored functional redundancy through specialization in some functional groups (i.e. niche partitioning). Rapid radiation (allopatric speciation) of functionally equivalent species together with lower extinction rates would then facilitate the observed tropical patterns of higher taxonomic richness, richness-saturated functional evenness and minimum functional evenness. On the other hand, functional unevenness may be to some extent a byproduct of the natural intrinsic differences in origination rates among functional groups and the disproportionately higher origination rates in the tropics, as evidenced by the fossil record (Berke *et al.*, 2014). The expansion and radiation of tropical genera into higher latitudes (i.e. the "Out of the Tropics" hypothesis) would result in extratropical assemblages comprising nested subsets of a functionally uneven tropical genus pool, hence with comparatively greater evenness. This explanation also adds to the fact that marine bivalve assemblages showed latitudinal variation in beta FD driven by nestedness rather than turnover, indicating that assemblages at higher latitudes are functional subsets of tropical ones (Schumm *et al.*, 2019). Indeed, no functional group was found to be exclusively represented outside tropical latitudes.

Global analyses on marine PD are also scarce. Diversification rates estimated from phylogenies in marine fishes follow a consistent unimodal increasing pattern with latitude (Rabosky *et al.*, 2018), from tropical minimum to maximum at the poles, challenging the long-established relationship between energy, metabolism and speciation that see the tropics as an evolutionary "cradle" and anticipate a declining pattern of species origination with latitude. Contrary to these expectations, fastest rates of diversification concentrated in high-latitude regions characterized by low temperatures, low species richness and high endemism. Eme *et al.* (2020) used calibrated molecular phylogeny to examine communities of marine ray-finned fishes (percomorphs) along depth gradients (50–1200 m) across New Zealand. They found phylogenetic diversity (PD), a metric based on the cumulative branch distance of all species across the phylogenetic tree that captures the overall amount of evolutionary history of a community (Faith, 1992), to strongly vary with depth and, to a lesser extent, latitude. These patterns mirrored those observed for taxonomic species richness. Across latitude, both metrics displayed a significant unimodal gradient, with maxima reaching at intermediate latitudes along the studied latitudinal gradient (29–51°S). Whereas both metrics significantly decreased with depth following a bimodal pattern with maxima at shallower depths (50 m) and a secondary peak at 900 m, other phylogenetic metrics, sensitive to the history of speciation events and time of divergence, showed unimodal patterns with maxima at intermediate depths. Taken together, and considering the inverse relationship of decreasing energy and increasing stability with increasing depth, these results suggest that energy-rich shallow environments support high speciation rates of ray-finned fishes, hence higher PD. However, recent speciation also occurs in ancient fish lineages in the comparatively energy-poor deep-sea habitats, suggesting that environmental stability facilitated their preservation during the Cenozoic. The observed peaks at intermediate depths further suggest that they represent a zone of overlap for distinct shallow and deep-water lineages, hence maximizing phylogenetic structure complexity, which may have provided a refuge to paleoceanographic perturbations affecting both shallow (Pliocene–Pleistocene climatic events) and deep-sea (ancient Mesozoic extreme anoxic events) environments (Eme *et al.*, 2020).

In an analysis of PD patterns of Southern Hemisphere brittle stars (Ophiuroidea), a dominant and cosmopolitan clade of deep-sea benthic invertebrates, O'Hara *et al.* (2019) also found concordance between taxonomic richness and PD latitudinal and depth gradients and a reversed gradient in diversification rates. Whereas richness and PD were maximum at shallow to intermediate (upper bathyal) depths within tropical latitudes and decline sharply into the abyss and the Antarctica, diversification rates peaked at polar and subpolar latitudes consistently across all depth strata. This result highlights the relatively recent phylogenetic radiation of Antarctic ophiuroids, likely reflecting an ongoing rebound from regional episodic extinctions during the mid-Cenozoic linked to the rapid cooling of water masses and the formation of icecaps. Whereas high diversification and net exchange into low latitudes characterized the Antarctic region, the shallow-water tropical–temperate region was found to be driven by relatively high diversification rates in the tropical shallows, with exchanges of linages across latitude and depth. The tropical deep sea represents, therefore, a net importer of ancient ophiuroid lineages characterized by high species diversity and low diversification rates (acting as a "museum"). In between, the upper- to mid-bathyal (200–2,000 m) region acts as a long-term refuge for deep-sea fauna.

Together, although still limited to a few studies and marine groups, these results demonstrate how the combination of taxonomic, functional and PD can help to identify and understand the evolutionary and ecological mechanisms that generate and support extant patterns of global marine diversity.

10.4. Conclusions

Marine biogeography is a relatively young but extremely active field that has experienced an enormous advance over the course of the last two centuries. The rapid growth of marine science, access to more extensive and better resolved data sets and multiple technological breakthroughs have catalyzed an increasing number of marine biogeographic studies at regional and global scales. The knowledge generated by these studies has allowed for the rapid development of fundamental hypotheses on the historical, evolutionary and ecological processes that generate and maintain biodiversity that were already tested and established in the terrestrial realm. More importantly, these developments vastly supported contemporary efforts in marine biodiversity conservation and ecosystem-based resource management (Sherman, 2005; Halpern *et al.*, 2008; Cochrane *et al.*, 2016; Costello and Chaudhary, 2017; Reygondeau and Dunn, 2019).

Nonetheless, potential research gaps and challenges in marine biogeography and biodiversity merit further attention. Progress in (marine) biogeography has often been hampered by a lack of "proof-of-concept" on the assumptions and verbal logic underpinning fundamental hypotheses. A clear example of this condition is the lack of consensus surrounding the plethora of hypotheses formulated to explain the mechanisms driving global diversity gradients, many of which offer partial, non-exclusive explanations. In an attempt to tackle this issue, Pontarp *et al.* (2019) recently proposed a more formal, integrative framework based on four key processes (selection, dispersal, ecological drift and speciation) underpinning the mechanisms linking variation in diversity to spatial and temporal environmental gradients. The authors demonstrate how the integration of this framework into mechanistic eco-evolutionary models can be used to empirically assess the relative importance of the various secondary processes that generate LDGs. In this and other fronts (Vichi *et al.*, 2011; Hill *et al.*, 2020), the development and application of quantitative assessments represent a crucial step to unify biogeographical theory, validate hypotheses and reduce uncertainty.

Although new marine species have been described at a higher pace than terrestrial species since the 1950s (Costello *et al.*, 2012), global taxonomic assessments in the marine realm remain highly incomplete relative to the terrestrial realm (Mora *et al.*, 2011; Costello and Chaudhary, 2017; Menegotto and Rangel, 2018). While the growth of open-access online repositories for marine species observations (e.g. OBIS) is rapidly allowing for the exploration of biodiversity patterns for an increasing number of taxonomic groups and larger spatial coverage, several caveats emerging from spatial and biological sampling biases as well as sampling effort efficiency have been highlighted (Webb *et al.*, 2010;

Chaudhary *et al.*, 2016; Menegotto and Rangel, 2018). In particular, polar and tropical waters are comparatively less sampled than the mid-latitude oceans at both hemispheres. This can potentially lead to interpretation errors and spurious patterns as suggested, for example, for the observed tropical dip in the LDG of many ocean taxa (Menegotto and Rangel, 2018). Likewise, deep-sea ecosystems are far less studied and sampled for physical, chemical and biological variables than shallower regions (Costello *et al.*, 2018). Future efforts should be directed towards strengthening international collaboration among research institutions, focusing on the development, exchange and promotion of biodiversity data standards and analytical tools, to improve further our knowledge of regional and global marine biodiversity and biogeography (Costello *et al.*, 2010, 2015; Saeedi *et al.*, 2019b).

Finally, impacts of multiple anthropogenic stressors on marine biodiversity are pervasive across regional and global scales and persisted for centuries (Gunderson *et al.*, 2016; Henson *et al.*, 2017). For instance, fisheries exploitation (e.g. overfishing and bycatch) and global climate change threaten some marine species with severe depletion or eventual extinction (Burgess *et al.*, 2013; Collins *et al.*, 2018; Gray and Kennelly, 2018). Other exacerbating anthropogenic impacts on marine biodiversity include eutrophication, hypoxia, ocean acidification and invasive species (Watson *et al.*, 2017; Beaury *et al.*, 2020; Guan *et al.*, 2020). The current overlap between marine biodiversity hotspots and regions of moderate or high human impact (Tittensor *et al.*, 2010), and the growing trend that many of these impacts are experiencing on a global scale (Halpern *et al.*, 2019) anticipate future implications for the redistribution of marine life and current biodiversity patterns that represent a clear research priority.

10.5. References

Adams, N.E., Inoue, K., Seidel, R.A., Lang, B.K., Berg, D.J. (2018). Isolation drives increased diversification rates in freshwater amphipods. *Molecular Phylogenetics and Evolution*, 127, 746–757.

Allen, A.P. and Gillooly, J.F. (2006). Assessing latitudinal gradients in speciation rates and biodiversity at the global scale. *Ecology Letters*, 9, 947–954.

Anderson, M.J., Tolimieri, N., Millar, R.B. (2013). Beta diversity of demersal fish assemblages in the North-Eastern pacific: Interactions of latitude and depth. *PLOS ONE*, 8, e57918.

Appeltans, W., Ahyong, S.T., Anderson, G., Angel, M.V., Artois, T., Bailly, N., Bamber, R., Barber, A., Bartsch, I., Berta, A. *et al.* (2012). The magnitude of global marine species diversity. *Current Biology*, 22, 2189–2202.

Bambach, R.K. (1999). Energetics in the global marinefauna: A connection between terrestrial diversification and change in the marine biosphere. *Geobios*, 32, 131–144.

Beaury, E.M., Fusco, E.J., Jackson, M.R., Laginhas, B.B., Morelli, T.L., Allen, J.M., Pasquarella, V.J., Bradley, B.A. (2020). Incorporating climate change into invasive species management: Insights from managers. *Biological Invasions*, 22, 233–252.

Benton, M.J. (2001). Biodiversity on land and in the sea. *Geological Journal*, 36, 211–230.

Benton, M.J. (2016). Origins of biodiversity. *PLOS Biology*, 14, e2000724.

Berke, S.K., Jablonski, D., Krug, A.Z., Valentine, J.W. (2014). Origination and immigration drive latitudinal gradients in marine functional diversity. *PLOS ONE*, 9, e101494.

Bloom, D.D., Weir, J.T., Piller, K.R., Lovejoy, N.R. (2013). Do freshwater fishes diversify faster than marine fishes? A test using state-dependent diversification analyses and molecular phylogenetics of New World silversides (Atherinopsidae). *Evolution*, 67, 2040–2057.

Brault, S., Stuart, C.T., Wagstaff, M.C., McClain, C.R., Allen, J.A., Rex, M.A. (2013). Contrasting patterns of α- and β-diversity in deep-sea bivalves of the eastern and western North Atlantic. *Deep Sea Research Part II: Topical Studies in Oceanography*, 92, 157–164.

Brayard, A., Escarguel, G., Bucher, H. (2005). Latitudinal gradient of taxonomic richness: Combined outcome of temperature and geographic mid-domains effects? *Journal of Zoological Systematics and Evolutionary Research*, 43, 178–188.

Briggs, J.C. (1995). *Global Biogeography*. Elsevier, Amsterdam, The Netherlands.

Briggs, D.E.G. (2015). The Cambrian explosion. *Current Biology*, 25, R864–R868.

Brown, J.H. (1995). *Macroecology*. University of Chicago Press, Chicago, IL.

Brown, J.H. (2014). Why are there so many species in the tropics? *Journal of Biogeography*, 41, 8–22.

Brown, J.H. and Maurer, B.A. (1989). Macroecology: The division of food and space among species on continents. *Science*, 243, 1145–1150.

Brown, J.H., Gillooly, J.F., Allen, A.P., Savage, V.M., West, G.B. (2004). Toward a metabolic theory of ecology. *Ecology*, 85, 1771–1789.

Burgess, M.G., Polasky, S., Tilman, D. (2013). Predicting overfishing and extinction threats in multispecies fisheries. *Proceedings of the National Academy of Sciences of the United States of America*, 110, 15943.

Bush, A.M., Hunt, G., Bambach, R.K. (2016). Sex and the shifting biodiversity dynamics of marine animals in deep time. *Proceedings of the National Academy of Sciences of the United States of America*, 113, 14073–14078.

de Candolle, A.P. (1820). Géographie botanique. *Dictionaire des Sciences Naturelles*, 18, 359–422.

Carrete Vega, G. and Wiens, J.J. (2012). Why are there so few fish in the sea? *Proceedings: Biological Sciences*, 279, 2323–2329.

Chaudhary, C., Saeedi, H., Costello, M.J. (2016). Bimodality of latitudinal gradients in marine species richness. *Trends in Ecology and Evolution*, 31, 670–676.

Chaudhary, C., Saeedi, H., Costello, M.J. (2017). Marine species richness is bimodal with latitude: A reply to Fernandez and Marques. *Trends in Ecology and Evolution*, 32, 234–237.

Clarke, A. and Gaston, K.J. (2006). Climate, energy and diversity. *Proceedings of the Royal Society B: Biological Sciences*, 273, 2257–2266.

Clarkson, M.O., Lenton, T.M., Andersen, M.B., Bagard, M.-L., Dickson, A.J., Vance, D. (2021). Upper limits on the extent of seafloor anoxia during the PETM from uranium isotopes. *Nature Communications*, 12, 399.

Cochrane, S.K.J., Andersen, J.H., Berg, T., Blanchet, H., Borja, A., Carstensen, J., Elliott, M., Hummel, H., Niquil, N., Renaud, P.E. (2016). What is marine biodiversity? Towards common concepts and their implications for assessing biodiversity status. *Frontiers in Marine Science*, 3, ea248.

Collins, K.S., Edie, S.M., Hunt, G., Roy, K., Jablonski, D. (2018). Extinction risk in extant marine species integrating palaeontological and biodistributional data. *Proceedings of the Royal Society B: Biological Sciences*, 285, 20181698.

Colwell, R.K. and Hurtt, G.C. (1994). Nonbiological gradients in species richness and a spurious Rapoport effect. *The American Naturalist*, 144, 570–595.

Costello, M.J. and Chaudhary, C. (2017). Marine biodiversity, biogeography, deep-sea gradients, and conservation. *Current Biology*, 27, R511–R527.

Costello, M.J., Coll, M., Danovaro, R., Halpin, P., Ojaveer, H., Miloslavich, P. (2010). A census of marine biodiversity knowledge, resources, and future challenges. *PLOS ONE*, 5, e12110.

Costello, M.J., Wilson, S., Houlding, B. (2012). Predicting total global species richness using rates of species description and estimates of taxonomic effort. *Systematic Biology*, 61, 871–871.

Costello, M.J., Vanhoorne, B., Appeltans, W. (2015). Conservation of biodiversity through taxonomy, data publication, and collaborative infrastructures. *Conservation Biology*, 29, 1094–1099.

Costello, M.J., Basher, Z., Sayre, R., Breyer, S., Wright, D.J. (2018). Stratifying ocean sampling globally and with depth to account for environmental variability. *Scientific Reports*, 8, 11259.

Dana, J.D. (1853). On an isothermal oceanic chart, illustrating the geographical distribution of marine animals. *The American Journal of Science and Arts*, 16(153–167), 314–327.

Dana, J.D. (1875). *Corals and Coral Islands*. Sampson Low, Marston, Low and Searle, London, UK.

Danovaro, R., Gambi, C., Lampadariou, N., Tselepides, A. (2008). Deep-sea nematode biodiversity in the Mediterranean basin: Testing for longitudinal, bathymetric and energetic gradients. *Ecography*, 31, 231–244.

Darwin, C.R. (1842). *The Structure and Distribution of Coral Reefs. Being the First Part of the Geology of the Voyage of the Beagle, Under the Command of Capt. Fitzroy, R.N. During the Years 1832 to 1836*. Smith Elder and Co, London, UK.

Dawson, M.N. (2012). Species richness, habitable volume, and species densities in freshwater, the sea, and on land. *Frontiers of Biogeography*, 4, 105–116.

Egerton, F.N. (2019). History of ecological sciences, Part 61C: Marine biogeography, 1690s–1940s. *The Bulletin of the Ecological Society of America*, 100, e01486.

Eme, D., Anderson, M.J., Myers, E.M.V., Roberts, C.D., Liggins, L. (2020). Phylogenetic measures reveal eco-evolutionary drivers of biodiversity along a depth gradient. *Ecography*, 43, 689–702.

Eschmeyer, W.N., Fricke, R., Fong, J.D., Polack, D.A. (2010). Marine fish diversity: History of knowledge and discovery (Pisces). *Zootaxa*, 2525, 19–50.

Faith, D.P. (1992). Conservation evaluation and phylogenetic diversity. *Biological Conservation*, 61, 1–10.

Field, C.B., Behrenfeld, M.J., Randerson, J.T., Falkowski, P. (1998). Primary production of the biosphere: Integrating terrestrial and oceanic components. *Science*, 281, 237–240.

Forbes, E. (1856). Map of the distribution of marine life. In *The Physical Atlas of Natural Phenomena*, Johnson, W. and Johnson, A.K. (eds). Published by the editors, Edinburgh and London.

Forbes, E. and Godwin-Austen, R. (1859). *The Natural History of European Seas*. John Van Voorst, London, UK.

Foster, L.C., Schmidt, D.N., Thomas, E., Arndt, S., Ridgwell, A. (2013). Surviving rapid climate change in the deep sea during the Paleogene hyperthermals. *Proceedings of the National Academy of Sciences of the United States of America*, 110, 9273–9276.

Fuhrman, J.A., Steele, J.A., Hewson, I., Schwalbach, M.S., Brown, M.V., Green, J.L., Brown, J.H. (2008). A latitudinal diversity gradient in planktonic marine bacteria. *Proceedings of the National Academy of Sciences of the United States of America*, 105, 7774–7778.

García Molinos, J., Halpern, B.S., Schoeman, D.S., Brown, C.J., Kiessling, W., Moore, P.J., Pandolfi, J.M., Poloczanska, E.S., Richardson, A.J., Burrows, M.T. (2016). Climate velocity and the future global redistribution of marine biodiversity. *Nature Climate Change*, 6, 83–88.

Gray, C.A. and Kennelly, S.J. (2018). Bycatches of endangered, threatened and protected species in marine fisheries. *Reviews in Fish Biology and Fisheries*, 28, 521–541.

Grosberg, R.K., Vermeij, G.J., Wainwright, P.C. (2012). Biodiversity in water and on land. *Current Biology*, 22, R900–R903.

Guan, Y., Hohn, S., Wild, C., Merico, A. (2020). Vulnerability of global coral reef habitat suitability to ocean warming, acidification and eutrophication. *Global Change Biology*, 26(10), 5646–5660.

Gunderson, A.R., Armstrong, E.J., Stillman, J.H. (2016). Multiple stressors in a changing world: The need for an improved perspective on physiological responses to the dynamic marine environment. *Annual Review of Marine Science*, 8, 357–378.

Halpern, B.S., McLeod, K.L., Rosenberg, A.A., Crowder, L.B. (2008). Managing for cumulative impacts in ecosystem-based management through ocean zoning. *Ocean and Coastal Management*, 51, 203–211.

Halpern, B.S., Frazier, M., Afflerbach, J., Lowndes, J.S., Micheli, F., O'Hara, C., Scarborough, C., Selkoe, K.A. (2019). Recent pace of change in human impact on the world's ocean. *Scientific Reports*, 9, 11609.

Hedgpeth, J.W. (1957). Chapter 13: Marine biogeography. *Geological Society of America Memoir*, 67, 1, 359–382.

Henson, S.A., Beaulieu, C., Ilyina, T., John, J.G., Long, M., Séférian, R., Tjiputra, J., Sarmiento, J.L. (2017). Rapid emergence of climate change in environmental drivers of marine ecosystems. *Nature Communications*, 8, 14682.

Hernández-Ávila, I., Guerra-Castro, E., Bracho, C., Rada, M., Ocaña, F.A., Pech, D. (2018). Variation in species diversity of deep-water megafauna assemblages in the Caribbean across depth and ecoregions. *PLOS ONE*, 13, e0201269.

Hill, N., Woolley, S.N.C., Foster, S., Dunstan, P.K., McKinlay, J., Ovaskainen, O., Johnson, C. (2020). Determining marine bioregions: A comparison of quantitative approaches. *Methods in Ecology and Evolution*, 11(10), 1258–1272.

Hillebrand, H. (2004). Strength, slope and variability of marine latitudinal gradients. *Marine Ecology Progress Series*, 273, 251–267.

Hou, Z., Sket, B., Fišer, C., Li, S. (2011). Eocene habitat shift from saline to freshwater promoted Tethyan amphipod diversification. *Proceedings of the National Academy of Sciences of the United States of America*, 108, 14533–14538.

Jablonski, D., Roy, K., Valentine, J.W. (2006). Out of the tropics: Evolutionary dynamics of the latitudinal diversity gradient. *Science*, 314, 102–106.

Jacobs, D.K. and Lindberg, D.R. (1998). Oxygen and evolutionary patterns in the sea: Onshore/offshore trends and recent recruitment of deep-sea faunas. *Proceedings of the National Academy of Sciences of the United States of America*, 95, 9396–9401.

Kaschner, K., Tittensor, D.P., Ready, J., Gerrodette, T., Worm, B. (2011). Current and future patterns of global marine mammal biodiversity. *PLOS ONE*, 6, e19653.

Kennett, J.P. and Stott, L.D. (1991). Abrupt deep-sea warming, palaeoceanographic changes and benthic extinctions at the end of the Palaeocene. *Nature*, 353, 225–229.

Kiessling, W., Simpson, C., Foote, M. (2010). Reefs as cradles of evolution and sources of biodiversity in the Phanerozoic. *Science*, 327, 196–198.

Kröger, B. (2017). Changes in the latitudinal diversity gradient during the great Ordovician Biodiversification event. *Geology*, 46, 127–130.

Leibold, M.A., Holyoak, M., Mouquet, N., Amarasekare, P., Chase, J.M., Hoopes, M.F., Holt, R.D., Shurin, J.B., Law, R., Tilman, D., Loreau, M., Gonzalez, A. (2004). The metacommunity concept: A framework for multi-scale community ecology. *Ecology Letters*, 7, 601–613.

Lévêque, C., Oberdorff, T., Paugy, D., Stiassny, M.L.J., Tedesco, P.A. (2008). Global diversity of fish (Pisces) in freshwater. In *Freshwater Animal Diversity Assessment*, Balian, E.V., Lévêque, C., Segers, H., Martens, K. (eds). Springer, Dordrecht, The Netherlands.

MacArthur, R.H. (1972). *Geographical Ecology: Patterns in the Distribution of Species*. Princeton University Press, Princeton, NJ.

Machac, A., Zrzavý, J., Smrckova, J., Storch, D. (2012). Temperature dependence of evolutionary diversification: Differences between two contrasting model taxa support the metabolic theory of ecology. *Journal of Evolutionary Biology*, 25, 2449–2456.

Maddison, W.P. (2006). Confounding asymmetries in evolutionary diversification and character change. *Evolution*, 60, 1743–1746.

Mannion, P.D., Upchurch, P., Benson, R.B.J., Goswami, A. (2014). The latitudinal biodiversity gradient through deep time. *Trends in Ecology and Evolution*, 29, 42–50.

May, R.M. (1994). Biological diversity: Differences between land and sea. *Philosophical Transactions of the Royal Society of London. Series B: Biological Sciences*, 343, 105–111.

McClain, C.R. and Rex, M.A. (2015). Toward a conceptual understanding of β-diversity in the deep-sea benthos. *Annual Review of Ecology, Evolution, and Systematics*, 46, 623–642.

Menegotto, A. and Rangel, T.F. (2018). Mapping knowledge gaps in marine diversity reveals a latitudinal gradient of missing species richness. *Nature Communications*, 9, 4713.

Meseguer, A.S. and Condamine, F.L. (2020). Ancient tropical extinctions at high latitudes contributed to the latitudinal diversity gradient. *Evolution*, 74, 1966–1987.

Mittelbach, G.G., Schemske, D.W., Cornell, H.V., Allen, A.P., Brown, J.M., Bush, M.B., Harrison, S.P., Hurlbert, A.H., Knowlton, N., Lessios, H.A., McCain, C.M., McCune, A.R., McDade, L.A., McPeek, M.A., Near, T.J., Price, T.D., Ricklefs, R.E., Roy, K., Sax, D.F., Schluter, D., Sobel, J.M., Turell, M. (2007). Evolution and the latitudinal diversity gradient: Speciation, extinction and biogeography. *Ecology Letters*, 10, 315–331.

Mora, C., Tittensor, D.P., Adl, S., Simpson, A.G.B., Worm, B. (2011). How many species are there on Earth and in the ocean? *PLoS Biology*, 9, e1001127.

Mouquet, N., Devictor, V., Meynard, C.N., Munoz, F., Bersier, L.-F., Chave, J., Couteron, P., Dalecky, A., Fontaine, C., Gravel, D., Hardy, O.J., Jabot, F., Lavergne, S., Leibold, M., Mouillot, D., Münkemüller, T., Pavoine, S., Prinzing, A., Rodrigues, A.S.L., Rohr, R.P., Thébault, E., Thuiller, W. (2012). Ecophylogenetics: Advances and perspectives. *Biological Reviews*, 87, 769–785.

Muir, P.R., Wallace, C.C., Done, T., Aguirre, J.D. (2015). Limited scope for latitudinal extension of reef corals. *Science*, 348, 1135–1138.

O'Hara, T.D., Hugall, A.F., Woolley, S.N.C., Bribiesca-Contreras, G., Bax, N.J. (2019). Contrasting processes drive ophiuroid phylodiversity across shallow and deep seafloors. *Nature*, 565, 636–639.

Owen, A.W. and Crame, J.A. (2002). Palaeobiogeography and the Ordovician and Mesozoic-Cenozoic biotic radiations. *Geological Society, London, Special Publications*, 194, 1–11.

Petchey, O.L. and Gaston, K.J. (2006). Functional diversity: Back to basics and looking forward. *Ecology Letters*, 9, 741–758.

Pinsky, M.L., Eikeset, A.M., McCauley, D.J., Payne, J.L., Sunday, J.M. (2019). Greater vulnerability to warming of marine versus terrestrial ectotherms. *Nature*, 569, 108–111.

Poloczanska, E.S., Brown, C.J., Sydeman, W.J., Kiessling, W., Schoeman, D.S., Moore, P.J., Brander, K., Bruno, J.F., Buckley, L.B., Burrows, M.T., Duarte, C.M., Halpern, B.S., Holding, J., Kappel, C.V., O'Connor, M.I., Pandolfi, J.M., Parmesan, C., Schwing, F., Thompson, S.A., Richardson, A.J. (2013). Global imprint of climate change on marine life. *Nature Climate Change*, 3, 919–925.

Pontarp, M., Bunnefeld, L., Cabral, J.S., Etienne, R.S., Fritz, S.A., Gillespie, R., Graham, C. H., Hagen, O., Hartig, F., Huang, S., Jansson, R., Maliet, O., Münkemüller, T., Pellissier, L., Rangel, T.F., Storch, D., Wiegand, T., Hurlbert, A.H. (2019). The latitudinal diversity gradient: Novel understanding through mechanistic eco-evolutionary models. *Trends in Ecology and Evolution*, 34, 211–223.

Powell, M.G., Moore, B.R., Smith, T.J. (2015). Origination, extinction, invasion, and extirpation components of the brachiopod latitudinal biodiversity gradient through the Phanerozoic Eon. *Paleobiology*, 41(330–341), 12.

Rabosky, D.L., Chang, J., Title, P.O., Cowman, P.F., Sallan, L., Friedman, M., Kaschner, K., Garilao, C., Near, T.J., Coll, M., Alfaro, M.E. (2018). An inverse latitudinal gradient in speciation rate for marine fishes. *Nature*, 559, 392–395.

Rex, M.A. (1973). Deep-sea species diversity: Decreased gastropod diversity at abyssal depths. *Science*, 181, 1051–1053.

Rex, M.A., McClain, C.R., Johnson, N.A., Etter, R.J., Allen, J.A., Bouchet, P., Warén, A. (2005). A source-sink hypothesis for abyssal biodiversity. *The American Naturalist*, 165, 163–178.

Reygondeau, G. and Dunn, D. (2019). Pelagic biogeography. In *Encyclopedia of Ocean Sciences*, 3rd edition, Cochran, J.K., Bokuniewicz, H.J., Yager, P.L. (eds). Academic Press, Oxford.

Righetti, D., Vogt, M., Gruber, N., Psomas, A., Zimmermann, N.E. (2019). Global pattern of phytoplankton diversity driven by temperature and environmental variability. *Science Advances*, 5, eaau6253.

Ronquist, F. and Sanmartín, I. (2011). Phylogenetic methods in biogeography. *Annual Review of Ecology, Evolution, and Systematics*, 42, 441–464.

Roy, K., Jablonski, D., Valentine, J.W., Rosenberg, G. (1998). Marine latitudinal diversity gradients: Tests of causal hypotheses. *Proceedings of the National Academy of Sciences of the United States of America*, 95, 3699–3702.

Rutherford, S., D'Hondt, S., Prell, W. (1999). Environmental controls on the geographic distribution of zooplankton diversity. *Nature*, 400, 749–753.

Saeedi, H., Dennis, T.E., Costello, M.J. (2017). Bimodal latitudinal species richness and high endemicity of razor clams (Mollusca). *Journal of Biogeography*, 44, 592–604.

Saeedi, H., Costello, M.J., Warren, D., Brandt, A. (2019a). Latitudinal and bathymetrical species richness patterns in the NW Pacific and adjacent Arctic Ocean. *Scientific Reports*, 9, 9303.

Saeedi, H., Reimer, J.D., Brandt, M.I., Dumais, P.-O., Jażdżewska, A.M., Jeffery, N.W., Thielen, P.M., Costello, M.J. (2019b). Global marine biodiversity in the context of achieving the Aichi Targets: Ways forward and addressing data gaps. *PeerJ*, 7, e7221.

Saeedi, H., Simões, M., Brandt, A. (2020). Biodiversity and distribution patterns of deep-sea fauna along the temperate NW Pacific. *Progress in Oceanography*, 183, 102296.

Schumm, M., Edie, S.M., Collins, K.S., Gómez-Bahamón, V., Supriya, K., White, A.E., Price, T.D. and Jablonski, D. (2019). Common latitudinal gradients in functional richness and functional evenness across marine and terrestrial systems. *Proceedings of the Royal Society B: Biological Sciences*, 286, 20190745.

Scotese, C.R. (2016). PALEOMAP PaleoAtlas for GPlates and the PaleoData Plotter Program, PALEOMAP Project [Online]. Available at: https://www.earthbyte.org/paleomap-paleoatlas-for-gplates/.

Sepkoski, J.J.J. (1984). A kinetic model of Pharenozoic taxonomic diversity III. Post-Paleozoic families and mass extinctions. *Paleobiology*, 10, 246–267.

Sherman, K. (2005). The large marine ecosystem approach for assessment and management of ocean coastal waters. In *Sustaining Large Marine Ecosystems: The Human Dimension*, Hennessey, T. and Suti-Nen, J. (eds). Elsevier, Amsterdam, The Netherlands.

Smith, K.F. and Brown, J.H. (2002). Patterns of diversity, depth range and body size among pelagic fishes along a gradient of depth. *Global Ecology and Biogeography*, 11, 313–322.

Smith, A.B. and Stockley, B. (2005). The geological history of deep-sea colonization by echinoids: Roles of surface productivity and deep-water ventilation. *Proceedings of the Royal Society B: Biological Sciences*, 272, 865–869.

Song, H., Wignall, P.B., Song, H., Dai, X., Chu, D. (2019). Seawater temperature and dissolved oxygen over the past 500 million years. *Journal of Earth Science*, 30, 236–243.

Song, H., Huang, S., Jia, E., Dai, X., Wignall, P.B., Dunhill, A.M. (2020). Flat latitudinal diversity gradient caused by the Permian–Triassic mass extinction. *Proceedings of the National Academy of Sciences of the United States of America*, 117, 17578–17583.

Strugnell, J.M., Rogers, A.D., Prodöhl, P.A., Collins, M.A., Allcock, A.L. (2008). The thermohaline expressway: The Southern Ocean as a centre of origin for deep-sea octopuses. *Cladistics*, 24, 853–860.

Stuart-Smith, R.D., Bates, A.E., Lefcheck, J.S., Duffy, J.E., Baker, S.C., Thomson, R.J., Stuart-Smith, J.F., Hill, N.A., Kininmonth, S.J., Airoldi, L., Becerro, M.A., Campbell, S.J., Dawson, T.P., Navarrete, S.A., Soler, G.A., Strain, E.M.A., Willis T.J., Edgar, G.J. (2013). Integrating abundance and functional traits reveals new global hotspots of fish diversity. *Nature*, 501, 539–542.

Sunday, J.M., Bates, A.E., Dulvy, N.K. (2012). Thermal tolerance and the global redistribution of animals. *Nature Climate Change*, 2, 686–690.

Teske, P.R., Sandoval-Castillo, J., Rao Golla, T., Emami-Khoyi, A., Tine, M., von der Heyden, S., Beheregaray, L.B. (2019). Thermal selection as a driver of marine ecological speciation. *Proceedings of the Royal Society B: Biological Sciences*, 286, 20182023.

Thuy, B., Gale, A.S., Kroh, A., Kucera, M., Numberger-Thuy, L.D., Reich, M., Stöhr, S. (2012). Ancient origin of the modern deep-sea fauna. *PLOS ONE*, 7, e46913.

Thuy, B., Kiel, S., Dulai, A., Gale, A.S., Kroh, A., Lord, A.R., Numberger-Thuy, L.D., Stöhr, S., Wisshak, M. (2014). First glimpse into lower Jurassic deep-sea biodiversity: *In situ* diversification and resilience against extinction. *Proceedings of the Royal Society B: Biological Sciences*, 281, 20132624.

Tittensor, D.P., Mora, C., Jetz, W., Lotze, H.K., Ricard, D., Berghe, E.V., Worm, B. (2010). Global patterns and predictors of marine biodiversity across taxa. *Nature*, 466, 1098–1101.

Valiela, I. (1995). Long-term and large-scale change in marine ecosystems. *Marine Ecological Processes*. Springer, New York.

Vermeij, G.J. (1977). The Mesozoic marine revolution: Evidence from snails, predators and grazers. *Paleobiology*, 3, 245–258.

Vermeij, G.J. (1978). *Biogeography and Adaptation: Patterns of Marine Life*. Harvard University Press, Cambridge, MA.

Vermeij, G.J. (1987). *Evolution and Escalation. An Ecological History of Life*. Princeton University Press, Princeton, NJ.

Vermeij, G.J. (2011). The energetics of modernization: The last one hundred million years of biotic evolution. *Paleontological Research*, 15(54–61), 8.

Vermeij, G.J. and Grosberg, R.K. (2010). The great divergence: When did diversity on land exceed that in the sea? *Integrative and Comparative Biology*, 50, 675–682.

Vichi, M., Allen, J.I., Masina, S., Hardman-Mountford, N.J. (2011). The emergence of ocean biogeochemical provinces: A quantitative assessment and a diagnostic for model evaluation. *Global Biogeochemical Cycles*, 25, GB2005.

Watson, A.J., Lenton, T.M., Mills, B.J.W. (2017). Ocean deoxygenation, the global phosphorus cycle and the possibility of human-caused large-scale ocean anoxia. *Philosophical Transactions of the Royal Society A: Mathematical, Physical and Engineering Sciences*, 375, 20160318.

Webb, T.J., Vanden Berghe, E., O'Dor, R. (2010). Biodiversity's big wet secret: The global distribution of marine biological records reveals chronic under-exploration of the deep pelagic ocean. *PLOS ONE*, 5, e10223.

Wiens, J.J. (2015). Faster diversification on land than sea helps explain global biodiversity patterns among habitats and animal phyla. *Ecology Letters*, 18, 1234–1241.

Wiens, J.J. and Donoghue, M.J. (2004). Historical biogeography, ecology and species richness. *Trends in Ecology and Evolution*, 19, 639–644.

Willig, M.R., Kaufman, D.M., Stevens, R.D. (2003). Latitudinal gradients of biodiversity: Pattern, process, scale, and synthesis. *Annual Review of Ecology, Evolution, and Systematics*, 34, 273–309.

Wilson, G. (1999). Some of the deep-sea fauna is ancient. *Crustaceana*, 72, 1019.

Witman, J.D. and Roy, K. (2009). *Marine Macroecology*. University of Chicago Press, Chicago, IL.

Woolley, S.N.C., Tittensor, D.P., Dunstan, P.K., Guillera-Arroita, G., Lahoz-Monfort, J.J., Wintle, B.A., Worm, B., O'Hara, T.D. (2016). Deep-sea diversity patterns are shaped by energy availability. *Nature*, 533, 393–396.

Worm, B. and Tittensor, D.P. (2018). *A Theory of Global Biodiversity*. Princeton University Press, Princeton, NJ.

Yasuhara, M., Hunt, G., Dowsett, H.J., Robinson, M.M., Stoll, D.K. (2012). Latitudinal species diversity gradient of marine zooplankton for the last three million years. *Ecology Letters*, 15, 1174–1179.

Zaffos, A., Finnegan, S., Peters, S.E. (2017). Plate tectonic regulation of global marine animal diversity. *Proceedings of the National Academy of Sciences of the United States of America*, 114, 5653–5658.

Zhang, Z.-Q. (2013). Animal biodiversity: An update of classification and diversity in 2013. *Zootaxa*, 3703, 5–11.

Zhang, Q.-G., Lu, H.-S., Buckling, A. (2018). Temperature drives diversification in a model adaptive radiation. *Proceedings of the Royal Society B: Biological Sciences*, 285, 20181515.

Zintzen, V., Anderson, M.J., Roberts, C.D., Harvey, E.S., Stewart, A.L. (2017). Effects of latitude and depth on the beta diversity of New Zealand fish communities. *Scientific Reports*, 7, 8081.

11

Biogeography of Diseases

Jesús OLIVERO
University of Málaga, Spain

11.1. Introduction

11.1.1. *The need of disease mapping for management and prevention policies*

Many studies have outlined the need of improving methods and conceptual approaches for disease mapping (Smith and Guégan 2010; Hay *et al.* 2013; Kraemer *et al.* 2016; Stensgaard *et al.* 2017). The tremendous experience people from around the world lived during the COVID-19 pandemic demonstrated how useful the quick surveillance of disease-risk propagation is for decision-making. From the most experienced epidemiologists to people outside the scientific and sanitary domains, many of us, confined at home, explored in the media looking for maps, seeking reasons why the virus spread, with the expectation of estimating whether we were safe. The SARS-COV-2 virus that monopolized our attention for months is only one among lots of pathogens that are currently gaining ground – just as AIDS in the 1980s, and microbial diseases like dengue, Zika, yellow fever, Lyme disease and West Nile virus have. Some of them, such as the Ebola virus, are emergent, totally unknown before the 1970s (Pourrut *et al.* 2005). Others, like yellow fever, were already spreading three to four centuries ago, probably reaching America on-board slave ships sailing from Africa (Bryant *et al.* 2007). Long before, plague had expanded along main trade routes (Maher *et al.* 2010). Past and present globalization episodes, defined by a high economic, political and social relevance of long-distance communications, are behind all these events, and, if there is a period across the

Biogeography,
coordinated by Eric GUILBERT.

history that can be marked by globalization, that is the time we are living in (Tatem *et al.* 2006).

Descriptive, explanatory and predictive disease maps are fundamental tools for health management. Foreign health departments, or health centers specializing in travel and tropical diseases, need descriptive maps in order to provide travelers with prophylactic, social distancing and vaccination advice (Hay *et al.* 2013). However, maps based on case reports rarely have fine enough spatial resolution for users in the most economically developed regions (Stengaard *et al.* 2017). In this context, biogeography provides tools for completing knowledge gaps, mostly through the application of distribution modeling (Peterson 2006; Kraemer *et al.* 2016), and through the search of biogeographic patterns associated with pathogens and diseases (Just *et al.* 2014; Murray *et al.* 2015). The analysis of distribution patterns involving disease co-occurrences in time and in space facilitates the search for causes of disease outbreak, severity and spread. Thus, the understanding of factors involved in disease dynamics might guide decision-making of health authorities who, for example, could locally and temporally restrict consumption of certain animal species in meteorologically conditioned zoonotic-transmission hotspots (Kamins *et al.* 2015). Finally, prevention against vectors and infectious diseases that are subject to spatio-temporal dynamics require predictive mapping. Sanitary action protocols in the face of epidemics are more efficient, often simply viable, if logistic and economic efforts can be concentrated in specific hotspots of the geography (Allen *et al.* 2017). Thus, in order to optimize medical resources (i.e. sanitary professionals, vaccines, drugs and survey efforts), we need to anticipate where new outbreaks could occur, and where a virus might establish new areas of endemic transmission. In short, maps are the needed bases of early warning systems.

11.1.2. *Hypotheses on which biogeography sustains the analysis of infectious diseases*

As much as 61% of pathogens involved in human diseases are zoonotic, direct or indirectly transmitted from animal reservoirs (Taylor *et al.* 2001). This makes biogeographical approaches to infectious diseases complex by definition. The "One Health" international initiative (Gibbs 2014) encourages the study of zoonotic diseases through the analysis of interconnections between animals, ecosystems and human health. Because of this, biogeographic studies on pathogens need to be multidisciplinary, requiring methods and concepts from diverse fields of knowledge (Ramette and Tiedje 2007). Biogeography addresses the spatio-temporal analysis of infectious diseases on the basis of the following hypotheses:

– Worldwide microbe communities are subject to biogeographical patterns that are shaped by the influence of environmental factors (Guernier *et al.* 2004; Martiny *et al.* 2006; Dunn *et al.* 2010; Murray *et al.* 2015).

– The extent to which specific environments are favorable to the emergence of a disease can be modeled according to spatial relations between disease-case reports and a range of predictor variables (Peterson 2006; Pigott *et al.* 2014).

– Depending on the microbe, infections may occur at any location where the environment favors the presence of hosts (i.e. "host-niche hypothesis"), or may also require conditions suitable for pathogen transmission and maintenance (i.e. "disease-niche hypothesis") (Maher *et al.* 2010).

– In zoonotic-disease modeling, when zoonotic-cycle components are complex or uncertain, animal types of distribution (i.e. chorotypes) can be used as predictor variables representing reservoir/vector distributions (Olivero *et al.* 2017a).

– The distribution of disease outbreaks may depend on socioeconomic factors determining the way humans interact with the environment (Patz *et al.* 2004; Bett 2011; Jones *et al.* 2013; Wallace and Wallace 2014; Mackenstedt *et al.* 2015).

– The occurrence of disease outbreaks can be influenced by great-distance connections between macroclimatic events and local phenomena favoring infections (Pinzon *et al.* 2004; Lowen *et al.* 2007).

– Climate change could modify the risk of zoonotic disease through its effects on land-use and the distribution of animal species (Patz *et al.* 1996, 2005; Epstein 2001; Hoberg and Brooks 2015).

The growing interest in the geography of diseases (Peterson 2008; Kraemer *et al.* 2016; Stephens *et al.* 2016; Murray *et al.* 2018) is defining the specialization of a biogeographical branch called "pathogeography" (Murray *et al.* 2018), recovering the term coined by Reicher and Palti (1968). The following sections of this chapter are a journey across the different aspects considered by this approach, not intending to be an exhaustive bibliographic revision but rather a conceptual guide based on examples.

11.2. Do microbes have their own biogeography?

Until recently, the existence of microbial biogeographical patterns was neglected and even denied (Whitfield 2005). Now, we know that biogeography can help build a context for the understanding of the microbe diversity and of the factors influencing this diversity across time and space (Ramette and Tiedje 2007; Fierer 2008). Different habitats contain different microbial communities, and the microbe diversity also varies within the same habitat (Green and Bohannan 2006; Fierer 2008). If we assume that plant and animal distribution patterns derive from mechanisms that regulate the biodiversity, ignoring these patterns for microbial species would mean a serious omission, as micro-organisms are a great part of biodiversity on Earth (Green and Bohannan 2006).

A main reason for regarding the microbe biogeography as irrelevant was the assumption that microbes can potentially occur anywhere provided that hosts are available. This is the "host-niche hypothesis". An alternative to this postulate, the "disease-niche hypothesis", was formulated for explaining the geographic distribution of plague (Maher *et al.* 2010) and suggests that the presence of the pathogen distribution may depend on other factors than the occurrence of hosts. Biogeographic analyses in North America based on plague have provided support for the latter, as infections by the bacterium *Yersinia pestis* occur in areas characterized by ecological conditions that are different from those explaining the host distributions (Maher *et al.* 2010). These conditions are expected to influence both the ecology of fleas (i.e. of vectors) and the survival of the pathogen outside vectors and hosts. Another study, focused on the determinants of environmental suitability for transmission of human malaria (Gething *et al.* 2011), suggested that the occurrence of infections by the protist *Plasmodium* may be influenced by temperature. This parameter modifies the performance of *Anopheles* mosquito vectors by influencing the distribution of environments suitable for oviposition, the rates of larval development and survival, and the adult life span and feeding rates, which ultimately affect the vector capacity to infect a person. Moreover, adult mosquitoes must survive long enough to accumulate sufficient *Plasmodium* sporozoites for a successful transmission, and the duration of sporogony (i.e. the production of sporozoites) is also determined by temperature.

Probably, there is not a single answer for the question of what factors shape the geographic distribution of microbes and disease. There may be cases in which simple transmission cycles lead to correlations between pathogen and host distributions, while in other cases, the disease-niche hypothesis might provide better predictions. The challenge is to formulate pathways for a robust testing of biogeographical hypotheses consistent with the observations, and able to integrate the complex biological relationships often shown by the zoonotic cycles. Microbiologists and biogeographers should work together for this purpose, taking advantage of pre-existing concepts originally developed for biogeographic studies on animals and plants (Fierer 2008). Significant biogeographic patterns in microbes have already been described in free-living micro-organisms (Martiny *et al.* 2006) and human pathogens (Guernier *et al.* 2004).

11.3. Historical biogeography and disease

Historical biogeography may have been one of the first approaches to the spatio-temporal analysis of disease. The spatial patterns of plague were analyzed in 1994 from a nearly phylogeographic point of view (Guiyoule *et al.* 1994), providing insights on how *Yersinia pestis* may have diversified for centuries. Phylogeography has been proposed as a framework for testing hypotheses on disease transmission (Gray and Salemi 2012). However, until very recently, explicit references to the phylogeographic discipline were not generalized in the medical literature. For example, Ebola-virus phylogenies based on georeferenced genetic sequences have been frequently addressed (e.g. Grard *et al.* 2011), providing hypotheses on the origin, evolution and spatio-temporal dynamics of the dreadful epidemic that killed more than 11,000 people between 2014 and 2016 in West Africa (Gire *et al.* 2014; Carroll *et al.* 2015). These studies helped to confirm the location of the initial outbreak in Guéckédou (Guinea) in December 2013, and described in detail the virus spread between countries and districts. Nevertheless, it was only when the phylogeographical framework became explicit in the analysis that the studies started to deepen the factors shaping the spatial dynamic of the disease (Dudas *et al.* 2017). Phylogeography has provided empiric evidence that the virus spread in West Africa following a gravity-model dynamic favored by dense human populations and short distances between villages and towns, that intranational transmissions were more frequent than international dispersals and that cultural links, urbanization and climate might have contributed to the maintenance of the virus in a region, but not as much to the virus spread. This is definitely useful information for delineating strategies for epidemic management and prevention.

Phylogeography has been applied to the analysis of disease vectors (e.g. Conn and Mirabello 2007; Monteiro *et al.* 2018). The invasive mosquitoes *Aedes aegypti* and *Aedes albopictus* are main vectors of dengue, yellow fever, chikungunya, Zika and other viral diseases. *A. aegypti* initially was an African species, from where it invaded other continents at some time between the 15th and the 18th centuries (Tabachnick 1991). In contrast, the origin of *A. albopictus* is Asian and the start of its geographical drift dates back to the second half of the 20th century (Kraemer *et al.* 2015). Now, both species have extensive distribution ranges with populations in all continents, and their spread still continues worldwide. *A. aegypti* and *A.*

albopictus belong to the same genus, transmit the same diseases, show roughly similar distributions that have been shaped by anthropogenic forces, and have similar anthropophilic and larval-development behaviors (Kraemer *et al.* 2015). This might suggest the existence of parallel genetic population structures and biogeographical histories, but phylogeography reveals a different scenario. The analyses confirmed the African origin of *A. aegypti*, and the human contribution to expand its distribution first in America, and subsequently in South-West Asia and in the Pacific (Brown *et al.* 2014). Before the species left Africa, the ancestral zoophilic populations evolved into a domestic form that surely experienced a genetic bottleneck. A relatively pure lineage would thus be in the origin of all the *A. aegypti* populations outside Africa, variations of which configured the current genetic diversity that still reflects the history of the species spread. In contrast, there is no evidence of genetic bottleneck in *A. albopictus*. Instead, a high propagule pressure may have been in the origin of this species spread, suggesting that repeated and probably massive introductions could have occurred (Goubert *et al.* 2016). There is no track of a single wave front for *A. albopictus*, but a stochastic dissemination along transportation infrastructures, and there is no genetic structure able to explain the high variability and the geographic pattern of invasive and native populations.

The study of pathogen phylogeographic dynamics has been used as an alternative to ecologic surveillance in animal reservoirs. A successful example of this was applied to the understanding of the spatiotemporal dispersal of the yellow-fever virus in South America (Mir *et al.* 2017). Human infections in Brazil since 1950 used to be sporadic and caused by contacts with non-human primates but, in 2017, non-endemic regions of South Brazil were affected by the largest yellow-fever outbreak recorded in the country over the last 50 years. In order to design and optimize vaccination and other prevention strategies, it is important to discover the drivers of the ongoing virus expansion, that is, how the evolution and spatio-temporal replacement of virus lineages have occurred, and to what extent animal hosts are contributing to the virus circulation. However, surveillance is extremely resource demanding in remote tropical forests. Using genome sequences available in databases, phylogeographic analyses concluded that the recent outbreak in South Brazil was the result of a new virus lineage, which probably emerged in the jungles of Trinidad and Tobago and was carried to the continent by infected humans. Thus, the mechanism explaining the recent epidemiological dynamic of the yellow fever in South America seems to be the *in situ* evolution of the virus, rather than the long-distance epizootic spread of infection waves (Mir *et al.* 2017). A brief but extremely complete review on the history of human pathogens since the origin of our species can be found in Smith and Guégan (2010).

11.4. Disease distribution patterns

Guided by the host-niche hypothesis, some scientists have studied the spatial patterns of disease through the analysis of animal distributions. A nice example is the analysis of diseases transmitted by *Aedes* mosquitoes, where the country-level risk of arboviral diseases (i.e. of diseases transmitted by arthropods acting as vectors) has been deduced from every country's potential to have populations of main vectors (Leta *et al.* 2018). A spatial correlation between the global pattern of pathogen diversity and that of birds and mammals (i.e. potential reservoir species) has been proved (Dunn *et al.* 2010). Moreover, areas with similar mammal communities have shown common assemblages of infectious diseases (Murray *et al.* 2015). However, as human pathogens follow biogeographic patterns that can be ecologically explained (Guernier *et al.* 2004), the analysis of disease-case records or of pathogen distributions is a more straightforward way of pattern detection.

Although predicting disease outbreaks is a main challenge to humanity, little is known about their global patterns and determinants at subnational scales (Steengard *et al.* 2017). Nonetheless, through the analysis of disease gamma-diversity hotspots, country-level resolution studies have helped to suggest where new diseases could emerge (Guernier *et al.* 2004; Jones *et al.* 2008; Smith and Guégan 2010). The richness of human pathogens, just as that of many animals and plants, seems to be correlated to latitude. The number of infectious microbes in the tropics is higher than in cold and temperate regions (Guernier *et al.* 2004). Climate is the argued reason for this pattern. However, the linear latitudinal correlation disappears when only agents of emerging disease (i.e. 13% of human pathogens, Smith and Guégan 2010) are considered. Pathogen emergence events are mainly concentrated in temperate countries (Jones *et al.* 2008), which is thought to be a result of socioeconomic factors, such as the higher reporting effort and the drug-resistance of pathogens in developed nations. Another gamma-diversity pattern that is classical in biogeography has been detected here: there is a significant agreement of insular pathogen patterns with the theory of island biogeography by MacArthur and Wilson (1967). Higher number of pathogen species can thus be found in islands with higher size and at lower distance from the continent (Cliff *et al.* 2000).

The patterns of pathogen-community composition are also associated with latitude (Guernier *et al.* 2004), as the pools of pathogen species in high-latitude countries tend to be subsets of the rich tropical pathogen assemblages. Just as with animals and plants, regionalization based on pathogen species composition informs about the probable history of these organisms, and so of disease. Just *et al.* (2014) demonstrated the existence of pathogeographic regions resulting from the presence of climatic differences, diverging biogeographic histories (i.e. vicariant and dispersal events), and human cultural and historical idiosyncrasies. Vector-borne disease communities showed similar regionalizations to those detected in vertebrates (Holt *et al.* 2013), and the analysis of non-vector-borne pathogens reproduced the classical Wallace's (1876) regionalization, suggesting that, despite human migrations and efforts to control pathogens and vectors, climate and the zoogeographical history still influence the biogeography of disease (Just *et al.* 2014). However, the human efforts against disease, and also the colonial history, have left a trace in pathogeographic regionalizations that is deduced from the similar pathogen assemblages shown by Australia and parts of the Holarctic Region (Just *et al.* 2014). Murray *et al.* (2015) quantified and mapped the extent of change in pathogen-community composition between countries through the analysis of beta-diversity patterns, showing again that infectious diseases are not randomly distributed in the global scale. Instead, some countries tend to share certain disease assemblages, following a pattern that is mostly explained by the diversity of mammals. Interestingly, this study provided a tool for developing hypotheses on "pathogeographic regions at risk of a given disease". For example, Murray *et al.* (2015) delimited the "Ebola co-zones", that is, the set of countries that are pathogeographically similar to those showing records of Ebola-virus disease outbreaks, suggesting that, in countries such as Nigeria, where zoonotic Ebola-virus transmissions are not reported, it is worth surveying for the presence of the virus in the wild. This approach could be helpful for strengthening biosecurity capacities and optimizing pathogen discovery efforts in regions with scarcity of data and resources.

11.5. Disease distribution modeling

11.5.1. *Mechanistic versus empirical modeling*

One of the biogeographic approaches with more pathogeographic potential is species distribution modeling (Guisan and Zimmermann 2000). Understanding the spatial and temporal distribution of infectious diseases is of central importance in order to understand the factors conditioning the risk of new outbreaks (Peterson 2008; Hay *et al.* 2013). A model is a hypothesis on the degree to which certain environments are favorable to the occurrence of infections, even in areas in which the disease has never been reported (Escobar and Craft 2016; Kraemer *et al.* 2016). Traditionally, modeling exercises in epidemiology have been based on mechanistic approaches, that is, predictive models using mathematic formulations of explicit processes limiting species distributions. These processes normally incorporate physiological constraints and specific responses of species to environmental conditions (i.e. vegetation, soil and physiography), and components from population-dynamics (i.e. growth and mortality rates, parameters quantifying competence and other interactions) (e.g. Patz *et al.* 1998; Ogden *et al.* 2005; Carver *et al.* 2015). Based on microenvironmental resolutions, mechanistic models detail the complexity of disease transmission systems at the ecological scale; but they face the challenge of projecting their results to coarse-resolution scales involving macroclimatic phenomena and geographic variations in the physiological functions of pathogens, vectors and hosts (Peterson 2014; but see Redding *et al.* 2019). Empirical models, also known as correlative models, are often seen as the alternative to mechanistic approaches in the macroecological scales. The spatial and temporal relations between the presence of pathogens, vectors or disease and a set of predictor variables is analyzed by correlative means, employing a variety of statistical and machine-learning routines (e.g. Hirzel *et al.* 2002; Phillips *et al.* 2006; Prasad *et al.* 2006; Real *et al.* 2006). This empirical approach is usually sustained on spatial units whose resolution varies from hundreds of square meters to countries, so it's trying to account for the presence of disease according to average environmental values and not microenvironmental conditions. Thus, the practice of empirical models is also subject to limitations, such as the fact that they do not capture the physiological variation of a pathogen or a vector within a spatial unit (Peterson 2014). Nevertheless, it has been demonstrated that both mechanistic and empirical models are able to produce robust forecasts when applied to the analysis of the same problem (Kearney *et al.* 2010).

Empirical distribution modeling has already produced robust predictions related to a number of different pathogens (e.g. Peterson *et al.* 2004; Peterson 2006; Reed *et al.* 2008; Neerinckx *et al.* 2009; Samy *et al.* 2014; Zhu and Peterson 2014) and to their animal vectors and reservoirs (e.g. Sweeney *et al.* 2006; Oliveira *et al.* 2013; Giles *et al.* 2014; Kraemer *et al.* 2015; Gossner *et al.* 2019). It is worth outlining a model that was developed for defining the global risk of disease emergence (Jones *et al.* 2008), based on diseases caused by 335 pathogens. The analysis reported that countries with the least risk of experiencing the emergence of important diseases accumulate, in contrast, most of the scientific and surveillance efforts. Jones *et al.* (2008) mapped the global hotspots of emerging disease, showing that new wildlife-zoonotic and vector-borne pathogens are most likely to originate in lower-latitude developing countries. These hotspots were characterized by ecological and socio-economic factors, concluding that conservation efforts in rich-wildlife-diversity areas can reduce the probability of future emergence of zoonotic disease.

11.5.2. *The search for risk factors in time and space*

Empirical distribution models are inferential, that is, they express, in the form of mathematic functions, patterns that can be generalized beyond the observations.

These patterns are significant relationships between the presence of the modeled entity (i.e. a pathogen, a vector, a reservoir species or disease reports) and a set of environmental, historical and/or socioeconomic variables representing predefined and tested hypotheses on the factors of disease. The current scientific production is prolific in the search of explanatory factors on which the risk of disease can be defined and predicted.

11.5.2.1. *Land-cover change*

More than 30 pathogens and infectious diseases are suspected to be linked to changes in landscape (Patz *et al.* 2004) such as forest loss and fragmentation, extractive industries, agriculture, urbanization, and the degradation of wetlands and coasts. The agricultural intensification around tropical forests is associated with the increased risk of zoonotic-disease, as it brings humans closer to zoonotic cycles, causes the migration of wildlife species and creates environments favorable to pathogens, vectors and hosts (Jones *et al.* 2013). Besides, the availability of food and shelter in urban areas attracts a range of animal species that can be hosts of pathogens and vector parasites (Mackenstedt *et al.* 2015).

Until 2016, the influence of deforestation on the local risk of Ebola-virus disease (EVD) outbreaks had been suggested (Wolfe *et al.* 2005; Wallace *et al.* 2014). This relation was supported in 2017, when two studies testing for the first time the link between Ebola and forest loss (Olivero *et al.* 2017b; Rulli *et al.* 2017). Rulli *et al.* (2017) showed that the rate of forest fragmentation, around the areas where Ebola-virus transmissions from wildlife took place from 2004 to 2014, was significantly higher than the average across Central and West Africa. The question was also addressed using a spatio-temporal distribution modeling approach (Olivero *et al.* 2017b). The rate of forest loss was quantified in a set of locations that were referenced in spatial and temporal coordinates. The resulting model pointed to some factors favoring the occurrence of EVD outbreaks: (1) the presence of humans and the Ebola virus; (2) a spatio-temporal structure of outbreak waves whose [probably meteorological] drivers need to be explored and (3) forest loss affecting some outbreaks. The analyses determined that, in the period 2006–2014, forest loss made a 27% contribution to explain the variation in favorability for the occurrence of new outbreaks, accounting for seven events of Ebola-virus spillover from wildlife (Figure 11.1). These events were restricted to areas at the limits of the rainforest domain where forest loss had principally affected dense forests. The influence of deforestation on the risk of disease was estimated to last a maximum of two years.

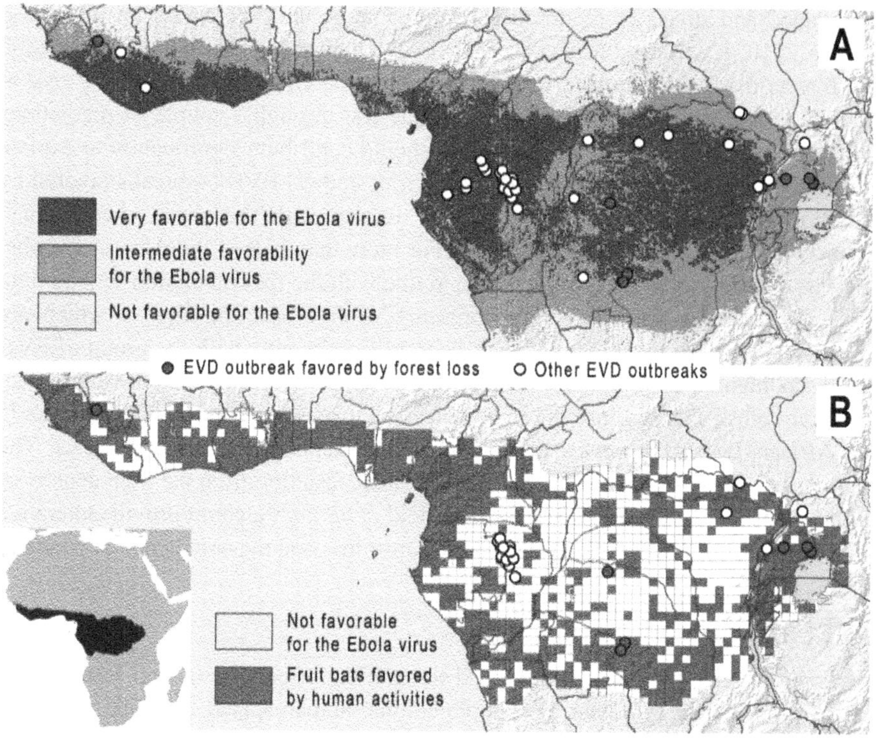

Figure 11.1. *Ebola-virus disease (EVD) outbreaks (events of zoonotic transmission), forest loss and fruit bats. A. The risk of Ebola-virus zoonotic transmission influenced by deforestation tends to be higher in the margins of the rainforest biome (Olivero* et al. *2017b), coinciding with the area with intermediate environmental/zoogeographic favorability for the presence of the virus (Olivero* et al. *2017a) (all EVD outbreaks reported between 1976 and 2021 are shown). B. EVD outbreaks linked to forest loss (Olivero* et al. *2017b) overlap significantly with the areas where the presence of five fruit-bat species* (Eidolon helvum, Epomops franqueti, Megaloglossus woermanni, Micropteropus pusillus *and* Rousettus aegyptiacus) *is favored by human activities (Olivero* et al. *2020) (only outbreaks between 2001 and 2014, i.e. those analyzed in Olivero* et al. *2017b, are shown). For a color version of this figure, see www.iste.co.uk/guilbert/biogeography.zip*

The following challenge was to explore which mechanisms explain the link between forest loss and EVD. The focus was put on the role of fruit-bat species, which are the most probable reservoirs of the virus (Leroy *et al.* 2005). Fruit-eater bats (family Pteropodidae) have been linked to forest fragmentation, logging, secondary succession forest and agriculture, through positive relations with abundance and diversity recurrently recorded in Asian and American sites (e.g. Luskin 2010; Castro and Michalski 2014). The hypothesis that human impacts on the forests favor the presence of fruit bats and, subsequently, increase the risk of new EVD outbreaks in Africa, was put to the test through a couple of predictions (Olivero *et al.* 2020): (1) human activities should contribute significantly to explain fruit-bat distributions in Central and West Africa and (2) EVD outbreaks favored by forest loss (Olivero *et al.* 2017b) should overlap with the areas where human impacts favor the presence of these bats. For more than half of the African fruit-bat species, distribution-model outputs were in line with the first prediction. In addition, the areas where humans favor the presence of five species, four of which are serologically linked to the Ebola virus, overlap significantly with the group of EVD outbreaks favored by forest loss (Figure 11.1). This biogeographic approach did not definitely prove that bats are the linkage between forest loss and EVD outbreaks in the African forests; however, it provided robust support for this hypothesis. The increase of contacts between humans and wildlife, resulting from the high density of forest edges in fragmented forests (Bloomfield *et al.* 2020), could provide alterative hypotheses for the spatial overlap between forest loss and the emergence of zoonotic diseases.

11.5.2.2. *Globalization and mobility*

Long-distance movements have had devastating effects on human health since early in the history of humanity. The Paleolithic human spread, the Silk and Spice Routes, the Crusades and the slave trade between Africa and Asia, might have contributed significantly to the expansion of diseases such as plague, malaria, yellow fever and their vectors (Bryant *et al.* 2007; Maher *et al.* 2010; Smith and Guégan 2010). In 2020, the whole world was witness to and victim of the likely correlation between the immediacy of intercontinental communications and the rapid spread of the virus SARS-CoV-2. However, sometimes, species and pathogens leave their original territories following their own biological dynamics, led by evolutionary, behavioral or environmental changes (e.g. the West Nile virus, Hadfield *et al.* 2019). Therefore, it is necessary to demonstrate in which cases the changing biogeography of disease is really a result of the increasing global connectivity. One of the most relevant cases of contemporary vector spread is that of the tiger mosquito *Aedes albopictus*, which has expanded from the forests of Asia to cities in all continents since the 1980s (Kraemer *et al.* 2015). Tatem *et al.* (2006) tested the influence of international communications on this insect's spread through the definition of "climatical suitability envelopes" around seaports and airports within the species' historical distribution. Traffic volumes from the historical distribution to ports located in areas with newly established populations resulted in twice as high traffic volumes to ports in climatically similar locations. The conclusion of this study was that the traffic volume between ports explained the

A. albopictus spread better than climatic similarities between regions. When the climatic suitability is equal in two areas, shipping volumes could determine the invasion likelihood.

11.5.2.3. *Atmosphere–ocean oscillations*

The local risk of zoonotic-disease outbreaks can be influenced by atmospheric events taking place in remote areas (Anyamba *et al.* 2001; Pinzon *et al.* 2004; Lowen *et al.* 2007). In 1999, the World Health Organization published a report in which disasters caused by the weather were related to the public health (Kovats *et al.* 1999). Some disasters are linked to atmospheric-oceanic phenomena such as the El Niño Southern Oscillation (ENSO). With a relative periodicity of two to seven years, changes in sea–surface temperatures and in air pressure occur in the Pacific basin, and these changes cause droughts and heavy rainfall in surrounding, and also remote, countries. This phenomenon is called "teleconnection" and is associated with the risk of rodent-borne diseases (such as those ones caused by hantaviruses) and of diseases transmitted by mosquitoes (such as malaria and dengue). Extreme weather conditions may produce, for example, a decrease in the protective immunity of people, and also an increase in the abundance of vectors. Interestingly, teleconnections are subject to temporal gaps between the detection of atmosphere–ocean changes and the occurrence of weather events in remote areas, and this involves a high potential for risk forecasting.

Distribution modeling is frequently used for assessing the relations between disease and atmospheric events with predictive purposes. Although it is located at the antipodes of the Pacific Ocean, Africa is subject to interannual meteorological changes responding to the ENSO. Anyamba *et al.* (2001) showed that, in the years with negative values of the Southern Oscillation Index (SOI, which is used to represent the phases of the ENSO), there is a sea–surface temperature increase in the West Indian Ocean that causes abnormally intense rainfall in the arid zones of East Africa. The effects of this singularity are seen, for example, as anomalously high values of greenness (which are detected by remote-sensing vegetation indices), in the flood of shallow wetlands called "dambos", and, ultimately, in the proliferation of hatching sites for *Aedes* mosquitoes that are vectors of phleboviruses causing Rift-Valley Fever (RVF). Anyamba *et al.* (2001) demonstrated that more than three out of four RVF outbreaks recorded in Kenia in the second half of the 20th century occurred during negative-SOI phases. These evidences provided support for the proposal of a predictive model (Redding *et al.* 2017), which concluded that the risk of RVF depends on interactions between time and space. Risk areas vary geographically depending on the ENSO. Thus, the highest risk values in South and East Africa occur in the earlier months of years with extreme ENSO behaviors (i.e.

in El Niño and La Niña events); whereas the risk in West Africa is concentrated in the second half of non-ENSO years.

There are multiple examples of the model-based estimation of disease-risk temporality considering meteorological patterns (e.g. Poveda *et al.* 2000; Mabaso *et al.* 2007; Báez *et al.* 2019). Events related to abnormal meteorological conditions could explain the spatio-temporal patterns of critical diseases. For example, the outbreak dynamics of the Ebola-virus disease in Africa seem to be associated with irregularities in the rainfall seasonality (Schmidt *et al.* 2017), specifically to exceptionally drier conditions at the end of the rainy season (Pinzon *et al.* 2004). Therefore, there is urgent need for the development of predictive models that can be useful as the basis for early warning systems, for which the predictive potential of teleconnections based on atmospheric–oceanic oscillations should be exploited.

11.5.2.4. *Climate change*

If short-term meteorological oscillations are able to influence the risk of disease, long-term climate changes surely have severe effects on the human health geography. According to the World Health Organization, around 150,000 people could be dying every year as a consequence of health problems derived from the global warming, which could be increasing the risk of cardiovascular and respiratory illness, of malnutrition and of infectious-disease transmission (Patz *et al.* 2005). In biogeography, forecasts on probable changes in species distributions are frequently obtained through the projection of distribution models to future climatic and socio-economic scenarios. Although some researchers are of the opinion that this kind of extrapolations is of limited value for the assessment of health risk (Patz *et al.* 2005), some studies have tested successfully the predictive potential of disease distribution models. For example, analyzing human cases of plague and tularemia, Nakazawa *et al.* (2007) trained models based on the disease and climatic records of the period between 1965 and 1969. Then, they projected these models to the climatic conditions of the 1970s, the 1980s and the 1990s, and demonstrated that there was consistency between predictions and disease-case observations regarding the same period. Supported on the validated predictive capacity of their models, Nakazawa *et al.* (2007) performed extrapolations based on climate-change scenarios for 2055 (Cubasch *et al.* 2001), and anticipated the risk of a little northward shift in the northern and southern limits of the distribution of plague and tularemia cases in North America.

A paradigmatic example of infectious disease recurrently analyzed in the face of climate changes is dengue. Using a mechanistic model, Patz *et al.* (1998) expressed the epidemic potential of this disease through an equation expressing that the higher the potential, the lower the number of mosquitoes needed for maintaining a state of dengue endemicity after the virus introduction in an area. The dengue epidemic potential depends on parameters based on the mosquito population dynamic, such as the incubation-period length and the survival rate; as well as on the virus transmission probability, which depends on the frequency of bites per person, the probability that a mosquito acquires the virus, and the probability that the mosquito infects the bitten person. According to known relationships between these parameters and temperature, Patz *et al.* (1998) projected the expected values of dengue epidemic potential for 2050 on a global grid of 250-km-side squares, finding that this potential could increase an average 34–47%, principally affecting temperate regions that were free from dengue by that time. An empirical-modeling approach to the influence of global warming on the risk of dengue was later applied by Hales *et al.* (2002). Based on country-level information on dengue cases, they produced models with the aim of determining the probability of dengue transmission according to a combination of climatic and human-population variables. A model focused on the period 1975–1990 suggested that around 30% of the total human population used to live in areas with high risk of dengue. This model was then projected to the future according to climate-change scenarios (Cubasch *et al.* 2001), suggesting that the percentage could increase up to 44% in 2055, and to 52%–70% in 2085. The availability of high-resolution georeferenced information on dengue cases, vector distributions, climate and socioeconomic variables has enabled the production of more complex empirical models. Messina *et al.* (2019) projected to 2020, 2050 and 2080 a model that was based on case reporting prior to 2015. This projection supported Hales *et al.*'s (2002) prediction of a nearly 60% increase in global population at risk of dengue by 2080, and suggested that outbreaks could soon reach regions in all continents (i.e. Australia, USA, Argentina, Japan, Eastern China and Southern Europe) that are currently outside the dengue endemic distribution area. Some worthy examples of disease-spread forecasting based on climate change are Fischer *et al.* (2011), with projections of climatic suitability for the mosquito vector *Aedes albopictus* in Europe, and Redding *et al.* (2019) on the epidemic potential of the Ebola-virus disease in Africa.

11.5.3. *Pathogeography: addressing the multifaceted analysis in disease mapping*

As suggested by the "One Health" initiative (Gibbs 2014), a balanced study of zoonotic infectious diseases requires scientific approaches able to integrate a great variety of factors in intelligible form, which involves reflecting the interactions between pathogens, vectors, reservoirs and final hosts. In addition, environmental, biological, economic and social dimensions of the emergence of zoonotic diseases must be taken into account in order to avoid a fragmented understanding of disease and inadequate policies in disease prevention (Jones *et al.* 2013). Determining how distribution patterns in infectious diseases are produced by the interactions among biotic and abiotic processes is the central objective of pathogeography (Murray *et al.* 2018). This subdiscipline of biogeography is a suitable framework from which empirical distribution models can capture the complexity of disease transmission by considering the processes behind correlational patterns, a characteristic that is usually associated with mechanistic approaches (Peterson 2014).

Empirical models have often neglected the potential contribution of animal distributions as explicit predictors of disease risk, assuming that it is represented by correlations with climate and other variables. However, biogeographers dealing with disease mapping have recently started to include reservoirs and/or vectors as predictor factors in empirical models. This is the case of Pigott *et al.* (2014), who included the distribution areas of three suspected-reservoir bat species as predictor variables in a model for the Ebola-virus disease, and Messina *et al.* (2019), who included environmental suitability values for two vector species as predictor variables in a model for dengue. This decision reinforced the explanatory power of the models, bringing empirical approaches, which are correlative by nature, closer to the hypothesis that sustains mechanistic models (i.e. the algorithm behind model predictions is based on the integrated effects of causative agents). Pigott *et al.* (2016) updated their first Ebola-virus disease model (Pigott *et al.* 2016) by considering a wide range of bat species as eligible predictors (and not only the three suspected reservoirs). This permitted avoiding the assumption that all is known about the group of Ebola-virus reservoirs. If taxonomically related species are potentially susceptible to similar pathogens (Plyusnin and Morzunov 2001), it is reasonable to test for biogeographical relations between the Ebola virus and bat species showing similar life history, physiology, ecology, morphology and demography to known reservoirs, so the range of meaningful predictors in the model could be expanded. However, the zoonotic cycle of a disease may be extremely complex or, what is worse for predictive purposes, unknown. In the case of the Ebola-virus disease, reservoirs have not been definitely identified (Leendertz 2016), and lots of other species might be forming part of the system as vectors, amplifier hosts or eventual victims capable of transmitting the virus to other species before it is transmitted to humans (Olival and Hayman, 2014). In the face of this complex scenario, in order to assess the biogeographic dimension of the Ebola-virus endemic distribution, Olivero *et al.* (2017a) adopted a multilevel integration approach. They selected a set of 96 mammal species living in Central Africa according to taxonomic proximity to the species that had been serologically linked to the Ebola virus. Then, they identified that these species were biogeographically arranged according to 16 types of distribution (i.e. chorotypes) and used the distribution of these chorotypes as predictor variables to build a distribution model for the Ebola virus, which was named a "zoogeographic model". It was argued that, whatever the specific taxa involved, the influence of the fauna on the probability of virus occurrence was exerted via these species distributions. Both the presence of species of the zoonotic cycle and of suitable environmental conditions for the virus transmission were finally considered to be limiting factors, which led to a final Ebola-virus model (Figure 11.1) that was conceived as the intersection of the zoogeographic model and an environmental model based on climate and landscape predictors, according to a fuzzy-logic approach (Romero *et al.* 2015).

In Olivero *et al.*'s (2017a) proposal for the Ebola virus, a range of conceptual and methodological resources from biogeography was put to the service of the empirical modeling of disease risk, as modeling algorithms and biogeographical patterns were integrated according to previous zoonotic hypotheses. This integration can be made more complex and specific in parallel to the detail of the available knowledge on a zoonotic cycle. For example, Figure 11.2 shows a methodological proposal for the empirical distribution modeling of a mosquito-borne disease that combines anthroponotic cycles (in which human-to-human transmission is mediated by mosquito vectors) with zoonotic cycles (in which undetermined tropical-forest primates are infected and can, in turn, transmit the virus to humans and be the substrate for the virus evolution). This is the case of the dengue and yellow-fever viruses (Vasilakis *et al.* 2011; Auguste *et al.* 2015). Reasonable starting hypotheses are as follows: (1) infections need the presence of at least one mosquito vector species (whose occurrence in the area should be, thus, considered to be a limiting factor); (2) the success of transmission to humans is limited by the vectorial capacity, which is in turn favored by certain environmental conditions (Gething *et al.* 2011) and (3) transmission rates can be regionally amplified by the presence of enzootic cycles in forest primates. Therefore, a vector model can be built through the union of individual mosquito-species models (i.e. at least one vector species must occur), a disease model will describe favorable areas for the occurrence of human infections through the combination of environmental/socioeconomic predictors and of zoogeographic variables (i.e. primate chorotypes) and a final transmission-risk model can be conceived as the intersection of both the vector model and the disease model (i.e. both representing limiting factors). An example of the application of this approach to the global risk of dengue transmission can be found in Aliaga-Samanez *et al.* (2021).

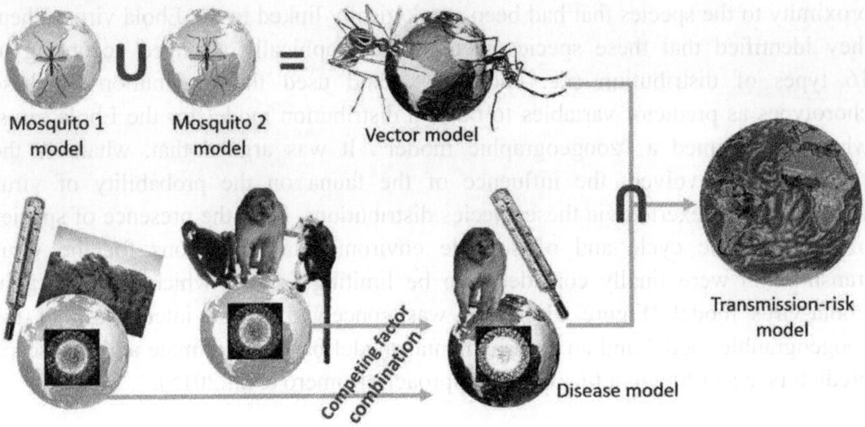

Figure 11.2. *Empirical modeling of transmission risk of a mosquito-borne disease combining anthroponotic cycles (i.e. human-to-human transmission) and zoonotic cycles (i.e. primates can transmit the virus to humans). The vector model results from the union of individual mosquito-species distribution models. The disease distribution model describes favorable areas for the occurrence of human infections, combining environmental variables and primate distribution types as predictor factors. The intersection of both models defines the transmission-risk model. Intersections and unions require commensurable model outputs and can be implemented according to fuzzy-logic operations. For a color version of this figure, see www.iste.co.uk/guilbert/biogeography.zip*

11.6. Concluding remarks

Much has been done in the last decades, from the biogeographical discipline, to improve the knowledge on disease distributions and broad-scale spatial dynamics. However, there is significantly much more that can be done through the application of biogeographic methods and theoretical concepts (which are usually committed to conservation and ecological purposes) to the research and management of human infectious diseases. This includes, for example, identifying processes of disease emergence, dispersal and establishment by exploring the organization of disease assemblages, and improving disease surveillance designs through the definition of contexts of pathogeographic representativeness (Murray *et al.* 2018). The availability of high-quality data on georeferenced disease reports (e.g. Messina *et al.* 2014), of databases providing structured access to this information (e.g. GIDEON: www.gideononline.com), and of predictor variables derived from remote sensing, big data and other cutting-edge technologies, provides a perfect scenario for addressing the high-resolution analysis of disease biogeography at the global scale.

In these times of health emergency, there is an urgent need to establish observatories from which temporal and spatial forecasts on disease risk are provided and updated on an ongoing basis. It is thus critical that biogeographers interested in pathogeography join efforts to predict where and when disease outbreaks are likely to occur imminently (Stephens *et al.* 2016), so they can contribute to the design of early warning systems with robust and meaningful geographical bases.

11.7. Acknowledgements

The project CGL2016-76747-R of the Spanish Ministerio de Economía, Industria y Competitividad and ERDF supported the writing of this chapter.

11.8. References

Aliaga-Samanez, A., Cobos-Mayo, M., Real, R., Segura, M., Fa, J.E., Olivero, J. (2021). Worldwide dynamic biogeography of zoonotic and anthroponotic dengue. *Plos Neglected Tropical Diseases*, 15, e0009496.

Allen, T., Murray, K.A., Zambrana-Torrello, C., Morse, S.S., Rondinini, C., Di Marco, M., Breit, N., Olival, K.J., Daszak, P. (2017). Global hotspots and correlates of emerging zoonotic diseases. *Nature Communications*, 8, 1124.

Anyamba, A., Linthicum, K.J., Tucker, C.J. (2001). Climate-disease connections: Rift Valley fever in Kenya. *Cadernos de Saúde Pública*, 17, 133–140.

Auguste, A.J., Lemey, P., Bergren, N.A., Giambalvo, D., Moncada, M., Morón, D., Hernandez, R., Navarro, J.-C., Weaver, S.C. (2015). Enzootic transmission of yellow fever virus, Venezuela. *Emerging Infectious Diseases*, 21, 99–102.

Báez, J.C., Olivero, J., Salazar-Aravena, L.E., Suazo-Galdanes, I.C. (2019). Effects of atmospheric oscillations on infectious diseases: The case of Chagas disease in Chile. *Memorias do Instituto Osvaldo Cruz*, 114, e:180569.

Bett, B. (2011). *Agriculture-associated Diseases Research at ILRI: Emerging Infectious Diseases*. Livestock Exchange Issue Brief 12. ILRI, Nairobi.

Bloomfield, L.S.P., McIntosh, T.L., Lambin, E.F. (2020). Habitat fragmentation, livelihood behaviors, and contact between people and nonhuman primates in Africa. *Landscape Ecology*, 35, 985–1000.

Brown, J.E., Evans, B.R., Zheng, W., Obas, V., Barrera-Martinez, L., Egizi, A., Zhao, H., Caccone, A., Powell, J.R. (2014). Human impacts have shaped historical and recent evolution in *Aedes aegypti*, the dengue and yellow fever mosquito. *Evolution*, 68, 514–525.

Bryant, J.E., Holmes, E.C., Barrett, A.D.T. (2007). Out of Africa: A molecular perspective on the introduction of yellow fever virus into the Americas. *PLoS Pathogens*, 3, 0668–0673.

Carroll, M.W., Matthews, D.A., Hiscox, J.A., Elmore, M.J., Pollakis, G., Rambaut, A., Hewson, R., García-Dorival, I., Bore, J.A., Koundouno, R., Abdellati, S., Afrough, B., Aiyepada, J., Akhilomen, P., Asogun, D. *et al.* (2015). Temporal and spatial analysis of the 2014–2015 Ebola virus outbreak in West Africa. *Nature*, 524, 97–101.

Carver, S., Mills, J.N., Parmenter, C.A., Parmenter, R.R., Richardson, K.S., Harris, R.L., Douglass, R.J., Kuenzi, A.J., Luis, A.D. (2015). Toward a mechanistic understanding of environmentally forced zoonotic disease emergence: Sin Nombre Hantavirus. *BioScience*, 65, 651–666.

Castro, I.J. and Michalski, F. (2014). Effects of logging on bats in tropical forests. *Brazilian Journal of Nature Conservation*, 12, 99–105.

Cliff, A.D., Haggett, A.D., Smallman-Raynor, M. (2000). *Island Epidemics*. Oxford University Press, Oxford.

Conn, J.E. and Mirabello, L. (2007). The biogeography and population genetics of neotropical vector species. *Heredity*, 99, 245–256.

Cubasch, U., Meehl, G.A., Boer, G.J., Stouffer, R.J., Dix, M., Noda, A., Senior, C.A., Raper, S., Yap, K.S., Abe-Ouchi, A., Brinkop, S., Claussen, M., Collins, M., Evans, J., Fischer-Bruns, I. *et al.* (2001). Projections of future climate change. In *Climate Change 2001: The Scientific Basis. Contribution of Working Group I to the Third Assessment Report of the Intergovernmental Panel on Climate Change*, Houghton, J.T., Ding, Y., Griggs, D.J., Noguer, M., van der Linden, P.J., Dai, X., Maskell, K., Johnson, C.A. (eds). Cambridge University Press, Cambridge.

Dudas, G., Carvalho, L.M., Bedford, T., Tatem, A.J., Gaele, G., Faria, N.R., Park, D.J., Ladner, J.T., Arias, A., Asogun, D., Bielejec, F., Caddy, S.L., Cotten, M., D'Ambrozio, J., Dellicour, S. *et al.* (2017). Virus genomes reveal factors that spread and sustained the Ebola epidemic. *Nature*, 544, 309–315.

Dunn, R.R., Davies, T.J., Harris, N.C., Gavin, M.C. (2010). Global drivers of human pathogen richness and prevalence. *Proceedings of the Royal Society of London B, Biological Sciences*, 277, 2587–2595.

Epstein, P.R. (2001). Climate change and emerging infectious diseases. *Microbes and Infection*, 3, 747–754.

Escobar, L.E. and Craft, M.E. (2016). Advances and limitations of disease biogeography using ecological niche modeling. *Frontiers in Microbiology*, 7, 1174.

Fierer, N. (2008). Microbial biogeography: Patterns in microbial diversity across space and time. In *Accessing Uncultivated Microorganisms: From the Environment to Organisms and Genomes and Back*, Zengler, K. (ed). ASM Press, Washington, DC.

Fischer, D., Thomas, S.M., Niemitz, F., Reineking, B., Beierkuhnlein, C. (2011). Projection of climatic suitability for *Aedes albopictus* Skuse (Culicidae) in Europe under climate change conditions. *Global and Planetary Change*, 78, 54–64.

Gething, P.W., van Boeckel, T.P., Smith, D.L., Guerra, C.A., Patil, A.P., Snow, R.W., Hay, S.I. (2011). Modelling the global constraints of temperature on transmission of *Plasmodium falciparum* and *P. vivax*. *Parasites and Vectors*, 4, 1–11.

Gibbs, E.P.J. (2004). The evolution of One Health: A decade of progress and challenges for the future. *Veterinary Record*, 174, 85–91.

Giles, J.R., Peterson, A.T., Busch, J.D., Olafson, P.U., Scoles, G.A., Davey, R.B, Pound, J.M., Kammlah, D.M., Lohmeyer, K.H., Wagner, D.M. (2014). Invasive potential of cattle fever ticks in the southern United States. *Parasites and Vectors*, 7, 189.

Gire, S.K., Goba, A., Andersen, K.G., Sealfon, R.S.G., Park, D.J., Kanneh, L., Jalloh, S., Momoh, M., Fullah, M., Dudas, G., Wohl, S., Moses, L.,M., Yozwiak, N.L., Winnicki, S., Matranga, C.B. *et al.* (2014). Genomic surveillance elucidates Ebola virus origin and transmission during the 2014 outbreak. *Science*, 345, 1369–1372.

Goubert, C., Minard, G., Vieira, C., Boulesteix, M. (2016). Population genetics of the Asian tiger mosquito *Aedes albopictus*, an invasive vector of human diseases. *Heredity*, 117, 125–134.

Gradr, G., Biek, R., Tamfum, J.-J.M., Fair, J., Wolfe, N., Formenty, P., Paweska, J., Leroy, E. (2011). Emergence of divergent Zaire Ebola virus strains in Democratic Republic of the Congo in 2007 and 2008. *The Journal of Infectious Diseases*, 204, S776–S784.

Gray, R.R. and Salemi, M. (2012). Integrative molecular phylogeography in the context of infectious diseases on the human–animal interface. *Parasitology*, 139, 1939–1951.

Green, J. and Bohannan, J.M. (2006). Spatial scaling of microbial biodiversity. *Trends in Ecology and Evolution*, 21, 501–507.

Guernier, V., Hochberg, M.E., Guégan, J.-F. (2004). Ecology drives the worldwide distribution of human diseases. *PLoS Bioloy*, 2, 740–746.

Guisan, A. and Zimmermann, N.E. (2000). Predictive habitat distribution models in ecology. *Ecological Modelling*, 135, 147–186.

Guiyoule, A., Grimont, F., Iteman, I., Grimont, P.A.D., Lefèvre, M., Carniel, E. (1994). Plague pandemics investigated by ribotyping of *Yersinia pestis* strains. *Journal of Clinical Microbiology*, 32, 634–641.

Hadfield, J., Brito, A.F., Swetnam, D.M., Vogels, C.B.F., Tokarz, R.E., Andersen, K.G., Smith, R.C., Bedford, R., Grubaugh, N.D. (2019). Twenty years of west Nile virus spread and evolution in the Americas visualized by Nextstrain. *PLoS Pathogens*, 15, e1008042.

Hales, S., de Wet, N., Maindonald, J., Woodward, A. (2002). Potential effect of population and climate changes on global distribution of dengue fever: An empirical model. *The Lancet*, 360, 830–834.

Hay, S.I., Battle, K.E., Pigott, D.M., Smith, D.L., Moyes, C.L., Bhatt, S., Brownstein, J.S., Collier, N., Myers, M.F., George, D.B., Gething, P.W. (2013). Global mapping of infectious diseases. *Philosophical Transactions of the Royal Society B*, 368, 20120250.

Hirzel, A.H., Hausser, J., Chessel, D., Perrin, N. (2002). Ecological-niche factor analysis: How to compute habitat-suitability maps without absence data? *Ecology*, 83, 2027–2036.

Hoberg, E.P. and Brooks, D.R. (2015). Evolution in action: Climate change, biodiversity dynamics and emerging infectious disease. *Philosophical Transactions of the Royal Society of London B, Biological Sciences*, 370, 20130553.

Holt, B.G., Lessard, J.-P., Borregaard, M.K., Fritz, S.A., Araújo, M.B., Dimitrov, D., Fabre, P., Graham, C.H., Graves, G.R., Jønsson, K.A. (2013). An update of Wallace's zoogeographic regions of the world. *Science*, 339, 74–78.

Jones, K.E., Patel, N.G., Levy, M.A., Storeygard, A., Balk, D., Gittleman, J.L., Daszak, P. (2008). Global trends in emerging infectious diseases. *Nature*, 451, 990–993.

Jones, B.A., Grace, D., Kock, R., Alonso, S., Rushton, J., Said, M.Y., McKeever, D., Mutua, F., Yound, J., McDermott, J., Pfeiffer, D.U. (2013). Zoonosis emergence linked to agricultural intensification and environmental change. *Proceedings of the National Academy of Sciences of the United States of America*, 110, 8399–8404.

Just, M.G., Norton, J.F., Traud, A.L., Antonelli, T., Poteate, A.S., Backus, G.A., Snyder-Beattie, A., Sanders, R.W., Dunn, R.R. (2014). Global biogeographic regions in a human-dominated world: The case of human diseases. *Ecosphere*, 5, 143.

Kamins, A.O., Rowcliffe, J.M., Ntiamoa-Baidu, Y., Cunningham, A.A., Wood, J.L.N., Restif, O. (2015). Characteristics and risk perceptions of Ghanaians potentially exposed to bat-borne zoonoses through bushmeat. *EcoHealth*, 12, 104–120.

Kearney, M.R., Wintle, B.A., Porter, W.P. (2010). Correlative and mechanistic models of species distribution provide congruent forecasts under climate change. *Conservation Letters*, 3, 203–213.

Kovats, R.S., Bouma, M.J., Haines, A. (1999). *El Niño and Health*. WHO Task Force on Climate and Health, Geneva.

Kraemer, M.U.G., Sinka, M.E., Duda, K.A., Mylne, A., Shearer, F.M., Barker, C.M., Moore, C.G., Carvalho, R.G., Coelho, G.E., Bortel, W.V., Hendrickx, G., Schaffner, F., Elyazar, I.R.F., Teng, H.-J., Brady, O.J. *et al.* (2015). The global distribution of the arbovirus vectors *Aedes aegypti* and *Ae. albopictus*. *eLife*, 4, e08347.

Kraemer, M.U.G., Hay, S.I., Pigott, D.M., Smith, D.L., Wint, G.R.W., Golding, N. (2016). Progress and challenges in infectious disease cartography. *Trends in Parasitology*, 32, 19–29.

Leendertz, S.A.J. (2016). Testing new hypotheses regarding Ebolavirus reservoirs. *Viruses*, 8, 30.

Leroy, E.M., Kumulungui, B., Pourrut, X., Rouquet, P., Hassanin, A., Yaba, P., Délicat, A., Paweska, J.T., Gonzalez, J.-P., Swanepoel, R. (2005). Fruit bats as reservoirs of Ebola virus. *Nature*, 438, 575–576.

Leta, S., Beyene, T.J., De Clercq, E.M., Amenu, K., Kraemer, M.U.G., Revie, C.W. (2018). Global risk mapping for major diseases transmitted by *Aedes aegypti* and *Aedes albopictus*. *International Journal of Infectious Diseases*, 67, 25–35.

Lowen, A.C., Mubareka, S., Steel, J., Palese, P. (2007). Influenza virus transmission is dependent on relative humidity and temperature. *PLoS Pathogens*, 3, e151.

Luskin, M.S. (2010). Flying foxes prefer to forage in farmland in a tropical dry forest landscape mosaic in Fiji. *Biotropica*, 42, 246–250.

Mabaso, M.L., Kleinschmidt, I., Sharp, B., Smith, T. (2007). El Niño Southern Oscillation (ENSO) and annual malaria incidence in Southern Africa. *Transactions of the Royal Society of Tropical Medicine and Hygiene*, 101, 326–30.

MacArthur, R.H. and Wilson, E.O. (1967). *The Theory of Island Biogeography*. Princeton University Press, Princeton, NJ.

Mackenstedt, U., Jenkins, D., Romig, T. (2015). The role of wildlife in the transmission of parasitic zoonoses in peri-urban and urban areas. *International Journal for Parasitology: Parasites and Wildlife*, 4, 71–79.

Maher, S.P., Ellis, C., Gage, K.L., Enscore, R.E., Peterson, A.T. (2010). Range-wide determinants of plague distribution in North America. *American Journal of Tropical Medicine and Hygiene*, 83, 736–742.

Martiny, J.B.H., Bohannam, B.J.M., Brown, J.H., Colwell, R.K., Fuhrman, J.A., Green, J.L., Horner-Devine, M.C., Kane, M., Krumins, J.A., Kuske, C.R., Morin, P.J., Naeem, S., Øvreås, L., Reysenbach, A.-L., Smith, V.H., Staley, J.T. (2006). Microbial biogeography: Putting microorganisms on the map. *Nature*, 4, 102–112.

Messina, J.P., Brady, O.J., Pigott, D.M., Brownstein, J.S., Hoen, A.G., Hay, S.I. (2014). A global compendium of human dengue virus occurrence. *Scientific Data*, 1, 140004.

Messina, J.P., Brady, O.J., Golding, N., Kraemer, M.U.G., Wint, G.R.W., Ray, S.E., Pigott, D.M., Shearer, F.M., Johnson, K., Earl, L., Marczak, L.B., Shjirude, S., Weaver, N.D., Gilbert, M., Velayudhan, R. *et al.* (2019). The current and future global distribution and population at risk of dengue. *Nature Microbiology*, 4, 1508–1515.

Mir, D., Delatorre, E., Bonaldo, M., Lourenço-de-Oliveira, R., Vicente, A.C., Bello, G. (2017). Phylodynamics of yellow fever virus in the Americas: New insights into the origin of the 2017 Brazilian outbreak. *Scientific Reports*, 7, 7385.

Monteiro, F.A., Weirauch, C., Felix, M., Lazoski, C., Abad-Franch, F. (2018). Evolution, systematics, and biogeography of the triatominae, vectors of chagas disease. *Advances in Parasitology*, 99, 265–344.

Murray, K.A., Preston, N., Allen, T., Zambrana-Torrelio, C., Hosseini, P.R., Daszak, P. (2015). Global biogeography of human infectious diseases. *Proceedings of the National Academy of Sciences of the United States of America*, 112, 12746–12751.

Murray, K.A., Olivero, J., Roche, B., Tiedt, S., Guégan, J.-F. (2018). Pathogeography: Leveraging the biogeography of human infectious diseases for global health management. *Ecography*, 41, 1411–1427.

Nakazawa, Y., Williams, R., Peterson, A.T., Mead, P., Staples, E., Gage, K.L. (2007). Climate change effects on plague and tularemia in the United States. *Vector Borne and Zoonotic Diseases*, 7, 529–540.

Neerinckx, S.B., Peterson, A.T., Gulinck, H., Deckers, J., Leirs, H. (2009). Geographic distribution and ecological niche of plague in sub-Saharan Africa. *International Journal of Health Geographics*, 7, 54.

Ogden, N.H., Bigras-Poulin, M., O'Callaghan, C.J., Barker, I.K., Lindsay, L.R., Maarouf, A., Smoyer-Tomic, K.E., Waltner-Toews, D., Charron, D. (2005). A dynamic population model to investigate effects of climate on geographic range and seasonality of the tick *Isodes Scapularis*. *International Journal for Parasitology*, 35, 375–389.

Olival, K.J. and Hayman, D.T.S. (2014). Filoviruses in bats: Current knowledge and future directions. *Viruses*, 6, 1759–1788.

de Oliveira, S.V., Escobar, L.E., Peterson, A.T., Gurgel-Gonçalves, R. (2013). Potential geographic distribution of hantavirus reservoirs in Brazil. *PLoS ONE*, 8, e85137.

Olivero, J., Fa, J.E., Real, R., Farfán, M.Á., Márquez, A.L., Vargas, J.M., González, J.P., Cunningham, A.A., Nasi, R. (2017a). Mammalian biogeography and the Ebla virus in Africa. *Mammal Review*, 47, 24–37.

Olivero, J., Fa, J.E., Real, R., Márquez, A.L., Farfán, M.A., Vargas, J.M., Gaveau, D., Salim, M.A., Park, D., Suter, J., King, S., Leendertz, S.A., Sheil, D., Nasi, R. (2017b). Recent loss of closed forests is associated with Ebola virus disease outbreaks. *Scientific Reports*, 7, 14291.

Olivero, J., Fa, J.E., Farfán, M.A., Márquez, A.L., Real, R., Juste, F.J., Leendertz, S.A., Nasi, R. (2020). Human activities link fruit bat presence to Ebola virus disease outbreaks. *Mammal Review*, 50, 1–10.

Patz, J.A., Epstein, P.R., Burke, T.A., Balbus, J.M. (1996). Global climate change and emerging infectious diseases. *Journal of the American Medicine Association*, 275, 217–223.

Patz, J.A., Martens, W.J.M., Focks, D.A., Jetten, T.H. (1998). Dengue fever epidemic potential as projected by general circulation models of global climate change. *Environmental Health Perspectives*, 106, 147–153.

Patz, J.A., Daszak, P., Tabor, G.M., Aguirre, A.A., Pearl, M., Epstein, J., Wolfe, N.D., Kilpatrick, A.M., Foufopoulos, J., Molyneux, D., Bradley, D.J., Members of the working group on land use change and disease emergence (2004). Unhealthy landscapes: Policy recommendations on land use change and infectious disease emergence. *Environmental Health Perspectives*, 112, 1092–1098.

Patz, J.A, Campbell-Lendrum, D., Holloway, T., Foley, J.A. (2005). Impact of regional climate change on human health. *Nature*, 438, 310–317.

Peterson, A.T. (2006). Ecological niche modeling and spatial patterns of disease transmission. *Emerging Infectious Diseases*, 12, 1822–1826.

Peterson, A.T. (2008). Biogeography of diseases: A framework for analysis. *Naturwissenschaften*, 95, 483–491.

Peterson, A.T. (2014). *Mapping Disease Transmission Risk: Enriching Models Using Biogeography and Ecology*. Johns Hopkins University Press, Baltimore, MA.

Peterson, A.T., Bauer, J.T., Mills, J.M. (2004). Ecologic and geographic distribution of filovirus disease. *Emerging Infectious Diseases*, 10, 40–47.

Phillips, S.J., Anderson, R.P., Schapire, R.E. (2006). Maximum entropy modeling of species geographic distributions. *Ecological Modelling*, 190, 231–259.

Pigott, D.M., Golding, N., Mylne, A., Huang, Z., Henry, A.J., Weiss, D.J., Brady, O.J., Kraemer, M.U.G., Smith, D.L., Moyes, C.L., Bhatt, S., Gething, P.W., Horby, P.W., Bogoch, I.I., Brownstein, J.S. *et al.* (2014). Mapping the zoonotic niche of Ebola virus disease in Africa. *eLife*, 3, e04395.

Pigott, D.M., Millear, A.I., Earl, L., Morozoff, C., Han, B.A., Shearer, F.M., Weiss, D.J., Brady, O.J., Kraemer, M.U.G., Moyes, C.L., Bhatt, S., Gething, P.W., Golding, N., Hay, S.I. (2016). Updates to the zoonotic niche map of Ebola virus disease in Africa. *eLife*, 5, e16412.

Pinzon, J.E., Wilson, J.M., Tucker, C.J., Arthur, R., Jahrling, P.B., Formenty, P. (2004). Trigger events: Enviroclimatic coupling of Ebola hemorrhagic fever outbreaks. *American Journal of Tropical Medicine and Hygiene*, 71, 664–674.

Plyusnin, A. and Morzunov, S. (2001). Virus evolution and genetic diversity of hantaviruses and their rodent hosts. *Current Topics in Microbiology and Immunology*, 256, 47–75.

Pourrut, X., Kumulungui, B., Wittmann, T., Moussavou, G., Délicat, A., Yaba, P., Nkoghe, D., González, J.-P., Leroy, E.M. (2005). The natural history of Ebola virus in Africa. *Microbes and Invection*, 7, 1005–1014.

Poveda, G., Graham, N.E., Epstein, P.R., Rojas, W., Quiñones, M.L., Vélez, I.D., Martens, W.J.M. (2000). Climate and ENSO variability associated with vector-borne diseases in Colombia. In *El Niño and the Southern Oscillation: Multiscale Variability and Global and Regional Impacts*, Diaz, H. and Markgraf, V. (eds). Cambridge University Press, Cambridge.

Prasad, A.M., Iverson, L.R., Liaw, A. (2006). Newer classification and regression tree techniques: Bagging and random forests for ecological prediction. *Ecosystems*, 9, 181–199.

Ramette, A. and Tiedje, J.M. (2007). Biogeography: An emerging cornerstone for understanding prokaryotic diversity, ecology and evolution. *Microbial Ecology*, 53, 197–207.

Real, R., Barbosa, A.M., Vargas, J.M. (2006). Obtaining environmental favourability functions from logistic regression. *Environmental and Ecological Statistics*, 13, 237–245.

Redding, D.W., Tiedt, S., Lo Iacono, G., Bett, B., Jones, K.E. (2017). Spatial, seasonal and climatic predictive models of Rift Valley fever disease across Africa. *Philosophical Transactions of the Royal Society B*, 372, 20160165.

Redding, D.W., Atkinson, P.M., Cunningham, A.A., Lo Iacono, G., Moses, L.M., Wood, J.L.N., Jones, K.E. (2019). Impacts of environmental and socio-economic factors on emergence and epidemic potential of Ebola in Africa. *Nature Communications*, 10, 4531.

Reed, K.D., Meece, J.K., Archer, J.R., Peterson, A.T. (2008). Ecologic niche modeling of Blastomyces dermatitidis in Wisconsin. *PLoS ONE*, 3, e2034.

Reichert, I. and Palti, J. (1967). Prediction of plant disease occurrence: A patho-geographical approach. *Mycopathologia et Mycologia Applicata*, 32, 337–355.

Romero, D., Olivero, J., Brito, J.C., Real, R. (2015). Comparison of approaches to combine species distribution models based on different sets of predictors. *Ecography*, 39, 561–571.

Rulli, M.C., Santini, M., Hayman, D.T.S., D'Odorico, P. (2017). The nexus between forest fragmentation in Africa and Ebola virus disease outbreaks. *Scientific Reports*, 7, 41613.

Samy, A.M., Fahal, A.H., Peterson, A.T. (2014). Mapping the potential risk of mycetoma infection in Sudan and South Sudan using ecological niche modeling. *PLoS Neglected Tropical Diseases*, 8, e3250.

Schmidt, J.P., Park, A.W., Kramer, A.M., Han, B.A., Alexander, L.W., Drake, J.M. (2017). Spatiotemporal fluctuations and triggers of Ebola virus spillover. *Emerging Infectious Diseases*, 23, 415–422.

Smith, K.F. and Guégan, J.-F. (2010). Changing geographic distributions of human pathogens. *Annual Reviews of Ecology, Evolution and Systematics*, 41, 231–250.

Stensgaard, A.-S., Dunn, R.R., Vennervald, B.J., Rahbek, C. (2017). The neglected geography of human pathogens and diseases. *Nature Ecology and Evolution*, 1, 0190.

Stephens, P.R., Altizer, S., Smith, K.F., Aguirre, A.A., Brown, J.H., Budischak, S.A., Byers, J.E., Dallas, T.A., Davies, T.J., Drake, J.M., Ezenwa, V.O., Farrell, M.J., Gittleman, J.L., Han, B.A., Huang, S. *et al.* (2016). The macroecology of infectious diseases: A new perspective on global-scale drivers of pathogen distributions and impacts. *Ecology Letters*, 19, 1159–1171.

Sweeney, A.W., Beebe, N.W., Cooper, R.D., Bauer, J.T., Peterson, A.T. (2006). Environmental factors associated with the distribution and range limits of the malaria vector *Anopheles farauti sensu stricto* in Australia. *Journal of Medical Entomology*, 43, 1068–1075.

Tabachnick, W.J. (1991). Evolutionary genetics and insect borne disease. The yellow fever mosquito, *Aedes aegypti*. *American Entomologist*, 37, 14–24.

Tatem, A.J., Hay, S.I., Rogers, D.J. (2006). Global traffic and disease vector dispersal. *Proceedings of the National Academy of Sciences of the United States of America*, 103, 6264–6247.

Taylor, L.H., Latham, S.M., Woolhouse, M.E. (2001). Risk factors for human disease emergence. *Philosophical Transsactions of the Royal Society of London B, Biological Sciences*, 356, 983–989.

Vasilakis, N., Cardosa, J., Hanley, K.A., Holmes, E.C., Weaver, S.C. (2011). Fever from the forest: Prospects for the continued emergence of sylvatic dengue virus and its impact on public health. *Nature Reviews Microbiology*, 9, 532–541.

Wallace, A.R. (1876). *The Geographical Distribution of Animals: With A Study of the Relations of living and Extinct Faunas as Elucidating the Past Changes of the Earth's Surface*. Cambridge University Press, Cambridge.

Wallace, R. and Wallace, R.G. (2014). Blowback: New formal perspectives on agriculturally driven pathogen evolution and spread. *Epidemiological Infection*, 143, 2068–2080.

Wallace, R.G., Gilbert, M., Wallace, R., Pittiglio, C., Mattioli, R., Koco, R. (2014). Did Ebola emerge in West Africa by a policy-driven phase change in agroecology? *Environment and Planning A*, 46, 2533–2542.

Whitfield, J. (2005). Biogeography: Is everything everywhere? *Science*, 310, 960–961.

Wolfe, N.D., Daszak, P., Kilpatrick, A.M., Burke, D.S. (2005). Bushmeat hunting, deforestation, and prediction of zoonotic disease emergence. *Emerging Infectious Diseases*, 11, 1822–1827.

Zhu, G. and Peterson, A.T. (2014). Potential geographic distribution of the novel avian-origin influenza A (H7N9) virus. *PLoS ONE*, 9, e93390.

12

Biogeography and Climate Change

Luisa Maria DIELE-VIEGAS
Federal University of Alagoas, Maceió, Brazil

12.1. Climate change

Climate change is any change in climatic patterns over time, due to natural variability or anthropogenic activities, that leads to changes in atmosphere composition (Cubasch *et al.* 2013). Among the natural forces responsible for climatic variation are astronomic phenomena, changes in solar activity, vulcanism, tectonism, geographic variation and terrestrial orbit. The sun is responsible for providing energy to the Earth, which is transformed and stored in the atmosphere, creating the greenhouse effect. This effect is natural and responsible for maintaining the Earth's temperature at acceptable levels for the occurrence of life.

The greenhouse effect is only possible due to the presence of gases composing the atmosphere, among which are water vapor, carbon dioxide (CO_2), methane, nitrous oxide and ozone (Schneider 1989). Fluctuations in the concentration of greenhouse gases (GHG) in the atmosphere were common in Earth's history (Huybers and Curry 2006), and paleoclimate data documents the transitions between different climate states at different time scales. This information allows us to describe multi-centennial to millennial baseline variability, against which the recent changes can be compared to assess whether they are unusual or unexpected (Masson-Delmotte *et al.* 2013). The Cenozoic climate changes contain relevant quantitative information on the assessment of human-made climate effects, and prior

Biogeography,
coordinated by Eric GUILBERT.

interglacial periods that were warmer than the Holocene can play a key role in assessing the dangerous levels of the current warming rate by anticipating future changes in the climate (Hansen and Sato 2012).

The dominant cause of long-term Cenozoic climate trends, as in other geological eras, was the total amount of CO_2 in the atmosphere, ocean, soil and biosphere, which changes over millions of years due to the imbalance of the volcanic source and weathering sinks (Hansen and Sato 2012). Global temperature change in the first half of the Cenozoic (65.5–35.5 mya) was consistent with expected effects of continental drift on atmospheric CO_2 (Hansen and Sato 2012). The influence of the Milankovitch oscillations is noticeable through the Pliocene and Pleistocene (5.33 mya), together with perturbations of Earth's orbital parameters, due to changes in the insolation distribution over Earth's surface (Berger 1978; Huybers and Curry 2006). These perturbations vary slowly, and climate oscillations occur at the same frequency as these orbital perturbations (Hays *et al.* 1976), demonstrating the sensitivity of climate to the weak orbital enforcing millennial time scales. Thus, large climate changes should be expected in response to other weak enforcing and climate noise (Hansen and Sato 2012).

Global surface temperatures have typically increased by 3°C to 8°C since the last ice age (about 10,000 years ago), with global surface temperatures probably fluctuating by maximum rates of change between 1°C and 1.5°C per 1,000 years (Masson-Delmotte *et al.* 2013). However, considering the more recent past, multiple lines of evidence have shown that anthropogenic drivers are the dominant cause of changes in the Earth's climate since the mid-20th century (Cubasch *et al.* 2013). Since the industrial revolution, humans are emitting increasing amounts of GHG due to economic and population growth. The current cumulative CO_2 emissions are unprecedented in at least the last 800 kyr (Stocker *et al.* 2013), with about half of the emissions from between 1750 and 2011 occurring between 1970 and 2010, and most of it between 2000 and 2010 (Stocker *et al.* 2013). These emissions changed the atmosphere composition, thus enhancing the greenhouse effect and creating the global warming.

Global warming has been of concern to scientists since 1975, when they first agreed that actions were needed to avoid a climate crisis (Broecker 1975). However, almost 50 years later, the GHG emissions are still rising, and the world is getting warmer at the fastest rate ever seen, despite a growing number of climate change mitigation policies (IPCC 2018). So, it is important to understand what is causing these climate changes and what their consequences are.

12.1.1. *Drivers of climate change*

Measurements of GHG emissions are split by sectors, considering carbon dioxide equivalents (total GHG terms, in tons; tCO_2e). By 2017, electricity and heat accounted for the highest emissions worldwide, reaching 15.01 billion tons of carbon dioxide equivalents (tCO_2e). It was followed by the transport sector (7.9 billion tCO_2e), manufacturing energy (6.11 billion tCO_2e), agriculture (5.8 billion tCO_2e), land-use change and forestry (3.22 billion tCO_2e), fugitive emissions from solid fuels, oil and gas, manufacturing industries, and construction (2.88 billion tCO_2e), industrial processes (2.77 billion tCO_2e), buildings (2.72 billion tCO_2e), waste (solid disposal, water handling, incineration and other waste handling; 1.56 billion tCO_2e) and other fossil fuel combustion (1.43 billion tCO_2e) (World Resources Institute, 2018).

Considering production-based CO_2, by 2017 Asia accounted for 53% of global emissions, followed by North America (18%), Europe (17%) and Africa (3.7%). Among countries, China is by far the world's largest emitter, followed by the United States and the 28 countries of the European Union (EU), which are grouped together since their GHG emission reduction targets are set on a collaborative basis. Together, these countries account for more than half of the annual global emissions. When comparing the countries' cumulative emissions over time, the United States is responsible for 25% of historical emissions, followed by the EU (22%) and China (12.7%) (World Resources Institute 2018).

12.1.2. *Observed changes in the climate system*

The increase of GHG emissions intensifies the greenhouse effect, leading to an increase in the world's average environmental, sea surface and marine air temperature, a phenomenon known as global warming (Allen *et al.* 2018). The period between 1983 and 2012 was considered the warmest period of the last 800 years in the Northern Hemisphere; by 2014, 2010–2020 was the warmest decade on record, and the global average combined land and ocean surface temperatures show a warming of 1.1°C from 1880 to 2017 (Stocker *et al.* 2013; IPCC 2018; McGrath 2020). Ocean temperature, in turn, increased at a rate of 0.11°C per decade from 1971 to 2010 (Cubasch *et al.* 2013). Also, surface solar radiation has increasing in many land-based sites for the last few decades, mainly in regions with high station densities, such as Europe and part of Asia (Hartmann *et al.* 2013). It is important to highlight that the terrestrial warming has been exacerbated by urban heat islands and changes in land use and land cover, since it changes the storage and transfer of heat, water and airflow by modifying surface characteristics and structure, the so-called albedo effect (Pincebourde *et al.* 2016).

Although very worrisome, the rising temperature is not the only consequence of the GHG emissions to the climate system (Hartmann *et al.* 2013). More than 90% of the heat from the greenhouse effect is absorbed by the oceans, leading to a thermal expansion that contributes to changes in ocean physical characteristics (Cubasch *et al.* 2013). Current estimates show a 0.1 decrease in ocean pH since 1975 and a rapid loss of sea ice loss (49%) since 1979. Losses of sea ice were 18% between 2007 and 2012 (Haas *et al.* 2008; Kwok *et al.* 2009). Sea level changes have decreased over the tropical Atlantic and increased over the Pacific and South Atlantic between 1979 and 2012 (Stocker *et al.* 2013). In addition, there was a considerable glacier mass loss and a sea-level rise varying from 1.7 ± 0.2 mm yr^{-1} to 3.2 ± 0.4 mm yr^{-1} since 1990 (Church and White 2011; Stocker *et al.* 2013).

Precipitation has changed worldwide, and extreme rain and drought events have become more frequent (Hartmann *et al.* 2013). In the Southern Hemisphere, an abrupt decline in overall precipitation was observed since the 2000s, which is consistent with an enhanced drying recently recorded. Changes in hydrological systems that affect runoff and water resources downstream were associated with changes in precipitation regimes and snow and ice melting (IPCC 2014). Also, snowfall events are decreasing worldwide, mainly where an increase in average air temperature during the winter has been observed, such as North America, Europe and Asia (IPCC 2018).

Evapotranspiration over land increased between the 1980s and 1990s, but a lack of moisture availability in the Southern Hemisphere after 1998 became a constraint to its further increase by decreasing soil moisture (Cubasch *et al.* 2013). Similarly, global near-surface and tropospheric air-specific humidity have also increased since the 1970s, but near-surface moisture and relative humidity have decreased in recent years (Stocker *et al.* 2013). Since ocean surface salinity depends on evaporation and precipitation, the water has become more saline in places where evaporation dominates, while in places where rainfall dominates, the water has become fresher (Stocker *et al.* 2013).

Changes in the intensity of daily temperature extremes had been observed worldwide, with decreases in the number of cold days and nights and increases in the length and frequency of heatwaves, especially in large parts of North America, Europe, Asia and Australia (Cubasch *et al.* 2013). Droughts, heavy precipitation events and floods, wildfires and cyclones are also more frequent (Cadenhead *et al.* 2016). Intense tropical hurricanes occur more often in the North Atlantic basin, while drought events are increasing in Mediterranean and West Africa and decreasing in central North America and northwest Australia (Stocker *et al.* 2013).

12.1.3. *Future projections of global climate change*

Projections of future climate change are uncertain for three main reasons: First, any projection depends on scenarios of future anthropogenic and natural forces that remain uncertain. Second, forecasts require an understanding of the complex climate system that is still being developed. Third, microclimatic variability exists (Collins *et al.* 2013). Given these caveats, current future predictions are based on a multi-model framework called the Coupled Model Intercomparison Project (CMIP), which is responsible for coordinating climate model experiments involving modeling teams worldwide (Eyring *et al.* 2016). The CMIP is now in its sixth phase, and consists of approximately 100 climate models, such as the global circulation models (GCMs), where a common set of future concentration of GHG, aerosols and other climate forcing factors are used as representations of important processes in Earth's climate system and simulate many climate aspects (Flato *et al.* 2013; Eyring *et al.* 2016, p. 6).

The GCMs use information from different scenarios of air pollutants, GHG emissions, biogeochemical cycles and land-use patterns to obtain climate change projections (Flato *et al.* 2013). In CMIP5, these different rates and magnitudes of climate change were measured through four representative concentration pathway (RCPs) scenarios, which describe specific emission trajectories and radiative forcing (i.e. the ability of a factor to change the energy balance of the Earth–atmosphere system) (Moss *et al.* 2010; Wayne, 2013).

The RCPs predict that warming will exceed up to 4.8°C by 2100, compared to the period between 1850 and 1900 (Collins *et al.* 2013). It also predicts that high latitudes and the equatorial Pacific will experience an increase in annual precipitation, while in mid-latitude and subtropical dry regions, it will likely decrease (Collins *et al.* 2013). Oceans are likely to become warmer, mainly in Northern Hemisphere subtropical regions, and more acid, with a decrease of about 0.06–0.32 in surface ocean pH. Finally, sea level is likely to rise at a rate of 8–16 mm/yr (Church *et al.* 2013).

In CMIP6, the RCPs were replaced by the shared socioeconomic pathways (SSPs), which are driven by different socioeconomic assumptions (Riahi *et al.* 2017). There are five SSPs offering a broader view of possibilities among which the societal choices can affect the GHG emissions and, consequently, lead to changes in the world climate system. Future SSP predictions indicate an increase in the world environmental temperature varying from 3.1°C to 5.1°C above pre-industrial levels by 2100 (Riahi *et al.* 2017).

12.1.3.1. *+1.5°C tipping point*

In 2018, the Intergovernmental Panel on Climate Change (IPCC), a United Nations body for assessing the science related to climate change, released a special report on the impacts of global warming of 1.5°C above pre-industrial levels (IPCC 2018). Human-induced warming reached 1.0°C in 2018, and, given current knowledge of the global climate response, it is likely to reach 1.5°C between 2030 and 2052 (Allen *et al.* 2018). This can lead to an increase in climate-related risks for both natural and human systems, which depends on the rate, peak and duration of warming. Some of these impacts may be long term and even irreversible, such as the loss of some ecosystems (IPCC 2018).

Some of the most worrisome impacts are not driven by the increase of temperature itself, but by multiple environmental drivers (Allen *et al.* 2018). Changes in rainfall patterns, for example, can affect the hydrological cycle and water availability, while changes in atmospheric composition can lead to a decrease in plant productivity and ocean acidification (Allen *et al.* 2018; Shukla *et al.* 2019). Moreover, ocean heat can lead to destabilization of coastal ice sheets and sea level rise, besides indirectly influencing economic activities, such as food production and tourism (Hoegh-Guldberg *et al.* 2018).

12.2. Impacts of climate change on biodiversity

Along the Earth's history, five mass extinctions have been recognized for different time periods considering the paleontological sense (i.e. when the pace of extinction seems to become faster than background extinctions) (Barnosky *et al.* 2011). The common features of these extinctions are all related to changes in the climate system, such as unusual climate dynamics, atmospheric composition and abnormally high-intensity ecological stressors (Barnosky *et al.* 2011). Thus, estimates of the magnitude and pace of paleoclimate change are useful tools to test ecological theories and identify how the biodiversity dynamics can be shaped by climatic shifts (Svenning *et al.* 2015).

The Late Pleistocene (12.6–11.7 ka) was characterized by intense and rapid climatic oscillations at regional levels, with variation in temperature reaching up to 16°C (Cooper *et al.* 2015). These rapid and extreme fluctuations, together with environmental shifts, were important factors promoting the extinction of several Pleistocene megafauna lineages. Broader scale metapopulation structure or processes (e.g. long-distance dispersal and refugia) were involved in maintaining ecosystem stability during the sudden climate change phases in the Pleistocene Holarctic, which was demonstrated by the rapid replacement of local genetic populations by congeners or conspecifics (Cooper *et al.* 2015). More recently, climatic fluctuations during the Quaternary (2.58 Mya to present) also promoted shifts in ecological conditions that lead past and present biota to modify their distribution across the planet, promoting the extinction of many lineages (Diele-Viegas and Rocha 2018; Fordham *et al.* 2017). Thus, the importance of climate change velocity for a species is related to different responses over time, depending on the species' ability to track suitable environments (Sandel *et al.* 2011).

The velocity of past climate change is also related to endemism patterns of several groups, such as amphibians, reptiles, birds and mammals (Araújo *et al.* 2008; Sandel *et al.* 2011). High rates of climate change are associated with variation in range sizes within assemblages, while low velocities are associated with high endemism rates and low dispersal capacities (Sandel *et al.* 2011). For example, in reptiles and amphibians, current distributions of widespread species are more constrained by current climate changes, while the distribution of narrow-ranging species is markedly constrained by past climate conditions (Araújo *et al.* 2008; Jansson, 2003). Thus, narrow-ranging species are at least partly restricted because of their poor ability to track current climate changes (Jansson 2003).

12.2.1. *Recent impacts*

While extreme climatic fluctuations were responsible for promoting the extinction of many lineages in the past, current rates of warming are already impacting natural systems worldwide. Climate change affects all biodiversity levels, from organisms to ecosystems (Bellard *et al.* 2012). Dramatic impacts of climate change have already been observed in several systems, such as the arctic and coral reef systems worldwide (Mooney *et al.* 2009). Ecosystem regime shifts resulting from climate change are triggered by different disturbance events, such as droughts, wind storms, fires and pest outbreaks (Settele *et al.* 2014). These changes may determine changes in species range and abundance rather than physiological tolerance for the mean thermal conditions (Thonicke *et al.* 2008).

Several aspects of terrestrial organismal biology are important to delimit the climate impacts on functional traits and the species' vulnerability to climate change (Settele *et al.* 2014). A species vulnerability to climate change is determined by the interaction between 1) the magnitude and frequency of exposure to environmental changes; 2) sensitivity to these changes; and 3) the degree to which it can respond to such changes and keep its structure, composition and functions (Dawson *et al.* 2011; Small-Lorenz *et al.* 2013; Settele *et al.* 2014). These responses are mainly driven by the potential for adaptation to local conditions, phenotypic plasticity and the ability to move to thermally suitable habitats (Bellard *et al.* 2012; Dawson *et al.* 2011).

If none of these are possible, the species are liable to be extirpated locally or extinguished globally (Sinervo *et al.* 2010; Dawson *et al.* 2011).

Climate change can be a significant selective agent acting on natural populations, affecting mainly thermal traits (Campbell-Staton *et al.* 2017). Phenotypic plasticity can improve the chances of adaptation to these stressful environments, but it depends on this change's fitness cost (Chevin *et al.* 2010). Also, adaptive responses can vary among different life stages, varying their vulnerability to environmental changes and affecting demographic parameters (Sinclair *et al.* 2016; Diele-Viegas *et al.* 2019). Some ectotherm species exhibit temperature-dependent sex determination, where the environmental temperature experienced during the embryonic/larval development determines the sex of the offspring (Bull 1980). Others might experience stressful environments affecting their population dynamics, leading to a population collapse within a few generations if no acclimatization or local adaptation is possible (Diele-Viegas *et al.* 2019).

Shifts in phenology of plants and animals have already occurred over recent decades, one of the most documented population being aspects influenced by the climate changes. The earlier occurrence of spring events, such as breeding, emergence from hibernation, onset of flowering and migration, has been observed in hundreds of species in many regions (e.g. Primack *et al.* 2009; Cook *et al.* 2012; Peñuelas *et al.* 2013). Thus, the disruption of ecological interactions, such as herbivory, predation, pollination networks and abundance of pest species, is also likely to occur (Tylianakis *et al.* 2008; Schweiger *et al.* 2012). Primary productivity, timber and pulp production are sensitive to most aspects of climate change, being indirectly affected by the effects of climate on pests and diseases (Seddon *et al.* 2016). Notably, evapotranspiration can be affected by multiple factors of climate change, such as temperature, drought, solar and thermal radiation (Luo *et al.* 2008), and transpiration can be additionally affected by CO_2 concentration of the atmosphere (Settele *et al.* 2014).

12.2.1.1. *Changes in species distribution*

Changes in species' geographic ranges, such as latitudinal and elevational shifts, are the most common response to stressful environments (Diele-Viegas and Rocha 2018), since it allows the species to track their climate niche when their original habitat lacks thermal adequacy (Dullinger *et al.* 2012; Urban *et al.* 2012; Bestion *et al.* 2015). However, it depends on the species' dispersal capacity to search for new thermally suitable environments, which is not a trivial decision made by an individual (McCarty 2001). Thus, species with slower dispersion rates are likely to be harmed, since changes in environment and ecosystem interactions are usually faster than their ability to move (Settele *et al.* 2014).

Suitable habitats may shift across different gradients of elevation, latitude and longitude, and several barriers can constrain dispersal, varying from natural geographic barriers, such as rivers and mountains, to human-made barriers, such as habitat fragmentation, roads, human settlements and dams (Settele *et al.* 2014). Thus, species that need to move long distances or from lower to higher elevation are likely to undergo more of these barriers, reducing the possibilities of these species to persist (Araújo *et al.* 2006; Ofori *et al.* 2017). For species with restricted distribution and unable to move, such as island or high-elevation endemics, this might be a threshold between survival and extinction (Bonino *et al.* 2014; Davis *et al.* 2015; Diele-Viegas and Rocha 2018).

Recent climate change is already noted as the main extinction cause of several species, such as the Costa Rican golden toad, *Bufo periglenes* (Amphibia, Anura) (Pounds and Crump 1994; Pounds *et al.* 1999), and the Australian melomys, *Melomys rubicola* (Mammalia, Rodentia) (Waller *et al.* 2017). In addition, indirect effects of climate change at local scales are likely to be more effective in dictating the pathway of evolutionary process than direct effects (Diele-Viegas *et al.* 2020). For example, snake diversity in tropical environments is collapsing due to the amphibian global decline crisis, which is caused by climate change, habitat loss and temperature-related pathogens, such as the fungal infection caused by *Batrachochytrium dendrobatidis* (Collins and Storfer, 2003; Greenberg and Palen, 2019; Zipkin *et al.* 2020).

12.2.2. *Future impacts*

Overall, substantial changes in structure and functioning of terrestrial and marine ecosystems are expected to occur in the near future (Settele *et al.* 2014). The interaction between the predicted changes with vulnerability and exposure of human and natural systems will increase the risk of severe, pervasive and sometimes irreversible detrimental impacts (Shukla *et al.* 2019). Thus, predicting the potential future risks to biodiversity caused by climate change has become an active field of research (Bellard *et al.* 2012). Some studies suggest that it could be the greatest global threat to biodiversity over the next few decades, which has been called the world's sixth mass extinction (Barnosky *et al.* 2011; Cahill *et al.* 2013).

Species distribution modeling (SDM) has been at the core of most estimates regarding the degree to which a species and/or population is vulnerable to climate changes (Sinervo *et al.* 2010, p. 201; Buckley *et al.* 2011; Carneiro *et al.* 2016;

Caetano *et al.* 2020). SDMs are also used to verify the relationship between hotspots of climate change vulnerability and threatened species richness, highlighting species that may become highly threatened over time (Böhm *et al.* 2016; Diele-Viegas *et al.* 2020; Gonçalves-Souza *et al.* 2020). The mechanism behind SDM predictions is the evaluation of present and potential future species ranges based on multiple predictor variables including climate, microclimate and others. SDMs have the potential for highlighting species that may be at risk or geographic areas that are likely to face changes in species composition (Elith and Leathwick 2009; Ackerly *et al.* 2010).

There are two broad methodological approaches to access species' vulnerability to climate change through SDMs. The correlative method correlates current climate data with historical species' presence/absence data and extrapolates the correlation to predict occurrence probabilities under different climate scenarios. Mechanistic niche models incorporate biological and physiological processes to directly model the fundamental niche of a species (Taylor *et al.* 2020). Despite the advantages of mechanistic niche models over correlative models, they require substantial amounts of experimental data to predict range shifts, which limits their applicability. Thus, alternative hybrid modeling frameworks that use both correlative and mechanistic approaches have been applied, producing more reliable predictions for how species will respond to climate change than purely correlative approaches (Taylor *et al.* 2020).

To be liable, changes in species distribution depend on a variety of factors, such as habitat quantity and quality and availability of suitable ecological conditions (Bellard *et al.* 2012) However, future range shifts transcend single systems or dimensions, impacting different interacting scales and leading some species to overlap their climate niche in the new occupied habitat, rearranging ecological interactions (including with humans) and increasing interspecific competition (Bombi *et al.* 2011; Pecl *et al.* 2017).

Species "on the move" have become a reality, so international agreements and global, regional and local policies must start to account for these effects (Pecl *et al.* 2017). Protected areas are a good strategy to keep thermal suitable environments and avoid local extirpation, but ecological corridors between these areas must be created and/or maintained (Caten *et al.* 2017). Besides, even when the species exhibit a good coverage and connectivity within the existing protected area, some can still be impacted by climate change by losing genetic variability through bottlenecks or decreasing their fitness through compromised physiological performance, demonstrating that the management of protected areas must consider all ecological and evolutionary processes threatening its biodiversity to improve its prospective conservation (Rödder and Schulte 2010; Diele-Viegas and Rocha 2018).

In conclusion, it is important to highlight that a species' vulnerability to climate change is expected to increase, considering the CMIP6 predictions of GHG emissions. Thus, to mitigate the further loss of species, human society must understand its role on Earth as part of the system, take responsibility for its actions and move towards a more sustainable economy, focusing on the preservation of species, biomes and ecosystems. After all, we are just another species among the almost 9 million known species in the world, and our actions are affecting not only their existence, but also ours.

12.3. References

Ackerly, D.D., Loarie, S.R., Cornwell, W.K., Weiss, S.B., Hamilton, H., Branciforte, R., Kraft, N.J.B. (2010). The geography of climate change: Implications for conservation biogeography: Geography of climate change. *Diversity and Distributions*, 16, 476–487.

Allen, M.R., Dube, O.P., Solecki, W., Aragón-Durand, F., Cramer, W., Humphreys, S., Kainuma, M., Kala, J., Mahowald, N., Mulugetta, Y., Perez, R., Wairiu, M., Zickfeld, K. (2018). Framing and context. *Global Warming of 1.5°C. An IPCC Special Report on the Impacts of Global Warming of 1.5°C Above Pre-Industrial Levels and Related Global Greenhouse Gas Emission Pathways, in the Context of Strengthening the Global Response to the Threat of Climate Change, Sustainable Development, and Efforts to Eradicate Poverty*. IPCC, Geneva.

Araújo, M.B., Thuiller, W., Pearson, R.G. (2006). Climate warming and the decline of amphibians and reptiles in Europe. *Journal of Biogeography*, 33, 1712–1728.

Araújo, M.B., Nogués-Bravo, D., Diniz-Filho, J.A.F., Haywood, A.M., Valdes, P.J., Rahbek, C. (2008). Quaternary climate changes explain diversity among reptiles and amphibians. *Ecography*, 31, 8–15.

Barnosky, A.D., Matzke, N., Tomiya, S., Wogan, G.O.U., Swartz, B., Quental, T.B., Marshall, C., McGuire, J.L., Lindsey, E.L., Maguire, K.C., Mersey, B., Ferrer, E.A. (2011). Has the Earth's sixth mass extinction already arrived? *Nature*, 471, 51–57.

Bellard, C., Bertelsmeier, C., Leadley, P., Thuiller, W., Courchamp, F. (2012). Impacts of climate change on the future of biodiversity: Biodiversity and climate change. *Ecology Letters*, 15, 365–377.

Berger, A.L. (1978). Long-term variations of daily insolation and quaternary climatic changes. *Journal of the Atmospheric Sciences*, 35, 2362–2367.

Bestion, E., Clobert, J., Cote, J. (2015). Dispersal response to climate change: Scaling down to intraspecific variation. *Ecol. Lett.*, 18, 1226–1233.

Böhm, M., Cook, D., Ma, H., Davidson, A.D., García, A., Tapley, B., Pearce-Kelly, P., Carr, J. (2016). Hot and bothered: Using trait-based approaches to assess climate change vulnerability in reptiles. *Biological Conservation*, 204, 32–41.

Bombi, P., Akani, G.C., Ebere, N., and Luiselli, L. (2011). Potential effects of climate change on high-and low-abundance populations of the Gaboon viper (*Bitis gabonica*) and the nose-horned viper (*B. nasicornis*) in southern Nigeria. *The Herpetological Journal*, 21(1), 59–64.

Bonino, M.F., Moreno Azócar, D.L., Schulte, J.A., Cruz, F.B. (2014). Climate change and lizards: Changing species' geographic ranges in Patagonia. *Regional Environmental Change*, 15, 1121–1132.

Broecker, W.S. (1975). Climatic change: Are we on the brink of a pronounced global warming? *Science*, 189, 460–463.

Buckley, L.B., Waaser, S.A., MacLean, H.J., Fox, R. (2011). Does including physiology improve species distribution model predictions of responses to recent climate change? *Ecology*, 92, 2214–2221.

Bull, J.J. (1980). Sex determination in reptiles. *The Quarterly Review of Biology*, 55(1), 3–21.

Cadenhead, N.C.R., Kearney, M.R., Moore, D., McAlpin, S., Wintle, B.A. (2016). Climate and fire scenario uncertainty dominate the evaluation of options for conserving the great desert skink: Resolving uncertainty in fire management. *Conservation Letters*, 9, 181–190.

Caetano, G.H., Santos, J.C., Godinho, L.B., Cavalcante, V.H., Diele-Viegas, L.M., Campelo, P.H., Martins, L.F., Oliveira, A.F., Alvarenga, J.M., Wiederhecker, H.C., de Novaes e Silva, V. (2020). Time of activity is a better predictor of the distribution of a tropical lizard than pure environmental temperatures. *Oikos*, 129(7), 953–963.

Cahill, A.E., Aiello-Lammens, M.E., Fisher-Reid, M.C., Hua, X., Karanewsky, C.J., Yeong Ryu, H., Sbeglia, G.C., Spagnolo, F., Waldron, J.B., Warsi, O., Wiens, J.J. (2013). How does climate change cause extinction? *Proceedings of the Royal Society B*, 280, 20121890.

Campbell-Staton, S.C., Cheviron, Z.A., Rochette, N., Catchen, J., Losos, J.B., Edwards, S.V. (2017). Winter storms drive rapid phenotypic, regulatory, and genomic shifts in the green anole lizard. *Science*, 357, 495–498.

Carneiro, L.R.A., Lima, A.P., Machado, R.B., Magnusson, W.E. (2016). Limitations to the use of species-distribution models for environmental-impact assessments in the Amazon. *PLoS ONE*, 11, e0146543.

Caten, C.T., Lima-Ribeiro, M.S., da Silva, N.J., Moreno, A.K., Terribile, L.C. (2017). Evaluating the effectiveness of Brazilian protected areas under climate change: A case study of *Micrurus brasiliensis* (Serpentes: Elapidae). *Tropical Conservation Science*, 10, 194008291772202.

Chevin, L.-M., Lande, R., Mace, G.M. (2010). Adaptation, plasticity, and extinction in a changing environment: Towards a predictive theory. *PLoS Biol*, 8, e1000357.

Church, J.A. and White, N.J. (2011). Sea-level rise from the late 19th to the early 21st century. *Surveys in Geophysics*, 32, 585–602.

Church, J.A., Clark, P.U., Cazenave, A., Gregory, J.M., Jevrejeva, S., Levermann, A., Merrifield, M.A., Milne, G.A., Nerem, R.S., Nunn, P.D., Payne, A.J., Pfeffer, W.T., Stammer, D., Unnikrishnan, A.S. (2013). Sea level change. *Climate Change 2013: The Physical Science Basis, Contribution of Working Group I to the Fifth Assessment Report of the Intergovernmental Panel on Climate Change.* Cambridge University Press, Cambridge, 1138.

Climate Change (2014). Synthesis report, 2014. Contribution of working groups I, II and III to the fifth assessment report of the intergovernmental panel on climate change. Intergovernmental Panel on Climate Change, Geneva, Switzerland.

Collins, J.P. and Storfer, A. (2003). Global amphibian declines: Sorting the hypotheses. *Diversity and Distributions*, 9, 89–98.

Collins, M., Knutti, R., Arblaster, J., Dufresne, J.-L., Fichefet, T., Friedlingstein, P., Gao, X., Gutowski, W.J., Johns, T., Krinner, G., Shongwe, M., Tebaldi, C., Weaver, A.J., Wehner, M. (2013). Long-term climate change: Projections, commitments and irreversibility. *Climate Change 2013: The Physical Science Basis., Contribution of Working Group I to the Fifth Assessment Report of the Intergovernmental Panel on Climate Change.* Cambridge University Press, Cambridge, 108.

Cook, B.I., Wolkovich, E.M., Davies, T.J., Ault, T.R., Betancourt, J.L., Allen, J.M., Bolmgren, K., Cleland, E.E., Crimmins, T.M., Kraft, N.J.B., Lancaster, L.T., Mazer, S.J., McCabe, G.J., McGill, B.J., Parmesan, C., Pau, S., Regetz, J., Salamin, N., Schwartz, M.D., Travers, S.E. (2012). Sensitivity of spring phenology to warming across temporal and spatial climate gradients in two independent databases. *Ecosystems*, 15, 1283–1294.

Cooper, A., Turney, C., Hughen, K.A., Brook, B.W., McDonald, H.G., Bradshaw, C.J.A. (2015). Abrupt warming events drove Late Pleistocene Holarctic megafaunal turnover. *Science*, 349, 602–606.

Cubasch, U., Wuebbles, D., Chen, D., Facchini, M.C., Frame, D., Mahowald, N., Winther, J.-G. (2013). Introduction. *Climate Change 2013: The Physical Science Basis., Contribution of Working Group I to the Fifth Assessment Report of the Intergovernmental Panel on Climate Change.* Cambridge University Press, Cambridge.

Davis, M.A., Douglas, M.R., Webb, C.T., Collyer, M.L., Holycross, A.T., Painter, C.W., Kamees, L.K., Douglas, M.E. (2015). Nowhere to go but up: Impacts of climate change on demographics of a short-range endemic (*Crotalus willardi obscurus*) in the Sky-Islands of Southwestern North America. *PLoS ONE*, 10, e0131067.

Dawson, T.P., Jackson, S.T., House, J.I., Prentice, I.C., Mace, G.M. (2011). Beyond predictions: Biodiversity conservation in a changing climate. *Science*, 332, 53–58.

Diele-Viegas, L.M. and Rocha, C.F.D. (2018). Unraveling the influences of climate change in Lepidosauria (Reptilia). *Journal of Thermal Biology*, 78, 401–414.

Diele-Viegas, L.M., Werneck, F.P., Rocha, C.F.D. (2019). Climate change effects on population dynamics of three species of Amazonian lizards. *Comparative Biochemistry and Physiology Part A: Molecular & Integrative Physiology*, 236, 110530.

Diele-Viegas, L.M., Figueroa, R.T., Vilela, B., and Rocha, C.F.D. (2020). Are reptiles toast? A worldwide evaluation of Lepidosauria vulnerability to climate change. *Climatic Change*, 159(4), 581–599.

Dullinger, S., Gattringer, A., Thuiller, W., Moser, D., Zimmermann, N.E., Guisan, A., Willner, W., Plutzar, C., Leitner, M., Mang, T., Caccianiga, M., Dirnböck, T., Ertl, S., Fischer, A., Lenoir, J., Svenning, J.-C., Psomas, A., Schmatz, D.R., Silc, U., Vittoz, P., Hülber, K. (2012). Extinction debt of high-mountain plants under twenty-first-century climate change. *Nature Climate Change*, 2, 619–622.

Elith, J. and Leathwick, J.R. (2009). Species distribution models: Ecological explanation and prediction across space and time. *Annual Review of Ecology, Evolution, and Systematics*, 40, 677–697.

Eyring, V., Bony, S., Meehl, G.A., Senior, C.A., Stevens, B., Stouffer, R.J., Taylor, K.E. (2016). Overview of the coupled model intercomparison project phase 6 (CMIP6) experimental design and organization. *Geoscientific Model Development*, 9, 1937–1958.

Flato, G., Marotzke, J., Abiodun, B., Braconnot, P., Chou, S.C., Collins, W., Cox, P., Driouech, F., Emori, S., Eyring, V., Forest, C., Gleckler, P., Guilyardi, E., Jakob, C., Kattsov, V., Reason, C., Rummukainen, M. (2013). Evaluation of climate models. *Climate Change 2013: The Physical Science Basis, Contribution of Working Group I to the Fifth Assessment Report of the Intergovernmental Panel on Climate Change*. Cambridge University Press, Cambridge.

Fordham, D.A., Saltré, F., Brown, S.C., Mellin, C., Wigley, T.M.L. (2017). Why decadal to century timescale palaeoclimate data are needed to explain present-day patterns of biological diversity and change. *Global Change Biology*, 24, 1371–1381.

Gonçalves-Souza, D., Verburg, P.H., Dobrovolski, R. (2020). Habitat loss, extinction predictability and conservation efforts in the terrestrial ecoregions. *Biological Conservation*, 246, 108579.

Greenberg, D.A. and Palen, W.J. (2019). A deadly amphibian disease goes global. *Science*, 363, 1386–1388.

Haas, C., Pfaffling, A., Hendricks, S., Rabenstein, L., Etienne, J.-L., Rigor, I. (2008). Reduced ice thickness in Arctic transpolar drift favors rapid ice retreat. *Geophysical Research Letters.*, 35, L17501.

Hansen, J.E. and Sato, M. (2012). Paleoclimate implications for human-made climate change. In *Climate Change*, Berger, A., Mesinger, F., Sijacki, D. (eds). Springer, Vienna.

Hartmann, D.L., Klein Tank, A.M.G., Rusticucci, M., Alexander, L.V., Brönnimann, S., Charabi, Y., Dentener, F.J., Dlugokencky, E.J., Easterling, D.R., Kaplan, A., Soden, B.J., Thorne, P.W., Wild, M., Zhai, P.M. (2013). Observations: Atmosphere and surface. *Climate Change 2013: The Physical Science Basis., Contribution of Working Group I to the Fifth Assessment Report of the Intergovernmental Panel on Climate Change.* Cambridge University Press, Cambridge.

Hays, J.D., Imbrie, J., Shackleton, N.J. (1976). Variations in the Earth's orbit: Pacemaker of the ice ages. *Science*, 194, 1121–1132.

Hoegh-Guldberg, O., Jacob, D., Taylor, M., Bindi, M., Brown, S., Camilloni, I., Diedhiou, A., Djalante, R., Ebi, K.L., Engelbrecht, F., Hijioka, Y., Mehrotra, S., Payne, A., Seneviratne, S.I., Thomas, A. *et al.* (2018). Impacts of 1.5°C of Global warming on natural and human systems. *Global Warming of 1.5°C. An IPCC Special Report on the Impacts of Global Warming of 1.5°C Above Pre-Industrial Levels and Related Global Greenhouse Gas Emission Pathways, in the Context of Strengthening the Global Response to the Threat of Climate Change, Sustainable Development, and Efforts to Eradicate Poverty.* IPCC, Geneva, 138.

Huybers, P. and Curry, W. (2006). Links between annual, Milankovitch and continuum temperature variability. *Nature*, 441, 329–332.

IPCC (2018). Summary for policymakers. In Global Warming of 1.5°C. An IPCC Special Report on the Impacts of Global Warming of 1.5°C Above Pre-Industrial Levels and Related Global Greenhouse Gas Emission Pathways, in the Context of Strengthening the Global Response to the Threat of Climate Change, Sustainable Development, and Efforts to Eradicate Poverty. IPCC, Geneva, 32.

Jansson, R. (2003). Global patterns in endemism explained by past climatic change. *Proceedings of the Royal Society of London. Series B*, 270, 583–590.

Kwok, R., Cunningham, G.F., Wensnahan, M., Rigor, I., Zwally, H.J., Yi, D. (2009). Thinning and volume loss of the Arctic Ocean sea ice cover: 2003–2008. *Journal of Geophysical Research*, 114, C07005.

Luo, Y., Gerten, D., Le Maire, G., Parton, W.J., Weng, E., Zhou, X., Keough, C., Beier, C., Ciais, P., Cramer, W., Dukes, J.S., Emmett, B., Hanson, P.J., Knapp, A., Linder, S., Nepstad, D., Rustad, L. (2008). Modeled interactive effects of precipitation, temperature, and [CO_2] on ecosystem carbon and water dynamics in different climatic zones. *Global Change Biology*, 14, 1986–1999.

Masson-Delmotte, V., Schulz, M., Abe-Ouchi, A., Beer, J., Ganopolski, A., Fidel, J., Rouco, G., Jansen, E., Lambeck, K., Luterbacher, J., Naish, T., Ramesh, R., Rojas, M., Shao, X., Anchukaitis, K. *et al.* (2013). Information from Paleoclimate Archives. *Climate Change 2013: The Physical Science Basis, Contribution of Working Group I to the Fifth Assessment Report of the Intergovernmental Panel on Climate Change.* Cambridge University Press, Cambridge, 82.

McCarty, J.P. (2001). Ecological consequences of recent climate change. *Conservation Biology*, 15, 320–331.

McGrath, M. (2020). Climate change: Last decade confirmed as warmest on record. *BBC News* [Online]. Available at: https://www.bbc.co.uk/news/science-environment-51111176.

Mooney, H., Larigauderie, A., Cesario, M., Elmquist, T., Hoegh-Guldberg, O., Lavorel, S., Mace, G.M., Palmer, M., Scholes, R., Yahara, T. (2009). Biodiversity, climate change, and ecosystem services. *Current Opinion in Environmental Sustainability*, 1, 46–54.

Moss, R.H., Edmonds, J.A., Hibbard, K.A., Manning, M.R., Rose, S.K., van Vuuren, D.P., Carter, T.R., Emori, S., Kainuma, M., Kram, T., Meehl, G.A., Mitchell, J.F.B., Nakicenovic, N., Riahi, K., Smith, S.J., Stouffer, R.J., Thomson, A.M., Weyant, J.P., Wilbanks, T.J. (2010). The next generation of scenarios for climate change research and assessment. *Nature*, 463, 747–756.

Ofori, B.Y., Stow, A.J., Baumgartner, J.B., Beaumont, L.J. (2017). Combining dispersal, landscape connectivity and habitat suitability to assess climate-induced changes in the distribution of Cunningham's skink, *Egernia cunninghami*. *PLoS ONE*, 12, e0184193.

Pecl, G.T., Araújo, M.B., Bell, J.D., Blanchard, J., Bonebrake, T.C., Chen, I.-C., Clark, T.D., Colwell, R.K., Danielsen, F., Evengård, B., Falconi, L., Ferrier, S., Frusher, S., Garcia, R.A., Griffis, R.B. *et al.* (2017). Biodiversity redistribution under climate change: Impacts on ecosystems and human well-being. *Science*, 355, eaai9214.

Peñuelas, J., Sardans, J., Estiarte, M., Ogaya, R., Carnicer, J., Coll, M., Barbeta, A., Rivas-Ubach, A., Llusià, J., Garbulsky, M., Filella, I., Jump, A.S. (2013). Evidence of current impact of climate change on life: A walk from genes to the biosphere. *Global Change Biology*, 19, 2303–2338.

Pincebourde, S., Murdock, C.C., Vickers, M., Sears, M.W. (2016). Fine-scale microclimatic variation can shape the responses of organisms to global change in both natural and urban environments. *Integrative and Comparative Biology*, 56, 45–61.

Pounds, J.A. and Crump, M.L. (1994). Amphibian declines and climate disturbance: The case of the golden toad and the harlequin frog. *Conservation Biology*, 8, 72–85.

Pounds, J.A., Fogden, M.P.L., Campbell, J.H. (1999). Biological response to climate change on a tropical mountain. *Nature*, 398, 611–615.

Primack, R.B., Ibáñez, I., Higuchi, H., Lee, S.D., Miller-Rushing, A.J., Wilson, A.M., Silander, J.A. (2009). Spatial and interspecific variability in phenological responses to warming temperatures. *Biological Conservation*, 142, 2569–2577.

Riahi, K., van Vuuren, D.P., Kriegler, E., Edmonds, J., O'Neill, B.C., Fujimori, S., Bauer, N., Calvin, K., Dellink, R., Fricko, O., Lutz, W., Popp, A., Cuaresma, J.C., Kc, S., Leimbach, M. *et al.* (2017). The shared socioeconomic pathways and their energy, land use, and greenhouse gas emissions implications: An overview. *Global Environmental Change*, 42, 153–168.

Rödder, D. and Schulte, U. (2010). Potential loss of genetic variability despite well established network of reserves: The case of the Iberian endemic lizard *Lacerta schreiberi*. *Biodiversity and Conservation*, 19, 2651–2666.

Sandel, B., Arge, L., Dalsgaard, B., Davies, R.G., Gaston, K.J., Sutherland, W.J., Svenning, J.-C. (2011). The influence of late quaternary climate-change velocity on species endemism. *Science*, 334, 660–664.

Schneider, S.H. (1989). The greenhouse effect: Science and policy. *Science*, 243(4892), 771–781.

Schweiger, O., Heikkinen, R.K., Harpke, A., Hickler, T., Klotz, S., Kudrna, O., Kühn, I., Pöyry, J., Settele, J. (2012). Increasing range mismatching of interacting species under global change is related to their ecological characteristics: Range mismatching of interacting species. *Global Ecology and Biogeography*, 21, 88–99.

Seddon, A.W.R., Macias-Fauria, M., Long, P.R., Benz, D., Willis, K.J. (2016). Sensitivity of global terrestrial ecosystems to climate variability. *Nature*, 531, 229–232.

Settele, J., Betts, R., Bunn, S., Leadley, P., Nepstad, D., Overpeck, J.T., Taboada, M.A. (2014). Terrestrial and inland water systems. *Climate Change 2014: Impacts, Adaptation, and Vulnerability, Part A: Global and Sectoral Aspects. Contribution of Working Group II to the Fifth Assessment Report of the Intergovernmental Panel on Climate Change*. Cambridge University Press, Cambridge, 271–359.

Shukla, P.R., Skea, J., Calvo Buendia, E., Masson-Delmotte, V., Pörtner, O., Roberts, D.C., Zhai, P., Slade, R., Connors, R., Van Diemen, R., Ferrat, M., Haughey, E., Luz, S., Neogi, S., Pathak, M. (2019). IPCC, 2019: Summary for policymakers. Climate Change and Land: An IPCC Special Report on Climate Change, Desertification, Land Degradation, Sustainable Land Management, Food Security, and Greenhouse Gas Fluxes in Terrestrial Ecosystems, 1–36.

Sinclair, B.J., Marshall, K.E., Sewell, M.A., Levesque, D.L., Willett, C.S., Slotsbo, S., Dong, Y., Harley, C.D.G., Marshall, D.J., Helmuth, B.S., Huey, R.B. (2016). Can we predict ectotherm responses to climate change using thermal performance curves and body temperatures? *Ecology Letters*, 19, 1372–1385.

Sinervo, B., Mendez-de-la-Cruz, F., Miles, D.B., Heulin, B., Bastiaans, E., Villagran-Santa Cruz, M., Lara-Resendiz, R., Martinez-Mendez, N., Calderon-Espinosa, M.L., Meza-Lazaro, R.N., Gadsden, H., Avila, L.J., Morando, M., de la Riva, I.J., Sepulveda, P.V. *et al.* (2010). Erosion of lizard diversity by climate change and altered thermal niches. *Science*, 328, 894–899.

Small-Lorenz, S.L., Culp, L.A., Ryder, T.B., Will, T.C., Marra, P.P. (2013). A blind spot in climate change vulnerability assessments. *Nature Climate Change*, 3, 91–93.

Stocker, T.F., Qin, D., Plattner, G.-K., Alexander, V.L., Allen, S.K., Bindoff, N.L., Bréon, F.-M., Church, J.A., Cubasch, U., Emori, S., Forster, P., Friedlingstein, P., Gillett, N., Gregory, J.M., Hartmann, D.L. *et al.* (2013). Technical summary. *Climate Change 2013: The Physical Science Basis., Contribution of Working Group I to the Fifth Assessment Report of the Intergovernmental Panel on Climate Change*. Cambridge University Press, Cambridge.

Svenning, J.-C., Eiserhardt, W.L., Normand, S., Ordonez, A., Sandel, B. (2015). The influence of paleoclimate on present-day patterns in biodiversity and ecosystems. *Annual Review of Ecology, Evolution, and Systematics*, 46, 551–572.

Taylor, E., Diele-Viegas, L.M., Ganglof, E., Hall, J., Halpern, B., Massey, M., Rödder, D., Rollinson, N., Spears, S., Sun, B., Telemeco, R. (2020). The thermal ecology and physiology of reptiles and amphibians: A user's guide. *Journal of Experimental Zoology Part A – Ecological and Integrative Physiology*. DOI: 10.1002/jez.2396.

Thonicke, K., Venevsky, S., Sitch, S., Cramer, W. (2008). The role of fire disturbance for global vegetation dynamics: Coupling fire into a dynamic global vegetation model: Fire disturbance and global vegetation dynamics. *Global Ecology and Biogeography*, 10, 661–677.

Tylianakis, J.M., Didham, R.K., Bascompte, J., Wardle, D.A. (2008). Global change and species interactions in terrestrial ecosystems. *Ecology Letters*, 11, 1351–1363.

Urban, M.C., Tewksbury, J.J., Sheldon, K.S. (2012). On a collision course: Competition and dispersal differences create no-analogue communities and cause extinctions during climate change. *Proceedings of the Royal Society B*, 279, 2072–2080.

Waller, N.L., Gynther, I.C., Freeman, A.B., Lavery, T.H., Leung, L.K.-P. (2017). The Bramble Cay melomys *Melomys rubicola* (Rodentia: Muridae): A first mammalian extinction caused by human-induced climate change? *Wildlife Research*, 44, 9.

Wayne, G.P. (2013). The beginner's guide to representative concentration pathways. *Sckeptical Science*, 25.

World Resources Institute. (2018). *Climate Watch*. World Resources Institute, Washington, DC.

Zipkin, E.F., Direnzo, G.V., Ray, J.M., Rossman, S., Lips, K.R. (2020). Tropical snake diversity collapses after widespread amphibian loss. *Science*, 367, 814–816.

13

Conservation Biogeography: Our Place in the World

Brett R. RIDDLE
University of Nevada, Las Vegas, United States

13.1. The emergence of conservation biogeography

Conservation biogeography was imagined as a distinct "applied and interdisciplinary science concerned with the conservation of nature" (Whittaker and Ladle, 2011, p. 4) less than two decades ago (Whittaker *et al.*, 2005). The researchers that argued for the need to think about biodiversity issues within an explicit biogeography framework sought to build on the legacy of the early- to mid-20th century disciplines of wildlife conservation and nature conservation, but at a "coarse landscape to geographic spatial scale" as an alternative to the "population scale" focus of more established biodiversity conservation research (Whittaker and Ladle, 2011). The first formal definition of Conservation Biology reflected these goals: "the application of biogeographical principles, theories, and analyses, being those concerned with the distributional dynamics of taxa individually and collectively, to problems concerning the conservation of biodiversity" (Whittaker *et al.*, 2005).

Here, I review the development of conservation biogeography since its inception a little over 1½ decades ago. My goals are to, first, provide a brief timeline of progress in the development of this subdiscipline, followed by an overview of the claimed and examined purview of the field. I then briefly examine several highlights emanating from this approach to date, and finish with a few speculations on how the subdiscipline might continue to develop in productive directions.

Biogeography,
coordinated by Eric GUILBERT.

13.2. Milestones in the development of conservation biogeography

Given that conservation biogeography was formally proposed as a distinct subdiscipline in conservation biology and biogeography in recent times (2005), a summary of milestones in the progressive maturation of the field is quite brief (Table 13.1). Notably, the same year of its inception, the journal *Diversity and Distributions* was rebranded as *A Journal of Conservation Biology* (Richardson, 2005) – a branding that was reinforced in 2016 (Franklin, 2016) and has persisted to this day.

In queries of the Web of Science Core Collection (search conducted on April 27, 2021), "Conservation Biogeography" has appeared in titles, abstracts, keywords or KeyWords Plus ("Topic" search) of some 346 peer-reviewed journal articles and in titles ("Title" search) of 43 peer-reviewed journal articles (2005–2021). Of those, 90 (26%) of "Topic" hits – but a larger proportion of "Title" hits (24: 56%) – come from articles published in *Diversity and Distributions*. The remaining "Topic" hits (256) appeared in a total of 81 different journals and "Title" hits (19) in 13 different journals. A textbook devoted to conservation biogeography, with edited chapters, appeared in 2011 (Ladle and Whittaker, 2011).

While the above survey suggests that conservation biogeography has yet to figure prominently in a large number of journal articles, the 43 with "Conservation Biogeography" in the title were cited by other articles a total of 4,167 times (Web of Science search, April 27, 2021), suggesting that these relatively small number of original articles are being accessed frequently by the research community (Table 13.2). The paper that introduced the subdiscipline to researchers (Whittaker *et al.*, 2005) has been cited over twice as frequently as the second most frequently cited article, and 16 articles with "Conservation Biology" in the title that have been cited 40 or more times (Table 13.2) represent 2,858 (69%) of the total number of citations; thus, the remaining 63% of these 43 articles represent only 31% of the citations. That a few seminal articles appear to have garnered the most interest across the research community perhaps is not surprising for a newer and still emerging discipline.

Next, I provide a synopsis of the range of topics that practitioners of conservation biogeography claim to be its core research themes, and address whether these are, in practice, the themes that appear to have featured most prominently in the field since 2005.

Date	Reference	Title	Summary of contributions
2005	Whittaker *et al.* (2005)	Conservation biogeography: assessment and prospect	Formal definition and conceptual delineation of new subdiscipline of conservation biology called "Conservation Biogeography".
2005	Richardson (2005)	Diversity, distributions and conservation biogeography	Formal rebranding by Editor-in-Chief of the journal *Diversity and Distributions* as *A Journal of Conservation Biogeography*.
2005	Second biennial meeting of the International Biogeography Society	Theme of conference: conservation biogeography	Twenty-five papers delivered in five symposia; five subsequently published as a Special Issue on Conservation Biogeography in the *Journal of Biogeography*.
2010	Richardson and Whittaker (2010)	Conservation biogeography – foundations, concepts and challenges	Introduction to Special Issue of *Diversity and Distributions*, summarizing contributions into six themes in conservation biogeography (notably, five years after its origination).
2011	Ladle and Whittaker, eds. (2011)	Conservation biogeography	Edited textbook providing overview of conservation biogeography over 10 chapters contributed by 17 authors.
2016	Franklin (2016)	*Diversity and Distributions* is (still) a journal of conservation biogeography	Reiteration by new Editor-in-Chief of journal *Diversity and Distributions* that it continues to be "a journal of conservation biogeography", with a focus on topics including species distribution modeling; conservation planning and risk assessment; and impacts of global change, climate change, land use change and invasive species on species distributions and range boundaries.
2019	Driscoll *et al.* (2019)	A critique of "countryside biogeography" as a guide to research in human-dominated landscapes	Essay arguing against relevance of several newer subdisciplines in conservation biology, focusing on "Countryside Biogeography" as an unnecessary and misleading "conceptual wrapper", and also criticizing "Conservation Biogeography" with similar arguments.
2019	Katinas and Ladle (2019)	Redundancy or progress? A response to Driscoll *et al.* (2019)	Essay providing a counter to Driscoll *et al.* (2019), arguing that evidence supports "Conservation Biogeography" as a productive emerging subdiscipline in conservation biology.

Table 13.1. *Chronology of milestones in the emergence of conservation biogeography as a subdiscipline of conservation biology (and biogeography)*

Citation	Title	Times cited
Whittaker *et al.* (2005)	Conservation biogeography: assessment and prospect	666
Olden (2006)	Biotic homogenization: a new research agenda for conservation biogeography	315
Ackerly *et al.* (2010)	The geography of climate change: implications for conservation biogeography	310
Franklin (2010)	Moving beyond static species distribution models in support of conservation biogeography	269
Devictor *et al.* (2010)	Beyond scarcity: citizen science programmes as useful tools for conservation biogeography	267
Olden *et al.* (2010)	Conservation biogeography of freshwater fishes: recent progress and future challenges	226
Franklin (2013)	Species distribution models in conservation biogeography: developments and challenges	135
Richardson and Whittaker (2010)	Conservation biogeography – foundations, concepts and challenges	116
Terauds *et al.* (2012)	Conservation biogeography of the Antarctic	113
Diniz *et al.* (2010)	Defying the curse of ignorance: perspectives in insect macroecology and conservation biogeography	101
Cumming *et al.* (2010)	Network analysis in conservation biogeography: challenges and opportunities	84
Devictor and Robert (2009)	Measuring community responses to large-scale disturbance in conservation biogeography	68
Real *et al.* (2009)	Conservation biogeography of ecologically interacting species: the case of the Iberian lynx and the European rabbit	67
Hu and Liu (2014)	Unveiling the conservation biogeography of a data-deficient endangered bird species under climate change	41
Liu *et al.* (2015)	The use of DNA barcoding as a tool for the conservation biogeography of subtropical forests in China	40
Fischer *et al.* (2008)	The role of landscape texture in conservation biogeography: a case study on birds in south-eastern Australia	40

Table 13.2. *Papers with "Conservation Biogeography" in the title that have been cited 40 or more times (Web of Science "Title" search, April 27, 2021)*

13.3. The purview of conservation biogeography: claimed and examined

What do practitioners of and advocates for the value of conservation biogeography claim to be its overarching goals? In an essay introducing a Special Issue of the journal *Diversity and Distributions*, Richardson and Whittaker (2010) provided a detailed synopsis of the purview of research areas in conservation biogeography (reproduced in a modified form here in Table 13.3). Across nine "research areas" – ranging from "pure" biogeographic research topics through disturbance-related and applied (management) topics to development and deployment of new methods – they listed some 61 "research themes". These read much like a long list of themes generally integrated into textbooks on conservation biology; for example, see "Contents" in Cardinale *et al.* (2019). This is not a particular surprise, however, given that the originators of the subdiscipline envisioned conservation biogeography not so much as a cauldron of new research themes – rather, they positioned it to emphasize more explicitly that subset of themes that tend to address patterns, processes and management issues already embedded within biodiversity conservation science, but targeting landscape to geographic spatial scales (Ladle and Whittaker, 2011).

A 2016 essay marking a transition at *Diversity and Distributions* to a new Editor-in-Chief highlighted the following focal topics for the journal: species distribution modeling; conservation planning and risk assessment; impacts of global change and climate change; and impacts of land use change and invasive species on species distributions and range boundaries (Franklin, 2016). Several of these topics are reflected in titles and subject matter of three of the top four cited articles with "Conservation Biogeography" in the title (Table 13.2), including "biotic homogenization" as a concern with invasive species (Olden, 2006), "the geography of climate change" (Ackerly *et al.*, 2010) and "species distribution models" (Franklin, 2010).

I used "Topic" searches of the Web of Science database (on April 27, 2021) to assess the number of hits in *Diversity and Distributions* (2005–2021) for each of the 55 terms listed in Table 13.4 (extracted from those 61 listed in Table 13.3). From a total of 9,154 hits, distribution* was most frequent (1,079: 12% of total), while range collapse* returned only two hits (other terms least frequently used include population viability analysis, Linnaean shortfall and Wallacean shortfall). All other terms ranged in hits from 817 to 10, with several of the more frequent including diversity*, climat*, scale*, invasi* and climate change*.

Research area	Prominent research themes
The biogeography of degradation	Habitat fragmentation, homogenization, urbanization and other human-induced impacts
Processes	Colonization and climate as fundamental determinants of distribution, dispersal, disturbance, extinction, persistence, range expansion, resilience, speciation
Inventory, mapping and data issues	Atlas data, breeding bird surveys, citizen science, detectability/discovery probabilities, herbaria and other collections, sampling intensity and biases
Species distribution modeling	Bioclimatic modeling, habitat-suitability analysis, model performance, niche-based models, presence-only data vs. presence–absence data, dispersal kernel analysis
Characterizing biotas	Conservation status, diversity indices and patterns, ecoregions, endemism, rarity, range size, species–area relationships, threatened species, identification of alternative baselines from long-term ecological data
Conservation planning	Complementarity, congruence, conservation units, ecosystem services, gap analysis, global conservation assessments, irreplaceability, reserve networks, surrogates
Methods	Molecular methods, paleoecology, remote sensing, scenario development
Related fields	Global change biology, invasion ecology, bioinformatics, molecular phylogenetics, network analysis, re-introduction ecology, risk analysis, behavioral ecology, population viability analysis
Overarching themes	Niche (fundamental vs. realized), novel climates/ecosystems, scale issues, uncertainty, Linnean shortfall, Wallacean shortfall

Table 13.3. *A comprehensive list of "prominent research areas in Conservation Biology", as outlined by Richardson and Whittaker (2010)*

Search term	"Topic" hits	Search term	"Topic" hits
distribution*	1,079	global change	76
diversity	817	molecular	72
climat*	765	population genetic*	60
scale*	688	conservation status	58
invasi*	602	rarity	56
climate change*	528	biotic homogeniz*	52
dispersal*	353	resilience	49
species distribution model*	342	congruen*	48
extinct*	330	ecosystem services	44
niche*	284	species–area	40
genetic*	251	citizen science	39
threaten*	229	museum*	29
disturbance*	216	speciation*	27
endemi*	187	gap analysis	26
phylogen*	179	remote sensing	26
vulnerab*	171	reserve network*	26
coloniz*	153	conservation assessment*	24
exotic*	147	genom*	23
endanger*	136	re-introduc* or reintroduc	17
persistence	126	bioclimatic model*	12
range shift*	106	irreplaceability	11
range size*	94	conservation unit*	10
niche model*	94	paleoecolog* or paleoecology*	10
habitat fragmentation	90	population viability analysis	7
complement*	87	Linnean shortfall	3
range expansion*	86	Wallacean shortfall	3
urban*	84	range collaps*	2
phylogeog*	80		

Table 13.4. *An array of 55 terms considered as viable "research themes" for conservation biogeography, drawn from Table 14.3 and modified from an original table in Richardson and Whittaker (2010). Terms are listed in descending order of number of hits returned through "Topic" searches in Web of Science (conducted on April 27, 2021) of the journal Diversity and Distributions (2005–2021). * = truncation of original term in order to capture full range of equivalent terms in the database (e.g. "invasi*" would return invasive, invasion, etc.)*

Research theme	Terms associated with theme	Total "Topic" hits
Global change and climate change	climate change, global change, climat*	1,369
Vulnerability	extinct*, threaten*, vulnerab*, endanger*, rarity	1,109
Species distribution characteristics	distribution*, endemic*, range size	1,098
Biotic diversity attributes	diversity, species-area	857
Range dynamics	dispersal*, coloniz*, range shift*, range expansion*, range collapse*	700
Scale issues	scale*	688
Genetics and evolution	genetic*, phylogen*, phylogeog*, molecular, population genetic*, speciation	669
Biotic invasion	invasi*, biotic homogeniz*	654
Conservation planning	persistence, complement*, conservation status, resilience, congruen*, gap analysis, reserve network*, conservation assessment*, re-introduc* or reintroduc* irreplaceability, conservation unit*	482
Species distribution modeling	species distribution model*, niche model*, bioclimatic model*	448
Land use change	disturbance*, habitat fragmentation, urban*,	390

Table 13.5. *Predominant "research themes" in conservation biogeography, as compiled from inspection and grouping of terms with overlapping or similar objectives from Table 13.4; listed here in descending order of total number of "Topic" hits from the Web of Science search of the journal Diversity and Distributions, conducted to generate numbers reported in that table*

Looked at a bit differently, we can readily group sets of terms from the list in Table 13.4 into a suite of more generalized categories of research themes. While somewhat subjective in nature, this exercise produced a set of 11 such themes (Table 13.5), from which a Web of Science survey similar to that conducted for terms in Table 13.4 might offer a more realistic comparison of relative popularity across several recognizable research themes as they have actually been addressed within the purview of conservation biogeography. In this case, terms likely to be most closely associated with "global change and climate change" (climate change, global change and perhaps most of climat*) returned 1,369 hits (15% of a total of

9,154 hits) and ranked highest among the 11 research themes. Other high-ranking themes included, in descending order, "vulnerability", grouping together terms that tend to describe levels of actual or perceived threats to vulnerable species (1,109 hits); "species distribution characteristics", describing attributes of distribution and the anatomy of geographic ranges (1,098 hits). Interestingly, within this more generalized thematic framework, "biotic invasion" drops to 8th (out of 11) in frequency of hits (from 5th for "invasi*" out of 55 terms in the original list), and "species distribution modeling" falls to 10th (out of 11) from 8th out of 55 in the original list.

What does this exercise suggest? First, while the lead developers of and advocates for the value of a distinct subdiscipline of conservation biogeography have rather grand aspirations for the breadth of research themes that could ultimately be incorporated within its conceptual purview (Table 13.3), the empirical usage of these concepts within the primary journal representing the field attenuates rather quickly following a handful of more popular themes (Table 13.4). This is not to say that the field has not achieved a respectable level of use and utility for biodiversity conservation researchers – indeed, total numbers of "hits" for 55 of these terms, as well as total numbers of citations of relatively few primary articles (Table 13.2) suggest that it has achieved a solid standing within this focal research community. Second, while the big research arenas that Franklin (2016) indicated to be, perhaps, among the predominant themes occupying subsequent volumes of *Diversity and Distributions* could well have come to be just so (Table 13.4), their "dominance" of the field perhaps depends on how one goes about defining the overarching research themes within the field (Table 13.5).

13.4. Has conservation biogeography provided unique contributions to biodiversity conservation?

What "value" has conservation biogeography brought to the realm of biodiversity conservation science since 2005? The relevance of conservation biogeography as a distinct subdiscipline of conservation biology has generated some recent controversy in the form of an exchange of essays in 2019. One perspective considered it to be either a diversionary or inaccurate approach (Driscoll *et al.*, 2019), while the counter-perspective was of a subdiscipline that represented productive advancement in conservation biology research (Katinas and Ladle, 2019). Does the assessment herein provide any clues in support of the latter view?

Stated differently, we might ask: where would conservation biology be today without an explicit conservation biogeography framework in which to conceptualize and implement research of value? While one way to address this question might be to design and conduct an exhaustive search of the conservation biology literature using terms such as those in Tables 14.4 and 14.5, both with and without connections with the term "conservation biogeography", I suspect that the effort required to do so would be disproportionate to the amount of information gained within the context of this contribution. However, if one views those same terms (55 total in Table 14.4; 11 general themes in Table 14.5) within an explicitly spatial and geographic context (many are done so in a straightforward fashion, others can be also thought of at the smaller population scale), and overlays the initial argument for a need to have a subdiscipline of conservation biogeography (i.e. landscape to geographic scale research themes), then an argument in favor of adding this subdiscipline to the arsenal of biodiversity conservation research can be visualized readily. Indeed, we might argue cogently that of the rather large numbers of studies that reference articles and papers within the conservation biogeography literature (Table 14.2), or that utilize terms of an explicitly geographic nature (within Tables 14.4 and 14.5), many might have had a rather different focus conceptually and emphasized somewhat different spatio-temporal scales in the absence of this focal, explicitly larger spatial (and hence, temporal) scale framework.

13.5. Future directions

Conservation biogeography appears to have made progress since 2005 in its attempt to delineate, conceptually and empirically, a portion of the spatial (and thereby, probably temporal) scale (landscape to geographic) within the vast realm of conservation biology that has not always been of a focal concern among biodiversity researchers. However, to continue to become even more relevant, practitioners would find value in making advances in several ways.

First, it appears that conservation biogeography has, to date, concentrated focus on a subset of the very large research themes that are conceivably nested within its purview. An array of very general research themes demonstrates a respectable frequency of topics "hits" in the subdiscipline's flagship journal (Table 14.5); however, there is a broad range of usage frequency of different topics embedded within each of these themes (Table 14.4). This result might indicate scope for a "fleshing out" of some concepts as researchers seek greater explanatory power in approaches to biodiversity conservation.

Second, as with all progressive disciplines in science, conservation biogeography will be well-served by continuing to incorporate advances in methods, theory and large datasets. Genomics, for example, is poised to make large and rapid strides in sequencing whole genomes for entire groups of major radiations (e.g. Earth BioGenome Project; https://www.earthbiogenome.org/). The wealth of data that could result from such endeavors was unimaginable in 2005 when conservation biogeography was conceived. Likewise, species distribution models continue to become more powerful and useful tools to predict biotic responses across geographic regions to, for example, global and climate change, and biotic invasions (Zurell *et al.*, 2020). Similar comments could be offered regarding the deployment of and data extracted from remote sensing technologies (Rocchini *et al.*, 2021).

In summary, given the wealth of themes, methods and approaches that appear to be available to conservation biogeographers, it will always to the case that future advances are likely to occur most rapidly to the degree that approaches to research themes become more synthetic and integrative, and therefore that researchers should seek when possible to form collaborative teams that bring a variety of skill sets together to focus on big, important and complex questions involving the geography of biodiversity conservation.

13.6. References

Ackerly, D.D., Loarie, S.R., Cornwell, W.K., Weiss, S.B., Hamilton, H., Branciforte, R., Kraft, N.J.B. (2010). The geography of climate change: Implications for conservation biogeography. *Diversity and Distributions*, 16, 476–487.

Cardinale, B.J., Primack, R.B., Murdoch, J.D. (2019). *Conservation Biology*. Sinauer Associates, New York.

Cumming, G.S., Bodin, O., Ernstson, H., Elmqvist, T. (2010). Network analysis in conservation biogeography: Challenges and opportunities. *Diversity and Distributions*, 16, 414–425.

Devictor, V. and Robert, A. (2009). Measuring community responses to large-scale disturbance in conservation biogeography. *Diversity and Distributions*, 15, 122–130.

Devictor, V., Whittaker, R.J., Beltrame, C. (2010). Beyond scarcity: Citizen science programmes as useful tools for conservation biogeography. *Diversity and Distributions*, 16, 354–362.

Diniz, J.A.F., De Marco, P., Hawkins, B.A. (2010). Defying the curse of ignorance: Perspectives in insect macroecology and conservation biogeography. *Insect Conservation and Diversity*, 3, 172–179.

Driscoll, D.A., Balouch, S., Burns, T.J., Garvey, T.F., Wevill, T., Yokochi, K., Doherty, T.S. (2019). A critique of "countryside biogeography" as a guide to research in human-dominated landscapes. *Journal of Biogeography*, 46(12), 2850–2859.

Fischer, J., Lindenmayer, D.B., Montague-Drake, R. (2008). The role of landscape texture in conservation biogeography: A case study on birds in south-eastern Australia. *Diversity and Distributions*, 14, 38–46.

Franklin, J. (2010). Moving beyond static species distribution models in support of conservation biogeography. *Diversity and Distributions*, 16, 321–330.

Franklin, J. (2013). Species distribution models in conservation biogeography: Developments and challenges. *Diversity and Distributions*, 19, 1217–1223.

Franklin, J. (2016). *Diversity and Distributions* is (still) a journal of conservation biogeography. *Diversity and Distributions*, 22, 1–2.

Hu, J.H. and Liu, Y. (2014). Unveiling the conservation biogeography of a data-deficient endangered bird species under climate change. *PLOS One*, 9, 11.

Katinas, L. and Ladle, R.J. (2019). Redundancy or progress? A response to Driscoll *et al.* (2019). *Journal of Biogeography*, 47(8), 1843–1845.

Ladle, R.J. and Whittaker, R.J. (2011). *Conservation Biogeography*. Wiley, Oxford, UK.

Liu, J., Yan, H.F., Newmaster, S.G., Pei, N.C., Ragupathy, S., Ge, X.J. (2015). The use of DNA barcoding as a tool for the conservation biogeography of subtropical forests in China. *Diversity and Distributions*, 21, 188–199.

Olden, J.D. (2006). Biotic homogenization: A new research agenda for conservation biogeography. *Journal of Biogeography*, 33, 2027–2039.

Olden, J.D., Kennard, M.J., Leprieur, F., Tedesco, P.A., Winemiller, K.O., Garcia-Berthou, E. (2010). Conservation biogeography of freshwater fishes: Recent progress and future challenges. *Diversity and Distributions*, 16, 496–513.

Real, R., Barbosa, A.M., Rodriguez, A., Garcia, F.J., Vargas, J.M., Palomo, L.J., Delibes, M. (2009). Conservation biogeography of ecologically interacting species: The case of the Iberian lynx and the European rabbit. *Diversity and Distributions*, 15, 390–400.

Richardson, D.M. (2005). Diversity, distributions and conservation biogeography. *Diversity and Distributions*, 11, 1–2.

Richardson, D.M. and Whittaker, R.J. (2010). Conservation biogeography – Foundations, concepts and challenges. *Diversity and Distributions*, 16, 313–320.

Rocchini, D., Salvatori, N., Beierkuhnlein, C., Chiarucci, A., De Boissieu, F., Forster, M., Garzon-Lopez, C.X., Gillespie, T.W., Hauffe, H.C., He, K.S., Kleinschmit, B., Lenoir, J., Malavasi, M., Moudry, V., Nagendra, H. *et al.* (2021). From local spectral species to global spectral communities: A benchmark for ecosystem diversity estimate by remote sensing. *Ecological Informatics*, 61, 10.

Terauds, A., Chown, S.L., Morgan, F., Peat, H.J., Watts, D.J., Keys, H., Convey, P., Bergstrom, D.M. (2012). Conservation biogeography of the Antarctic. *Diversity and Distributions*, 18, 726–741.

Whittaker, R.J. and Ladle, R.J. (2011). The roots of conservation biogeography. In *Conservation Biogeography*, Ladle, R.J. and Whittaker, R.J. (eds). Wiley, Oxford, UK.

Whittaker, R.J., Araujo, M.B., Paul, J., Ladle, R.J., Watson, J.E.M., Willis, K.J. (2005). Conservation biogeography: Assessment and prospect. *Diversity and Distributions*, 11, 3–23.

Zurell, D., Franklin, J., Konig, C., Bouchet, P.J., Dormann, C.F., Elith, J., Fandos, G., Feng, X., Guillera-Arroita, G., Guisan, A., Lahoz-Monfort, J.J., Leitao, P.J., Park, D.S., Peterson, A.T., Rapacciuolo, G. *et al.* (2020). A standard protocol for reporting species distribution models. *Ecography*, 43, 1261–1277.

List of Authors

Irene D. ALABIA
Arctic Research Center
Hokkaido University
Sapporo
Japan

Mohammad BAHRAM
Swedish University of
Agricultural Science
Uppsala
Sweden
and
University of Tartu
Estonia

Matthew BIDDICK
School of Life Sciences
Technical University of Munich
Germany

Luisa Maria DIELE-VIEGAS
Federal University of Alagoas
Maceió
Brazil

Malte C. EBACH
University of New South Wales
and
The Australian Museum
Sydney
Australia

Arnaud FAILLE
Stuttgart State Museum of
Natural History
Germany

Jorge GARCÍA MOLINOS
Arctic Research Center
Hokkaido University
Sapporo
Japan

Eric GUILBERT
UMR7179 MECADEV
National Museum of Natural History
Paris
France

Laurent HUSSON
ISTerre
CNRS
Université Grenoble Alpes
France

Battle KARIMI
UMR Agroécologie
INRAE
Dijon
France

Tarquin NETHERWAY
Swedish University of
Agricultural Science
Uppsala
Sweden

Violaine NICOLAS
Institut de Systématique, Evolution,
Biodiversité (ISYEB)
National Museum of Natural History
Paris
France

Anthi OIKONOMOU
Institute of Marine Biological Resources
and Inland Waters
Hellenic Centre for Marine Research
Anavyssos
Greece

Jesús OLIVERO
University of Málaga
Spain

Lionel RANJARD
UMR Agroécologie
INRAE
Dijon
France

Brett R. RIDDLE
University of Nevada
Las Vegas
United States

Isabel SANMARTÍN
Real Jardín Botánico, CSIC
Madrid
Spain

Julia SCHMACK
Centre for Biodiversity and Biosecurity
University of Auckland
New Zealand

Pierre SEPULCHRE
LSCE
IPSL
CNRS
Université Versailles Saint Quentin
France

Inessa VOET
Institut de Systématique, Evolution,
Biodiversité (ISYEB)
National Museum of Natural History
Paris
France

Index

A, B

C, D

E, F

G, H

I, J, L

M, N, O

P, R, S

T, V, Z